Thorsten Gressling
Data Science in Chemistry

Also of interest

Computational Chemistry Methods
Ramasami (Ed.), 2020
ISBN 978-3-11-046536-5, e-ISBN 978-3-11-063162-3

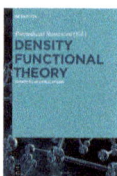

Density Functional Theory.
Advances in Applications
Ramasami (Ed.), 2018
ISBN 978-3-11-056675-8, e-ISBN 978-3-11-056819-6

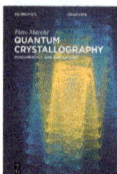

Quantum Crystallography.
Fundamentals and Applications
Macchi, 2021
ISBN 978-3-11-060710-9, e-ISBN 978-3-11-060716-1

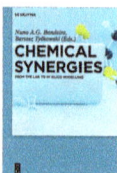

Chemical Synergies.
From the Lab to In Silico Modelling
Bandeira, Tylkowski, 2018
ISBN 978-3-11-048135-8, e-ISBN 978-3-11-048206-5

Thorsten Gressling

Data Science in Chemistry

Artificial Intelligence, Big Data, Chemometrics
and Quantum Computing with Jupyter

DE GRUYTER

Author
Dr. Thorsten Gressling
Am Glunzbusch 9
15741 Bestensee, Germany
mail@gressling.com

ISBN 978-3-11-062939-2
e-ISBN (PDF) 978-3-11-062945-3
e-ISBN (EPUB) 978-3-11-063053-4

Library of Congress Control Number: 2020942815

Bibliographic information published by the Deutsche Nationalbibliothek
The Deutsche Nationalbibliothek lists this publication in the Deutsche Nationalbibliografie; detailed
bibliographic data are available on the Internet at http://dnb.dnb.de.

© 2021 Walter de Gruyter GmbH, Berlin/Boston
Cover image: IBM Q – Dilution Refrigerator from the Zürich Quantum Computer,
taken at 2018 ASCE. Credit: Graham Carlow, License: Attribution-NoDerivs 2.0
Generic (CC BY-ND 2.0)
Typesetting: Integra Software Services Pvt. Ltd.
Printing and binding: CPI books GmbH, Leck

www.degruyter.com

Preface

The speed of digital transformation is increasing. Although the introduction of computing started in the middle of last century with the first wave of cheminformatics and the digitization of production plants, new fields of application are emerging with data science and artificial intelligence (AI). However, the new fields are a fusion of *three* domains:

- computer science,
- mathematics, and
- the specialist domain.

There are many excellent books on AI, cheminformatics, or data science, in general. But practicing data science *must* address the specialist domain or even take the perspective of the business expert. This is where *Data Science in Chemistry* starts.

Acknowledgments

The idea of writing this book came up by Karin Sora of De Gruyter at the ACS Spring Meeting 2019 in Orlando. I was complaining why there were no books on this important field, and she told me "Then write it!" I also thank Ms. Vivien Schubert of De Gruyter for guiding me in authoring this book and helping me with all my questions.

I thank my wife Isabella Lewandowski who endured a nerd writing even on Christmas Eve.

Credits to the coders

The book would not have been possible without the passion and the contributions of hundreds of scientists who had listed their programs on GitHub! I thank *Greg Landrum, Oliver Beckstein, Daniel Pelliccia, Björn Dahlgreen, and Jean-Baptiste Lamy,* five individuals who are part of the GitHub community.

Contact

In case of errors or comments, please open an issue on github.com/Gressling/examples.

I am available on LinkedIn or ResearchGate for further contact and discussion.

https://doi.org/10.1515/9783110629453-202

Contents

Data science as field of activity

Introduction to ML and AI

Part B: Jupyter in cheminformatics
Physical chemistry

Material science

Organic chemistry

Engineering, laboratory, and production

Part C: **Data science**
Data engineering in analytic chemistry

Applied data science and chemometrics

Applied artificial intelligence

Knowledge and information

Part D: Quantum computing and chemistry introduction

Quantum computing applications

Introduction

The book covers the classic definition of the data science process and how it is adopted in chemistry today. It discusses the state-of-the-art process and the possible directions this relatively young field of expertise might take.

The book also covers daily life questions in digital research:

– How to integrate data?
– How to manage the alignment of ingestion and experiment?
– What is the role of computational chemistry in daily work?
– How to do cheminformatics in Python?

Finally, the book contains and discusses emerging new technology areas like quantum computing and DevOps Toolchains for the chemist.

The new pillar *"data science"* in the world of chemistry comes along with the third wave of IT[1, 2]: Big data, cloud-centric computing, and machine learning are providing new building blocks for it. This is covered in part A of the book.

Inspiration

The book is a *stimulus and entertainment* for those who are already in the field of expertise. It contains a new bird's eye view of programming and chemistry. It aims to push forward the construction of an integrated environment covering *all aspects of chemistry* with Python.

Structure

In 100 small chapters, each of around three pages, the field of data science is covered. After theoretical abstracts, there are program code snippets based on Python and Jupyter notebooks. This book is written from the perspective of the domain knowledge of chemistry. So it avoids mathematical formulas and focuses on chemical structures.

The book has four sections. **Part A** introduces the fundamentals, without repeating from other textbooks on AI, IT, and mathematics, but with a strong focus on their latest developments and on chemistry. **Part B** is the domain knowledge part addressing Jupyter and Python in cheminformatics, along the outline of classic chemistry: OC, AC, and PC so that chemists feel to be at home. **Part C** contains various aspects of data science such as its use in analytical chemistry or in the form of chemometrics. The use of artificial intelligence in daily chemistry processes has its own section. Finally, **part D** includes the quantum part with the bandwidth from theoretical considerations up to the use of QM in AI.

https://doi.org/10.1515/9783110629453-204

The red thread of the book covers *all aspects of modern chemistry with some python program code*. But as data science in Python is a young discipline, there is a lot of "*alpha*" and "*beta*." Despite the various maturity levels of the libraries – and the ongoing migration from Python 2.x to Python 3 – most examples are usable in Jupyter 3 notebooks, and at least running in Python 2 CLI at the lowest maturity level.

How to use this book

This book is intended to provide the chemist a manual that serves both as a *reference* and as a *practice book*. It is also intended to work vice versa, that is, to provide the data scientist, the programmer, and the computational chemist with use cases. It is also intended that the code examples can be used as a source for exams in academic environments.

The chemist learns how to use data science for deep insights and pattern analyses. This book is intended to be a practical guide that gives a state-of-the-art overview of available processes and technologies.

Education

There must be changes to chemical sciences curricula to provide training in data science, statistics, machine learning, and programming. Experimentalists are not immune to changes in their field of expertise! The next generation of material scientists – whether experimentalists or computationalists – will be *data native*.

References

1. McKinsey. Digital in chemicals: From technology to impact https://www.mckinsey.com/industries/chemicals/our-insights/digital-in-chemicals-from-technology-to-impact.
2. Deloitte. Global Digital Chemistry https://www2.deloitte.com/content/dam/Deloitte/de/Documents/consumer-industrial-products/Deloitte%20Global%20Digital%20Chemistry%20Survey2016Extract.pdf.

Technical setup and naming conventions

In order to facilitate fluent reading, some basic layouts are used avoiding text frag-mentation and allowing the integration of code snippets to the red thread of each chapter.

Program code

All snippets are shortened so that just *the core and the working principle* are carved out. All listings have this form:

```
# Title
# author: Name, A
# License: MIT License        # code: URL to the original code
# activity: (status)          # index: (chapter)(number)

With (status):
    -   active (2020)          : Excellent
    -   single example         : (N/A, persistent)
    -   maintained (2019, 2018) : OK
    -   on hold? (<2018)        : Unclear
    -   outdated (2016)         : Be careful

# code is always in a box, comments are grey
this.isACall('With a parameter')
```

Citations

Bold text as well as text in *blue + italic* are citations and indicate intellectual property from other authors. There is always a reference at the end of each chapter.

Legal questions

All listings and citations in the book are based on license types that allow *commercial use*. Even the use of CC-BY-**SA** was avoided as the "SA/build upon material" is unclear from the perspective of a printed book. Only Wikipedia has enriched the SA license with a clear statement:

If you want to use Wikipedia's text materials in your own books/articles/web sites or other publications, you can generally do so, but you must comply with one of the licenses that Wikipedia's text is licensed under. If you make modifications or additions to the page

https://doi.org/10.1515/9783110629453-205

you re-use, you must license them under the Creative Commons Attribution-Share-Alike License 3.0 or later.[1]

Even if not necessary, I give references to the creator and author – (*please: avoid fantasy names* in GitHub!). Some authors added license terms after request, thank you!

There are four related groups of licenses: Apache, MIT and BSD, CommonCreative, and GPL. All citations and code listings in this book have one of these types:

License	Author	Version	Date	Link	Commercial use	Modify	Patent grant	TM grant
Apache License	Apache Software Foundation	2	2004	OK	OK	OK	Yes	No
BSD License	University of California	3.0		OK	OK	OK	(Part.)	(Part.)
BSD License	University of California	2-Clause "Simplified"						
MIT license	MIT		1988	OK	OK	OK	(Part.)	(Part.)
CC BY SA	Creative Commons	4		OK	OK	(Part.) Share alike*		
CC BY (without SA part)	Creative Commons	4	2002	OK	OK	OK	No	
CC0	Creative Commons			OK	OK	OK	OK	OK
GNU LGPL (not GPL)	GNU Lesser General Public License	3.0	2007	OK	OK	OK		
Copyleft	Like GPL or CC-BY-SA							
The Unlicense	Like CC0							

*Remix, transform, or build upon the material (with change indication).

Liability

All information and program listings are for informational use only. There is no product liability.

Technical setup: Windows, Linux, or cloud

A good platform for working is Anaconda with Python 3. Windows™ or Linux are equivalent, and Apple™ is not tested. On Windows, you can install a Linux subsystem and Ubuntu as an App.[2]

Sometimes there are citations of unported Python 2 code. Using `$ 2to3 -n -W --add-suffix=3 example.py`, can perform an semi-automatic transformation of[3] the code but mostly some minor tweaks in function call, syntax, and error handling may remain manual.

Working in the cloud/AWS

A cloud-centric solution is based on an instance on Amazon with C9 (e.g., if you use Chromebooks like the author). As a cloud solution, it is not a local system; sometimes a few tricks are necessary, for example, using PyMol (C9 does not support XWindows).

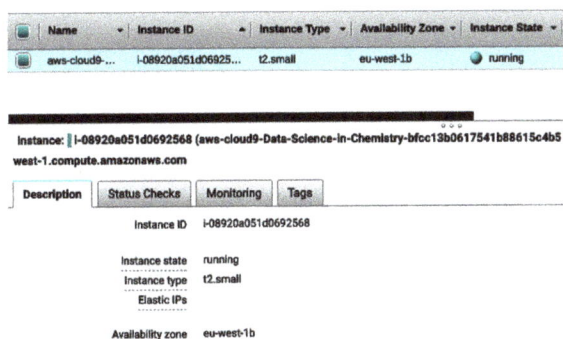

Name	Instance ID	Instance Type	Availability Zone	Instance State
aws-cloud9-...	I-08920a051d06925...	t2.small	eu-west-1b	running

Instance: I-08920a051d0692568 (aws-cloud9-Data-Science-in-Chemistry-bfcc13b0617541b88615c4b5 west-1.compute.amazonaws.com

| Description | Status Checks | Monitoring | Tags |

Instance ID	I-08920a051d0692568
Instance state	running
Instance type	t2.small
Elastic IPs	
Availability zone	eu-west-1b

Figure Setup 1: Amazon C9 environment.

Anaconda as a coding environment for the book

Conda already has fully open package management and is on a steady trend toward fully open packages; see installation and handling in Chapter 11. Of course, other environments like Google Colab or IBM Watson Studio can be used, but there are limitations in environment installation and file handling.

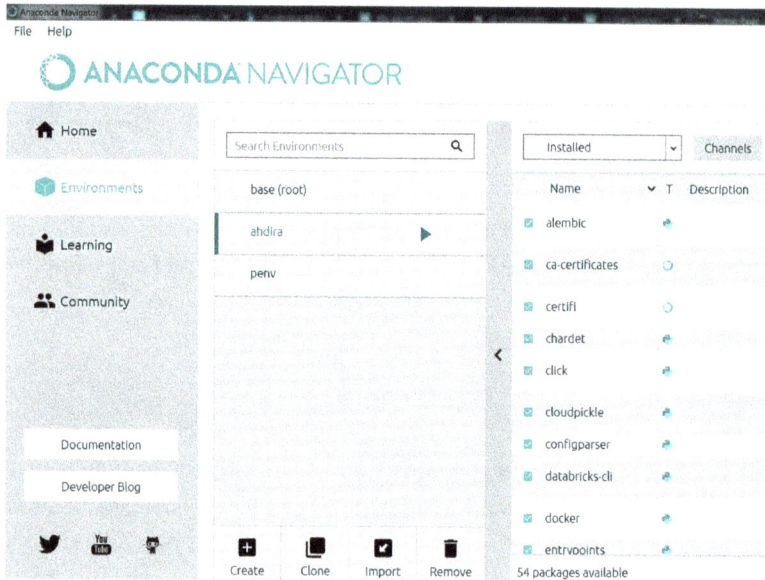

Figure Setup 2: Anaconda administration with environment and package management.

References

1. Reusing Wikipedia content (Wikipedia) https://en.wikipedia.org/wiki/Wikipedia:Reusing_Wikipedia_content.
2. Install Windows Subsystem for Linux (WSL) on Windows 10 (Microsoft) https://docs.microsoft.com/en-us/windows/wsl/install-win10.
3. Python Software Foundation. Porting Python 2 Code to Python 3 https://docs.python.org/3/howto/pyporting.html.

1 Data science: introduction

Data science[1] is an interdisciplinary field that uses *scientific* methods, processes, algorithms, and systems to *extract knowledge* and insights from structural and unstructured data. Data science is related to data mining and big data.

The keyword in "Data Science" is not Data, it is Science[2]

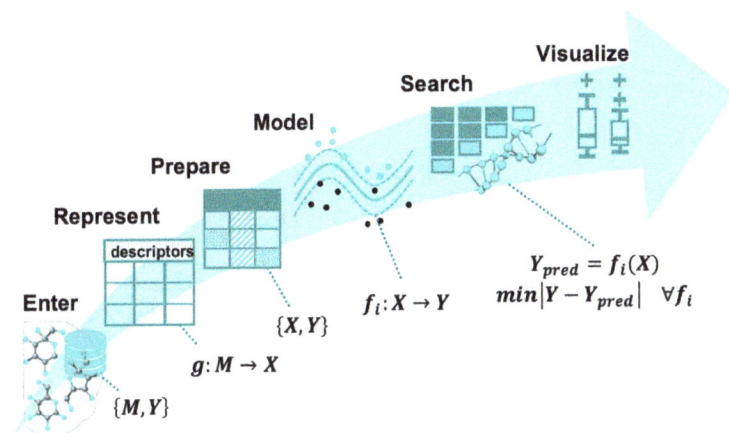

Figure 1.1: The major tasks and mathematical setup of a supervised machine learning workflow[3] (*Hachmann*).

*The first problem is that most great data scientists don't sufficiently understand busi-
ness and most great business leaders don't sufficiently understand data science.[4]*

Relation of science and digital research

Since the beginning of the 1980s, the term "data science" has appeared in various
contexts, but was never well defined in the scientific community.[5] Originally, it was
used as a substitute for computer science.[1]

Definitions

*Data science is performed to analyze, understand, and extract actual phenomena in
data. The challenge is to identify unique patterns and variables.*

https://doi.org/10.1515/9783110629453-001

The goal is to:
- understand
- extract insights

To do this, data science is a multidisciplinary field that brings together concepts from:
- computer science
- statistics/machine learning
- data analysis
- domain knowledge

Bringing together these fields of expertise in data science is also a concept of *unification*.

Discussion

Leek[2] discusses the relation of structure and the desired results:
- *It is easy to discover "structure" or "networks" in a data set. There will always be correlations for a thousand reasons if you collect enough data.*
- *Understanding whether these correlations matter for specific, interesting questions is much harder.*
- *Often the structure you found on the first pass is due to phenomena (measurement error, artifacts, and data processing) that do not answer an interesting question.*

The two paradigms of data research

1. Hypothesis driven
Given a problem, what kind
of data do we need to help solve it?

2. Data driven
Given some data, what interesting
problems can be solved with it?

The heart of data science is to always ask questions:
1. What can we learn from this data?
2. What actions can we take, once we find whatever it is we are looking for?

Main types of problems

Two problems arise repeatedly in data science. This is discussed in detail in Chapters 33 to 37. As a rule of thumb, these are:
- Classification: Assigning something to a discrete set of possibilities
- Regression: Predicting a numerical value

References

1. Data science (Wikipedia) https://en.wikipedia.org/wiki/Data_science.
2. Leek, J. Simply Statistics https://simplystatistics.org/2013/12/12/the-key-word-in-data-science-is-not-data-it-is-science/.
3. Haghighatlari, M.; Hachmann, J. Advances of Machine Learning in Molecular Modeling and Simulation https://www.researchgate.net/publication/330845218_Advances_of_Machine_Learning_in_Molecular_Modeling_and_Simulation.
4. Boyle, D. Data Science vs. the C Suite https://www.linkedin.com/pulse/data-science-vs-c-suite-solution-david-boyle/.
5. Che-Workshop. Framing the Role of Big Data and Modern Data Science in Chemistry https://www.nsf.gov/mps/che/workshops/data_chemistry_workshop_report_03262018.pdf.

2 Data science: the "fourth paradigm" of science

Statistics + computer science + domain knowledge = data science

Data science interrogates (scientific) data at scale. Additional success can be achieved when data science paradigms integrate tools with domain-specific knowledge and expertise.[1]
- *Transform* chemical sciences and engineering
- *Knowledge* discovery at scale
- *Interdisciplinary*: statistics, computer science, applied math, AI, and domain tools

The fourth paradigm: knowledge discovery from data (KDD)

The Fourth Paradigm: Data-Intensive Scientific Discovery[2] is a 2009 anthology of essays on the topic of data science-based on data-intensive computing.
- *Theory*
- *Experiment*
- *Simulation*
- *Data science*

Increase in the use of data is bringing a paradigm shift to the nature of science.[3] The *Fourth Paradigm* is *data science* and the first three paradigms, in order, are empirical evidence, scientific theory, and computational science as shown in figure 2.1.

New technologies and approaches are generating large, diverse data sets. Data science offers methods and tools that are needed to integrate, analyze, and manage these data sets. However, data science applications in the chemical sciences and engineering communities have been relatively limited and many opportunities for advancing the fields have gone unexplored.[5]

Data science life cycle

In general, the life cycle has phases of *exploration* and *production* as shown in figure 2.2. In science, the exploration part may be the main focus while the production part may not be as relevant as in commercial applications. But from the perspective of the different roles in a data science value chain, using a model in production can make the results of the exploration phase available to other scientists – to speed up the knowledge innovation cycle.[6]

https://doi.org/10.1515/9783110629453-002

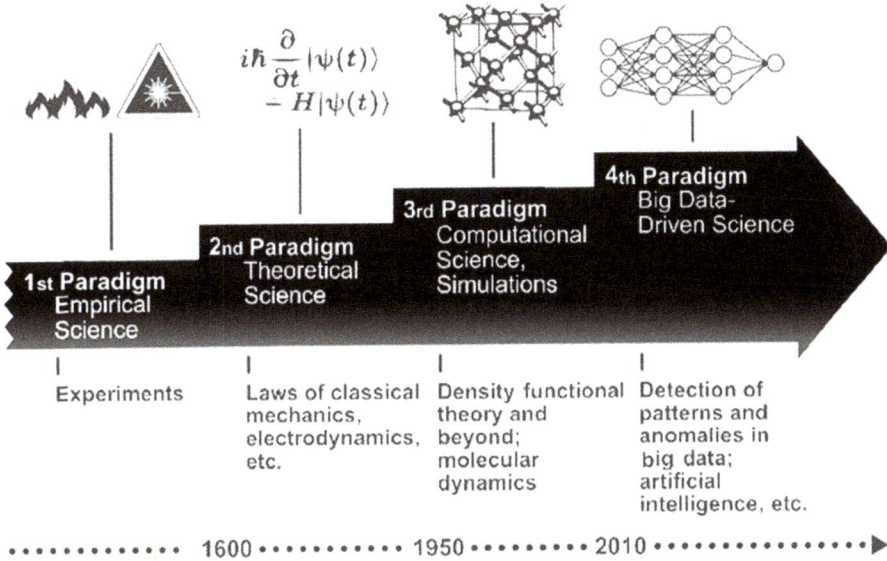

$$i\hbar \frac{\partial}{\partial t}|\psi(t)\rangle = H|\psi(t)\rangle$$

1st Paradigm
Empirical
Science

2nd Paradigm
Theoretical
Science

3rd Paradigm
Computational
Science,
Simulations

4th Paradigm
Big Data-
Driven Science

| Experiments | Laws of classical mechanics, electrodynamics, etc. | Density functional theory and beyond; molecular dynamics | Detection of patterns and anomalies in big data; artificial intelligence, etc. |

1600 · · · · · · · · · · 1950 · · · · · · · · 2010 · · · · · · · · · · · · · ▶

Figure 2.1: Development of the four (materials) science and engineering paradigms[4] *(Draxl, Scheffler)*.

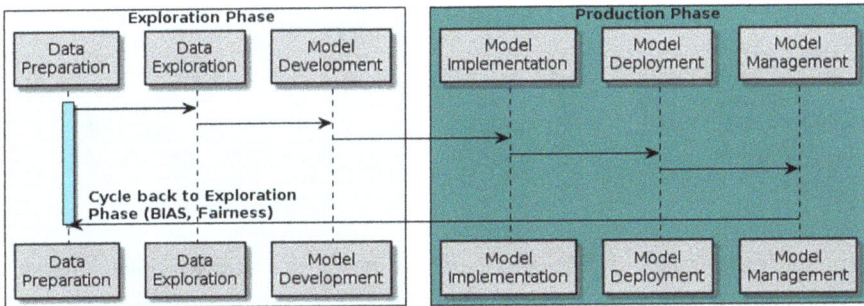

Figure 2.2: Data science life cycle *(Gressling)*.

Challenges

Although data science is a rapidly growing field, some of the building blocks still experience difficulties:

Big data

–   Data sets are so large that standard approaches and tools for storage, analysis, and sharing fail.
–   Data may be too large to fit in a 10 TB memory.

Data mining

–   Extracting and understanding *human-relevant* insight and predictions
–   *"Knowledge is a process of piling up facts; wisdom lies in their simplification"* – *M. H. Fischer (Physician)*

Machine learning/statistical learning

–   Algorithms that learn without explicit human instruction
–   Unlearning
–   Transfer learning

Artificial intelligence

–   Machine behavior that mimics human cognition
–   Turing test: "Intelligent behavior equivalent to or indistinguishable from that of a human" – *Alan Turing*

So, data science brings statistical methods to new scales and prioritizes approximation and uncertainty. It brings new challenges to IT with demand for computing power, memory, and hardware.

References

1. Ferguson, A. What is data science? http://nas-sites.org/csr/files/2018/04/2.-Ferguson.pdf.
2. Hey, T.; Tansley, S. The Fourth Paradigm: Data-intensive Scientific Discovery https://www.immagic.com/eLibrary/ARCHIVES/EBOOKS/M091000H.pdf.
3. The Fourth Paradigm (Wikipedia) https://en.wikipedia.org/wiki/The_Fourth_Paradigm.
4. Draxl, C.; Scheffler, M. NOMAD: The FAIR Concept for Big Data-Driven Materials Science. *MRS Bull.* **2018**, *43* (9), 676–682. https://doi.org/10.1557/mrs.2018.208.
5. Data Science: Opportunities to Transform Chemical Sciences & Engineering http://nas-sites.org/csr/data-science-opportunities-to-transform-chemical-sciences-engineering/.
6. Gressling, T. Classification of Software in Digital Chemistry https://www.researchgate.net/publication/338139891_Classification_of_Software_in_Digital_Chemistry. https://doi.org/10.13140/RG.2.2.21845.99044.

3 Relations to other domains and cheminformatics

Differentiation of computer science, mathematics, and the specialist domain:

- **Mathematical chemistry**[1] *is the area of research engaged in* **novel applications of mathematics to chemistry**; *it concerns itself principally with the mathematical modeling of chemical phenomena. Mathematical chemistry has also sometimes been called computer chemistry, but should not be confused with computational chemistry.*
- **Computational chemistry**[2] *is a branch of chemistry that uses* **computer simulation to assist** *in solving chemical problems. It uses methods of theoretical chemistry, incorporated into efficient computer programs, to calculate the structures and properties of molecules and solids. While computational results normally complement the information obtained by chemical experiments, it can, in some cases, predict hitherto unobserved chemical phenomena. It is widely used in the design of new drugs and materials.*
- **Cheminformatics** *is the use of computer and informational techniques applied to a range of problems in the field of chemistry. These in silico techniques are used, for example, in pharmaceutical companies and academic settings in the process of drug discovery. These methods can also be used in chemical and allied industries in various other forms.*[3] *Cheminformatics combines the scientific working fields of chemistry, computer science, and information science, for example, in the areas of topology, chemical graph theory, information retrieval, and data mining in the chemical space.*
- **Theoretical chemistry**[4] *is the branch of chemistry that develops theoretical generalizations that are part of the theoretical arsenal of modern chemistry: for example, the concepts of chemical bonding, chemical reaction, valence, the surface of potential energy, molecular orbitals, orbital interactions, and molecule activation.*

Domain expertise

Since the beginning of the 1980s, cheminformatics[5] emerged into a specialist domain of its own, working on questions like 2D and 3D molecule editors and generators and operating with data and structures like simplified molecular-input line-entry system (SMILES), Chemical Markup Language (CML), and chemical table file (MDL). Another focus was on working with structures like ring finding or substructure search. Also, features like fingerprint or force field calculations are calculated.

https://doi.org/10.1515/9783110629453-003

Why data science in chemical science?

Data science in chemistry can be:
- a chemical science solution of *forwarding problems* – measurement or prediction of materials' properties at unprecedented new scales;
- the solution of inverse problems like reverse engineering of molecular and materials properties (which is not simply to "invert" physical models and run in reverse);
- a way of problem search in chemical space that is astronomical.

Also, some chemical sciences methods (like statistical thermodynamics or molecular simulation) integrate well with data science/ML paradigms-like ensembles, Bayesian statistics, or non-Boltzmann (Monte Carlo) sampling. There is potential for chemical sciences to not just use and adapt ML algorithms but also inspire new computational tools, giving them back in the ML community.

References

1. Mathematical chemistry (Wikipedia) https://en.wikipedia.org/wiki/Mathematical_chemistry.
2. Computational chemistry (Wikipedia) https://en.wikipedia.org/wiki/Computational_chemistry.
3. Cheminformatics (Wikipedia) https://en.wikipedia.org/wiki/Cheminformatics.
4. Theoretical chemistry (Wikipedia) https://en.wikipedia.org/wiki/Theoretical_chemistry.
5. Engel, T.; Gasteiger, J. Wiley-VCH – Chemoinformatics: https://www.wiley-vch.de/de/fachgebiete/naturwissenschaften/chemoinformatics-978-3-527-33109-3.

4 Cheminformatics application landscape

Cheminformatic toolkits[1] are software development kits that allow cheminformaticians to develop custom computer applications for use in virtual screening, chemical database mining, and structure–activity studies. Toolkits are often used for experimentation with new methodologies.

All vendors share many core functions but with different approaches either in internal perspective with the implementation or external perspective (with the application programming interfaces[2] API). Hence, comparing and contrasting the functionalities of toolkits, for example, RDKit versus CDK or OpenBabel versus OEChem often show the same idea but with slightly different implementations.

Programming comparison: heavy atom counts from an SD file

For each record from the benzodiazepine file, print the total number of heavy atoms in each record (i.e., exclude hydrogens).

```
# Heavy atom counts code comparison (Chemistry Toolkit Rosetta Wiki³)
# author: Dahlke, A; Landrum,     G; Ihlenfeld, W (Et al.)
# license: CC-BY-SA             # code: wiki/Chemistry Toolkit Rosetta Wiki
# activity: single examples (2016) # index: 4-1
```

Cinfony

```
from cinfony import rdk

for mol in rdk.readfile("sdf", "benzodiazepine.sdf"):
    print mol.calcdesc(['HeavyAtomCount'])['HeavyAtomCount']
```

Indigo

```
from indigo import Indigo
indigo = Indigo()

for mol in indigo.iterateSDFile("benzodiazepine.sdf.gz"):
    print mol.countHeavyAtoms()
```

https://doi.org/10.1515/9783110629453-004

Openbabel/Pybel

```python
import pybel

for mol in pybel.readfile("sdf", "benzodiazepine.sdf.gz"):
    print mol.OBMol.NumHvyAtoms()
```

OpenEye

```python
from openeye.oechem import *

ifs = oemolistream()
ifs.open("benzodiazepine.sdf.gz")
for mol in ifs.GetOEGraphMols():
    print OECount(mol, OEIsHeavy())
```

RDKit

```python
from rdkit import Chem

suppl = Chem.SDMolSupplier("benzodiazepine.sdf")
heavy patt = Chem.MolFromSmarts("[!#1]")

for mol in suppl:
    print len(mol.GetSubstructMatches(heavy_patt))
```

Cactvs

```python
# reference Xemistry4
for eh in Molfile('benzodiazepine.sdf.gz'):
    print(eh.E_HEAVY_ATOM_COUNT)
```

References

1. Cheminformatics toolkits (Wikipedia) https://en.wikipedia.org/wiki/Cheminformatics_toolkits.
2. Application programming interface API (Wikipedia): https://en.wikipedia.org/wiki/Application_programming_interface.
3. Chemistry Toolkit Rosetta Wiki https://ctr.fandom.com/wiki/Chemistry_Toolkit_Rosetta_Wiki.
4. Xemistry Chemoinformatics https://www.xemistry.com/.

5 Cloud, fog, and AI runtime environments

- ***Cloud[1,2] computing*** *is the on-demand availability of computer system resources, especially data storage and computing power, without direct active management by the user. The term is generally used to describe data centers available to many users over the Internet. Large clouds, predominant today, often have functions distributed over multiple locations from central servers. If the connection to the user is relatively close, it may be designated an edge server.*
- ***Fog[3] computing*** *is an architecture that uses edge devices (see Chapter 9) to carry out a substantial amount of computation, storage, and communication locally and routed over the Internet backbone. Both cloud computing and fog computing provide storage, applications, and data to end-users. However, fog computing is closer to end-users and has a wider geographical distribution.*

Compared to the cycles of artificial intelligence, also in the world of IT architecture, there were three cycles with design paradigms. The first wave was monolithic and compilation was specific to computers where the programs ran. The second generation that came in the 1990s had client–server applications and was based on open frameworks like eclipse, which is the base of Knime. Now we have approached the third generation, which is cloud computing. REST APIs and calls over the web belong to the new architectural pattern.

AI runtime with Apache PredictIO

Apache PredictionIO®[4] is an open-source[5] machine learning server built on top of a state-of-the-art open-source stack for developers and data scientists to create *predictive engines* for any machine learning task. The event server can collect and store arbitrary events. Event server collects data in real time or in batch and can also unify the data. After data is collected, it mainly serves two purposes:
- Provide data to engine(s) for model training and evaluation
- Offer a unified view for data analysis

In the beginning, it is recommended to collect as much data as possible. Later on, data that are not relevant can be excluded from the predictive model in the data preparator.

https://doi.org/10.1515/9783110629453-005

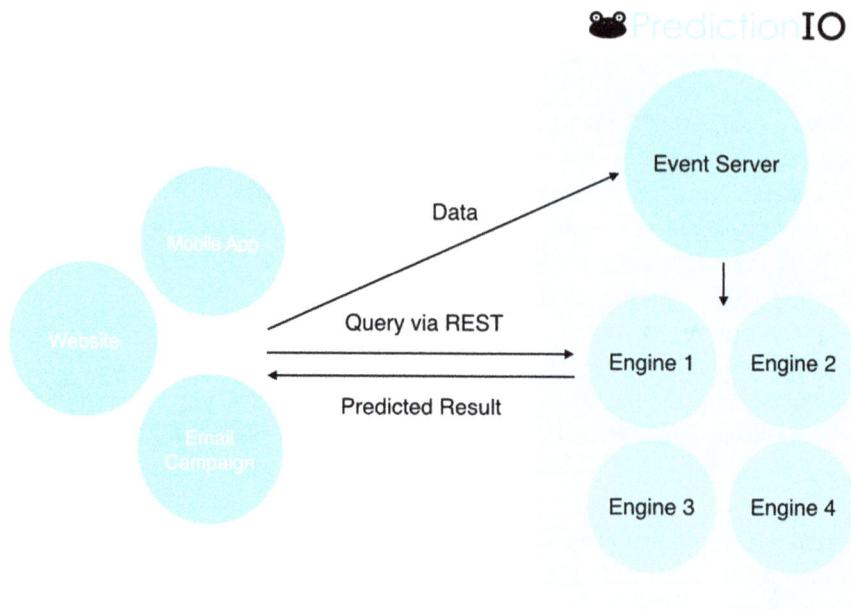

Figure 5.1: PredictionIO components *(Apache foundation)*.

Deploy

```
$ pio app new **your-app-name-here**
$ pio build # to update the engine
$ pio train # to train a predictive model with training data
$ pio deploy # to deploy the engine as a service
```

Call Apache PIO from Python

EventClient is for importing data into the PredictionIO platform.

```
# Interact with Apache PredictIO
# author: TappingStone, Inc (Et al.)
# license: Apache License 2.0  # code: pythonhosted.org/PredictionIO
# activity: active (2020)       # index: 5-1

from predictionio import EventClient, NotFoundError

access_key = "<YOUR_ACCESS_KEY>" # create key with 'pio app new'
client = EventClient(access_key=access_key, url="http://localhost:7070")
# Admin is on port 8000
```

```
first_event_properties = {
  "temp": 277.8,
  "worker": "Susan",
  "pH": [4.26, 4.15, 3.94],
  "stir": True
}
first_event_time = datetime(2020, 12, 13, 09, 39, pytz.timezone('US/Moun-
tain'))
first_event_response = client.create_event(
  event="sample A66 G39",
  entity_type="user",
  entity_id="susanb",
  properties=first_event_properties,
  event_time=first_event_time
)

# Get the first event from Event Server
first_event_id = first_event_response.json_body["eventId"]
event = client.get_event(first_event_id)
```

Docker

*Docker[6] is a set of platform as a service (PaaS) products that uses OS-level virtualiza-
tion to deliver software in packages called containers. Containers are isolated from one
another and bundle their own software, libraries, and configuration files; they can communicate with each other through well-defined channels.[7] All containers are run by a
single operating-system kernel and are thus more lightweight than virtual machines.*

It is very easy to set up services like an **MQTT broker** or a **Postgres database**:

```
$ docker pull eclipse-mosquitto
$ docker run -it -p 1883:1883 -p 9001:9001 eclipse-mosquitto
# Open port 1883 in the firewall to access mqtt broker,
# use MQTTBox (Chrome) or myMQTT (Android) to monitor

$ docker pull postgres
$ docker volume create my-vol # Persist volume
$ docker run --name postgres -e POSTGRES_PASSWORD=postgres -p 5432:5432
      -v my-vol:/var/lib/postgresql/data -d postgres
```

In this book, Docker is also used, for example, for installing `Gromacs`, `GROMACS-CUDA`,
`lammps`, `VASP/ESPRESSO`[7] and `ChemBL`.[8] It bypasses the `make` process and is a very easy
way to get up-to-date software.

References

1. Cloud computing (Wikipedia) https://en.wikipedia.org/wiki/Cloud_computing.
2. Crocker, C. Understanding Cloud Data Services: https://www.kdnuggets.com/2019/06/understanding-cloud-data-services.html.
3. Fog computing (Wikipedia) https://en.wikipedia.org/wiki/Fog_computing.
4. Apache PredictionIO http://predictionio.apache.org/.
5. predictionio-sdk-python (GitHub) https://github.com/apache/predictionio-sdk-python.
6. Docker (software) (Wikipedia) https://en.wikipedia.org/wiki/Docker_(software).
7. Hampel, A. VASP/ESPRESSO Dockerfiles: https://github.com/materialstheory/Dockerfiles.
8. Bachorz, R. A. The ChEMBL database of molecules in a Docker environment: https://medium.com/@rbachorz/the-chembl-database-of-molecules-in-a-docker-environment-6cdc64d4f0e.

6 DevOps, DataOps, and MLOps

Machine learning and its *operations* are called MLOps.[1] It helps manage the production of ML lifecycle and looks to increase automation and improve quality. MLOps introduce an approach to AI lifecycle management, applying to the entire lifecycle.[2]

This chapter does *not* contain:
- *Model optimization* which is part of data science in Chapter 31 (Model metrics)
- *xAI and BIAS* detection that is discussed in Chapter 40 (eXplainable AI).

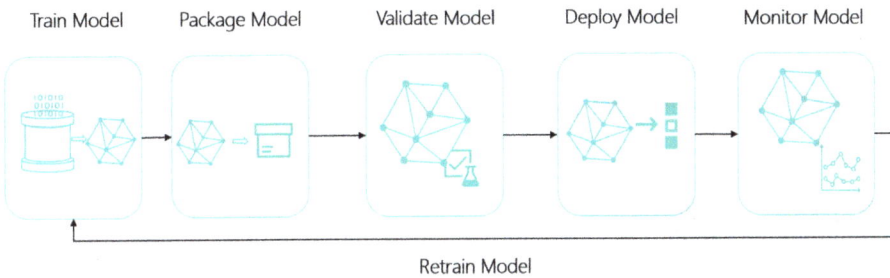

Figure 6.1: MLOps[3] *(Microsoft)*.

Kubeflow

Kubeflow[4] is an ML Toolkit for Docker/Kubernetes – an open-source system for automating the deployment, scaling, and management of containerized applications. The project makes machine learning on Kubernetes easy, portable, and scalable.

```
# MiniKF - deploy production-ready, full-fledged,
# local Kubeflow on a laptop
# https://www.kubeflow.org/docs/.../getting-started-minikf/

# Prerequisites:
# Install Vagrant - https://www.vagrantup.com/
# Install Virtual Box - https://www.virtualbox.org/

# Open a terminal on your laptop, create a new directory,
# switch into it, and run the following commands to install MiniKF:
```

https://doi.org/10.1515/9783110629453-006

```
vagrant init arrikto/minikf
vagrant up

# MiniKF will take a few minutes to boot.
# When this is done, navigate to http://10.10.10.10
# and follow the on-screen instructions to start Kubeflow and Rok.
```

Use Jupyter on Kubeflow

1. Click notebook servers in the left panel of the Kubeflow UI.
2. Choose the namespace corresponding to your Kubeflow profile.
3. Click NEW SERVER to create a notebook server.
4. When the notebook server provisioning is complete, click CONNECT.
5. Click Upload to upload an existing notebook or click New to create an empty note-book.

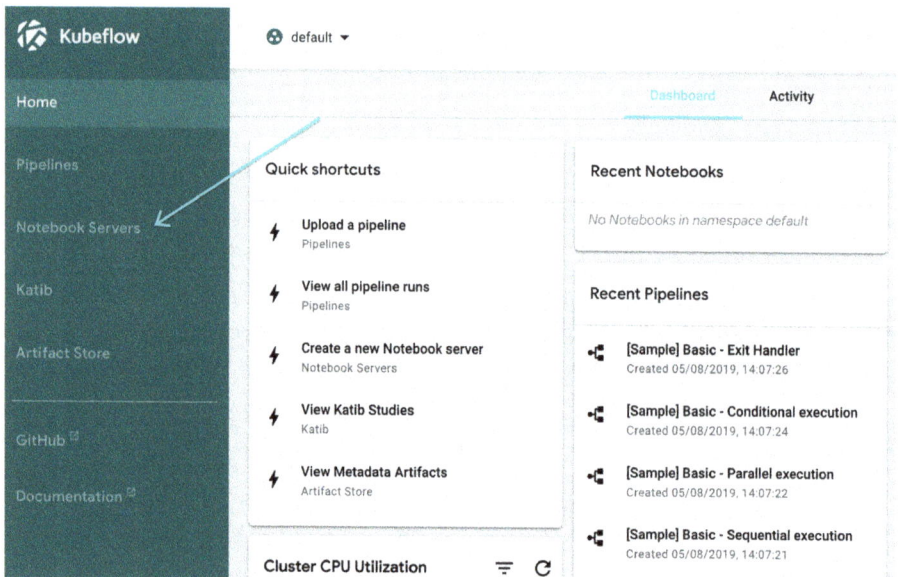

Figure 6.2: Kubeflow[5] and Jupyter *(Google)*.

Pipeline example

The following is an extract from the Python code that defines an xgboost-training pipeline.

```
# Pipeline in Kubeflow
# author: Kubeflow team
# license: CC BY 4.0        # code: kubeflow.org/.../pipelines-overview/
# activity: active (2020)   # index: 6-1

@dsl.pipeline(
    name='XGBoost Trainer',
    description='A trainer that does end-to-end distributed training for
    XGBoost models.'
)

def xgb_train_pipeline(
    output='gs://your-gcs-bucket',
    project='your-gcp-project',
    cluster_name='xgb-%s' % dsl.RUN_ID_PLACEHOLDER,
    train_data='gs://ml-pipeline-playground/sfpd/train.csv',
    eval_data='gs://ml-pipeline-playground/sfpd/eval.csv',
    target='resolution',
    rounds=200,
    workers=2):
...
    transform_output_train = os.path.join(output_template, 'train', 'part-*')
    transform_output_eval = os.path.join(output_template, 'eval', 'part-*')
    train_output = os.path.join(output_template, 'train_output')
    predict_output = os.path.join(output_template, 'predict_output')

dsl.get_pipeline_conf().add_op_transformer(
                gcp.use_gcp_secret('user-gcp-sa'))
```

Comparison between Kubeflow and MLFlow

Both are open-source projects and are supported by major players in the data analytics industry: Databricks and Google. Both address the tracking of ML experiments and supporting the production of ML lifecycle.[6] Both tools are enablers of data science and experimentation. Due to the ease of collaboration, MLflow might be a better enabler. It has a smaller footprint and a focus on tracking and archiving. It is a far more flexible tool. Both enable the serving of models at an API endpoint.

MLflow is an open-source machine learning platform for managing the ML lifecycle. MLflow is a single python package that covers key steps in model management. The API interface is built by Flask which is single-threaded. Therefore, it cannot handle a large number of requests at the same time.

Kubeflow is a combination of open-source libraries that depends on a Kubernetes cluster. Kubeflow pipelines capture the 'last mile' of the data pipeline after BigQuery, Dataproc, or containerized scripts. Kubeflow is an orchestration tool. Kubeflow addresses the company-wide orchestration more while MLFlow is more for the individual user. Some authors[6] feel a lack of maturity in Kubeflow (4/2020).

Databricks

Databricks is a company founded by the original creators of Apache Spark, related to a project of the University of California that was involved in making Apache Spark. Databricks develops a web-based platform for working with Spark, which provides automated cluster management and IPython-style notebooks.[7]

The *Managed MLflow* extension is built on top of MLflow and adds features like security and high availability, thus targeting the Kubeflow advantages. Also available is a Model Registry.[8]

Model registries

MLFlow Model Registry: The Model Registry is a collaborative hub where teams can share ML models, work together from experimentation to online testing and production, integrate with approval and governance workflows, and monitor ML deployments and their performance.[8] The Model Registry gives MLflow new tools to share, review, and manage ML models throughout their lifecycle.

Marketplaces are discussed in Chapter 30 as part of the data science process.

References

1. Gallatin, K. The Rise of the Term "MLOps" https://towardsdatascience.com/the-rise-of-the-term-mlops-3b14d5bd1bdb.
2. MLOps (Wikipedia) https://en.wikipedia.org/wiki/MLOps.
3. MLOps Python (Microsoft, MIT License) https://github.com/microsoft/MLOpsPython.
4. Kubeflow https://www.kubeflow.org/.
5. Set Up Your Notebooks https://www.kubeflow.org/docs/notebooks/setup/.
6. Allen, B. The Cheesy Analogy of MLflow and Kubeflow (Medium) https://medium.com/weareservian/the-cheesy-analogy-of-mlflow-and-kubeflow-715a45580fbe.
7. Databricks (Wikipedia) https://en.wikipedia.org/wiki/Databricks.
8. Databricks. Managed MLflow https://databricks.com/product/managed-mlflow.

7 High-performance computing (HPC) and cluster

Performance[1] is the amount of useful work accomplished by a computer system. Outside of specific contexts, computer performance is estimated in terms of accuracy, efficiency, and speed of executing computer program instructions. When it comes to high computer performance, one or more of the following factors might be involved:
- Short response time for a given piece of work
- High throughput (rate of processing work)

Jupyter notebooks[2] are suitable for various HPC usage patterns and workflows.

Interactive parallel computing with IPython

ipyparallel[3] is a Python package and collection of CLI scripts for controlling clusters for Jupyter. Serial code can be parallelized with only a few extra lines of code. The parallel code can be run without leaving the Jupyter's shell. Any data computed in parallel can be explored interactively through visualization or further numerical calculations.

ipyparallel contains a cluster engine and scheduler using the following CLI scripts:

```
$ ipcluster # start/stop a cluster
$ ipcontroller # start a scheduler
$ ipengine # start an engine
```

To install for all users on JupyterHub, as root:

```
$ jupyter nbextension install --sys-prefix --py ipyparallel
$ jupyter nbextension enable --sys-prefix --py ipyparallel
$ jupyter serverextension enable --sys-prefix --py ipyparallel
```

How to use it from Python

```
# Controlling clusters with ipyparallel
# author: IPython Development Team
# license: 3-Clause BSD        # code: ipyparallel.readthedocs.io
# activity: stable (2017)      # index: 7-1
```

https://doi.org/10.1515/9783110629453-007

```
import ipyparallel as ipp

rc = ipp.Client()
ar = rc[:].apply_async(os.getpid)
pid_map = ar.get_dict()
```

Example

Use case: Calculate frequencies of all two digit sequences (00–99) in PI.

1 – classic, without ipyparallel:

```
# Idea: calculate frequencies of all 1 digits sequences (0-9)
# using functions defined in the pidigits.py
import sympy
pi = sympy.pi.evalf(10000)
digits = (d for d in str(pi)[2:]) # from pidigits: create a sequence

freqs = one_digit_freqs(digits) # from pidigits: how many times each digit
occurs
plot_one_digit_freqs(freqs) # uses Matplotlib to plot the result
```

2 – with ipyparallel:

```
# Use ipcluster to start 15 engines
import ipyparallel as ipp
c = ipp.Client(profile='mycluster')
machines = c[:]
c.ids
> 15 machines:
> [0, 1, 2, 3, 4, 5, 6, 7, 8, 9, 10, 11, 12, 13, 14]

# The %px magic executes a single Python command
# on the engines specified by the targets attribute
```

```
%px print('test')
> Parallel execution on engine(s): all
> [stdout:0] hi
...
> [stdout:14] hi

# trick: precalculated 150 million digits of pi in ASCII file
filestring = 'pi200m.ascii.%(i)02dof20'
files = [filestring % {'i':i} for i in range(1,16)]

# some helper functions contained in parallelpi.py
machines.map(fetch_pi_file, files)
# final step the counts from each engine will be added up
freqs_all = machines.map(compute_two_digit_freqs, files)

freqs = reduce_freqs(freqs_all)
plot_two_digit_freqs(freqs)
```

HPC usage with SLURM

The *Simple Linux Utility for Resource Management* (SLURM) is a free and open-source job scheduler for Linux and Unix-like kernels, used by many of the world's super-computers.[4] To interact with SLURM through a notebook, a special Python package developed at NERSC[5] can be used, which implements SLURM commands as magic commands. SLURM can be combined with ipyparallel (see earlier).[6]

```
%sacct. Display accounting data for all jobs and job steps in the Slurm job
accounting log or Slurm database. Is modal.
%sacctmgr: View and modify Slurm account information. Is modal.
%salloc: Obtain a Slurm job allocation (a set of nodes), execute a command,
and then release the allocation when the command is finished.
%sattach (TBC): Attach to a Slurm job step.
%%sbatch: Submit a batch script to Slurm.
%srun: Run parallel jobs.
```

```
# PDC Center for High-Performance Computing[7]

# !pip install git+https://github.com/NERSC/slurm-magic.git
# In the notebook, we then need to load the IPython extension:
%load_ext slurm_magic
```

```
# use the %%sbatch cell magic to submit a GROMACS job
%%sbatch
#!/bin/bash -l
#SBATCH -A 20XX-YY-ZZ
#SBATCH -N 1
#SBATCH -t 00:05:00
#SBATCH -J gromacs
module load gromacs/2018.3
gmx_seq grompp -f npt.mdp -c start.gro -p topol.top
gmx_seq mdrun -s topol.tpr -deffnm npt

# Monitor your job with the %squeue line magic
%squeue -u username
```

Other batch queue packages

BatchSpawner[8] and WrapSpawner[9] are JupyterHub Spawners that let HPC users run notebooks on compute nodes supporting a variety of batch queue systems.

Architecture aspects

CUDA[10] (Compute Unified Device Architecture) is a parallel computing platform and application programming interface (API) model created by Nvidia. It allows software developers and software engineers to use a CUDA-enabled graphics processing unit (GPU) for general-purpose processing.

```
# Using CUDA in Python
# author: Gressling, T
# license: MIT License      # code: github.com/gressling/examples
# activity: single example  # index: 7-2

# !conda install cudatoolkit
from numba import cuda

len(cuda.gpus)
> 1
cuda.gpus[0].name
>b'GeForce GTX 980M'

@cuda.jit
def MD_trajectory_calculation():
    ...
```

References

1. Computer performance (Wikipedia) https://en.wikipedia.org/wiki/Computer_performance.
2. Using IPython for parallel computing https://ipyparallel.readthedocs.io/en/latest/.
3. ipyparallel (GitHub) https://github.com/ipython/ipyparallel.
4. Using Jupyter Notebooks to manage SLURM jobs https://www.kth.se/blogs/pdc/2019/01/using-jupyter-notebooks-to-manage-slurm-jobs/.
5. slurm-magic (GitHub) https://github.com/NERSC/slurm-magic.
6. Slurm Workload Manager (Wikipedia) https://en.wikipedia.org/wiki/Slurm_Workload_Manager.
7. PDC-support (GitHub) https://github.com/PDC-support/jupyter-notebook/blob/master/2-slurm-analysis.ipynb.
8. batchspawner (GitHub) https://github.com/jupyterhub/batchspawner.
9. wrapspawner (GitHub) https://github.com/jupyterhub/wrapspawner.
10. CUDA (Wikipedia) https://en.wikipedia.org/wiki/CUDA.

8 REST and MQTT

Representational State Transfer (REST) is a software architectural style that defines a set of constraints to be used for creating Web services. Web services that conform to the REST architectural style, called RESTful Web services, provide interoperability between computer systems on the Internet.[1]

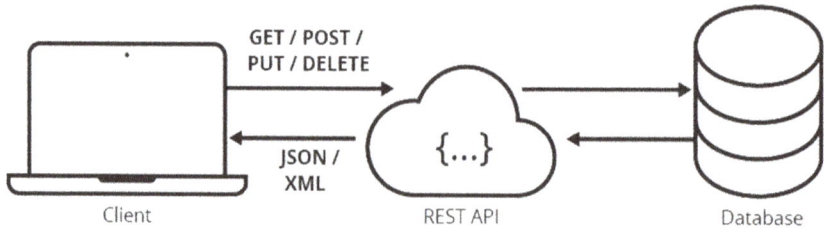

Figure 8.1: REST API[2] (Seobility).

Turn any Jupyter notebook into a REST API

You can use Jupyter Nnotebooks as REST backend:t

```
# Make Jupyter a REST Server
# author: Gressling, T
# license: MIT License        # code: github.com/gressling/examples
# activity: Single example    # index: 8-1
```

1. Setup

```
$ pip install jupyter_kernel_gateway
$ jupyter kernelgateway --generate-config
```

2. The notebook

```
import math
import json
if 'angle' not in args:
  print(json.dumps({'convertedAngle': None}))
else:
  angle = int(args['angle'][0])
  converted = math.radians(angle)
  print(json.dumps({'convertedAngle': converted}))
```

https://doi.org/10.1515/9783110629453-008

3. Run as REST service

```
$ jupyter kernelgateway
--KernelGatewayApp.api='kernel_gateway.notebook_http' --
KernelGatewayApp.seed_uri='/home/gressling/Notebook-08-1.ipynb'
> [KernelGatewayApp] Kernel started: 72515fd8-6314-4d65-9f48-8214756850d7
> [KernelGatewayApp] Registering resource: /convert, methods: (['GET'])
> [KernelGatewayApp] Registering resource: /_api/spec/swagger.json, methods: (GET)
> [KernelGatewayApp] Jupyter Kernel Gateway at http://*:8888
```

Call REST service

```
$ curl "http://serverIp:8888/convert?angle=180"
> {"convertedAngle": 3.141592653589793}
```

Jupyter notebook for retrieving from REST APIs

This section shows an example of data available from a REST API. After retrieving the data, it stores it in in a processable format.[3,4]

```
# Jupyter Notebook for retrieving from REST APIs
# author: Gressling, T
# license: MIT License        # code: github.com/gressling/examples
# activity: Single example    # index: 8-2

import requests

api_url = 'http://127.0.0.1:8081/predict'
data = {'name': 'reaction1', 'molecules': ['COCc1cc(C=O)ccc10', 'CN=C=O']}
r = requests.post(api_url + '/APIv1',
    headers={
            'Authorization': 'token %s' % <API KEY>,
        },
    json=data
)
r.raise_for_status()
r.json()
```

MQTT – message exchange in lab and production

MQTT[5] (Message Queuing Telemetry Transport) is an open ISO standard. It is a light-weight, publish-subscribe network protocol that transports messages between devices. Any network protocol that provides ordered, lossless, b-directional connections can support MQTT. It is designed for connections with remote locations where a "small code footprint" is required or the network bandwidth is limited.

Any device can publish or subscribe messages via one or more channels. In between, there is a broker. As MQTT is easy to implement. It has the potential to become a standard in connecting laboratory devices.sSee also Chapter 62.

You can use docker to set up a local broker $ `docker run -it -p 1883:1883 -p 9001:9001 eclipse-mosquitto` or use a public service.

```
# MQTT subscriber in Jupyter (paho)
# author: Gressling, T
# license: MIT License       # code: github.com/gressling/examples
# activity: Single example   # index: 8-3


import time, datetime
import paho.mqtt.client as mqttClient


Connected = False
# using public available service from hivemq
broker_address="broker.hivemq.com"
port = 1883
user = "roger" # default
password = "password" # default


def on_log(client, userdata, level, buf):
    print( str(datetime.datetime.now()) + ": ",buf)


def on_message(client, userdata, message):
    print ("Client received: "  + str(message.topic))
    print ("Message received: "  + str(message.payload))


def on_connect(client, userdata, flags, rc):
    if rc == 0:
        print("Connected to broker")
        global Connected
        Connected = True
    else:
        print("Connection failed")


client = mqttClient.Client("datascience")    #create new instance
# datascience will just subscribe in this example, but can also publish!
```

```
client.on_connect = on_connect          #attach function to callback
client.on_message = on_message          #attach function to callback
client.on_log = on_log                  #attach function to callback

client.username_pw_set(user, password=password)
client.connect(broker_address, port=port)
client.loop_start()

while Connected != True:
    time.sleep(0.1)

client.subscribe("someDeviceChannel/pH")
client.subscribe("someDeviceChannel/temp")
client.subscribe("otherchannel/onoff")

try:
    while True:
        time.sleep(1)
except KeyboardInterrupt:
    client.disconnect()
    client.loop_stop()
```

Capturing data with Arduino

Arduino[6] is an open-source hardware and software company, project, and user community that designs and manufactures single-board microcontrollers and microcontroller kits for building digital devices.

Board designs use a variety of microprocessors and controllers. The boards are equipped with sets of digital and analog input/output (I/O) pins that may be interfaced with various expansion boards 'shield') or breadboards (For prototyping) and other circuits.

For *prototyping* data collection in laboratory and pre-production, the Arduino platform is easy to use, see also Chapter 9. To develop and test, it is easy to use a mobile app such as MQTT Dash or similar applications.

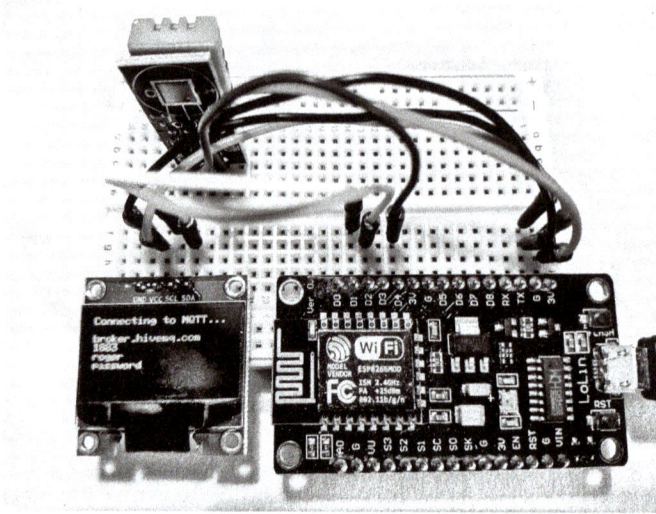

Figure 8.2: Logging data via MQTT with Arduino *(Gressling)*.

```
# MQTT publisher for Jupyter (paho)
# author: Gressling, T
# License: MIT License        # code: github.com/gressling/examples
# activity: Single example    # index: 8-4

#include <ESP8266WiFi.h>
#define WLAN_SSID        "*****"
#define WLAN_PASS        "*****"
WiFiClient client;

#include "Adafruit_MQTT.h"
#include "Adafruit_MQTT_Client.h"
#define mqttSERVER       "broker.hivemq.com"
#define mqttSERVERPORT   1883
#define mqttUSERNAME     "roger"
#define mqttKEY          "password"
#define mqttID           "someDeviceChannel"
Adafruit_MQTT_Client mqtt(&client, mqttSERVER, mqttSERVERPORT, mqttUSERNAME,
mqttKEY);
Adafruit_MQTT_Publish pH = Adafruit_MQTT_Publish(&mqtt, mqttID "/pH");
Adafruit_MQTT_Publish temp = Adafruit_MQTT_Publish(&mqtt, mqttID "/temp");

#include <stdio.h>
#define phSENSOR         12  // D0
#define tempSENSOR       14  // D1
```

```
void MQTT_connect();

void setup() {
    WiFi.begin(WLAN_SSID, WLAN_PASS);
    while (WiFi.status() != WL_CONNECTED) {
        delay(500);
    }
}

void MQTT_connect() {
    int8_t ret;
    if (mqtt.connected()) {
        return;
    }
    while ((ret = mqtt.connect()) != 0) { // return 0 for connected
        mqtt.disconnect();
        delay(5000);
    }
}

void loop() {
    MQTT_connect();
    pH.publish(digitalRead(phSENSOR));      // publish ph Value
    temp.publish(digitalRead(tempSENSOR)); // publish Temperature
    delay(10000);
}
```

References

1. Representational state transfer (Wikipedia) https://en.wikipedia.org/wiki/Representational_ state_transfer.
2. What is REST API? – Seobility Wiki (License: CC BY-SA 4.0) https://www.seobility.net/en/wiki/ REST_API.
3. Jellema, L. Jupyter Notebook for retrieving JSON data from REST APIs https://technology.amis.nl/ 2019/04/29/jupyter-notebook-for-retrieving-json-data-from-rest-apis/.
4. Jellema, L. DataAnalytics--IntroductionDataWrangling-JupyterNotebooks (GitHub) https://github.com/ lucasjellema/DataAnalytics--IntroductionDataWrangling-JupyterNotebooks.
5. MQTT (Wikipedia) https://en.wikipedia.org/wiki/MQTT.
6. Arduino (Wikipedia) https://en.wikipedia.org/wiki/Arduino.

9 Edge devices and IoT

Edge computing[1] is a distributed computing paradigm which brings computation and data storage closer to the location where it is needed, to improve response times and save bandwidth. A definition of edge computing is any type of computer program that delivers low latency nearer to the requests.

The Internet of things, or IoT, is a system of interrelated computing devices, mechanical and digital machines, objects, animals, or people that are provided with unique identifiers (UIDs) and the ability to transfer data over a network, without requiring human-to-human or human-to-computer interaction.

The definition of the Internet of things has evolved due to the convergence of multiple technologies, real-time analytics, machine learning, commodity sensors, and embedded systems.[2]

Arduino code in a Jupyter notebook cell

Integrating Jupyter notebooks and Arduino requires separate environments. Jupyter has dozens of language kernels, none of which support Arduino sketches. To bypass writing a new kernel, one idea is to use magic switches (jam – Jupyter notebook and Arduino Mash-Up using cell magics[3]). The Arduino IDE must be installed on your system. The switches are defined in the jam.py library.

```
%%jamcell # Saves the cell to a .ino Arduino file, compiles and
loads it.
```

```
# Arduino in a Jupyter Notebook Cell
# author: ylabrj
# License: GNU GPL 3        # code: github.com/ylabrj/jam
# activity: active (2020)   # index: 9-1


%%jamcell mysketch

int x;
void setup(){
    Serial.begin(9600);
    x = 1;
}
```

https://doi.org/10.1515/9783110629453-009

```
void loop(){
    Serial.println(x);
    x++;
}

%jam --serialports
> Arduino ports on system:
> ['Arduino Uno (COM6)']
```

Plot serial data read from Arduino

This example reads and plots temperature and humidity measured by Arduino Mega and DHT11 over the serial port in a Jupyter notebook.

```
# Plot serial data read from Arduino
# author: Gressling, T
# license: MIT License      # code: github.com/gressling/examples
# activity: single example  # index: 9-2
```

Arduino code

```
#include "DHT.h"
#define DHTPIN 2
#define DHTTYPE DHT11
DHT dht(DHTPIN, DHTTYPE);

void setup() {
  Serial.begin(9600);
  dht.begin();
}

void loop() {
  delay(2000);
  Serial.print(dht.readTemperature());
  Serial.print(" , ");
  Serial.println(dht.readHumidity());
}
```

Python

```python
import serial
import numpy
import matplotlib.pyplot as plt
from drawnow import *
%matplotlib auto

tempF=[]
humidity=[]

arduinoData=serial.Serial('COM5',9600)
plt.ion()

def makeFig():
    plt.plot(tempF,'ro-')
    plt.show

while True:
    while arduinoData.inWaiting()==0:
        pass
    arduinoString=arduinoData.readline()
    # readline result are bytes that have to be byte splitted
    dataArray=arduinoString.split(b',')
    temp=float(dataArray[1])
    H=float(dataArray[0])
    tempF.append(temp)
    humidity.append(H)
    drawnow(makeFig)
    plt.pause(.000001)
```

Jupyter notebooks on Raspberry Pi

```
$ sudo pip3 install jupyter
$ sudo ipython3 kernelspec install-self
$ sudo apt-get install rpi.gpio
$ jupyter notebook
```

Raspberry-read sensor data act on actuators with the Jupyter notebook[4] and also control Raspberry Pi GPIO's.[5]

Jupyter notebook

```
# Raspberry GPIO access to log laboratory data
# author: Gressling, T
# license: MIT License       # code: github.com/gressling/examples
# activity: single example  # index: 9-3

# DS18B20 1-Wire library (GPIO 4 is used)
from w1thermsensor import W1ThermSensor
ds18b20Sensor = W1ThermSensor()
import Adafruit_DHT
DHT22Sensor = Adafruit_DHT.DHT22
import Adafruit_BMP.BMP085 as BMP085
sensor = BMP085.BMP085()

import RPi.GPIO as GPIO
GPIO.setmode(GPIO.BCM)
GPIO.setwarnings(False)

temp = sensor.read_temperature()
pres = sensor.read_pressure()
alt =  sensor.read_altitude()
```

AI on edge devices/run the same ML model on multiple hardware platforms

Implementing machine learning on devices like Raspberry or Arduino makes no sense, as the performance is too bad even for the simplest use cases. For other devices like mobile phones (ARM) or NVIDIA GPU, there are frameworks optimizing, for example, TensorFlow models for the targets. For example, Amazon SageMaker Neo[6] allows training a model once and running it virtually anywhere with a single executable. Neo understands how to optimize the model for Intel, NVIDIA, ARM, and others.

Special AI chips

A neural processor or a neural processing unit (NPU) is a specialized circuit that implements all the necessary control and arithmetic logic necessary to execute machine learning algorithms, typically by operating on predictive models such as artificial neural networks (ANNs) or random forests (RFs). NPUs sometimes go by similar names such as tensor processing unit (TPU), neural network processor (NNP), and intelligence processing unit (IPU) as well as vision processing unit (VPU) and graph processing unit (GPU).[7]

Examples for this special hardware (commonly available) are the NVIDIA Jetson[8] series (nano, TX, or Xavier).

References

1. Edge computing (Wikipedia) https://en.wikipedia.org/wiki/Edge_computing.
2. Internet of things (Wikipedia) https://en.wikipedia.org/wiki/Internet_of_things.
3. jam – Arduino on Python (GitHub) https://github.com/ylabrj/jam.
4. Rovai, M. Physical Computing using Jupyter Notebook: https://towardsdatascience.com/physical-computing-using-jupyter-notebook-fb9e83e16760.
5. Rovai, M. Python4DS https://github.com/Mjrovai/Python4DS.
6. Amazon SageMaker Neo https://aws.amazon.com/sagemaker/neo/.
7. Neural Processor (WikiChip) https://en.wikichip.org/wiki/neural_processor.
8. Nvidia Jetson (Wikipedia) https://en.wikipedia.org/wiki/Nvidia_Jetson.

Programming

10 Python and other programming languages

Python is[1] an interpreted, high-level, general-purpose programming language. Created by Guido van Rossum and first released in 1991, Python's design philosophy emphasizes code readability with its notable use of significant whitespace. Its language constructs and object-oriented approach aim to help programmers write clear, logical code for small and large-scale projects.

Python is the de-facto standard language for scripting in cheminformatics. (Noel M O'Boyle)

Python helps the programmers to code in fewer steps as compared to Java or C++. There are advantages2 compared to other languages:

- Extensive Support Libraries
- Integration of COM, CORBA, XML, or languages
- Clean design

Limitations:

- Returning to other languages is hard
- Slow, due to interpretation (compensated by native C libraries like in TensorFlow)
- dynamically typed: design restrictions
- more testing time

The Zen of Python

Code conventions and coding style are fundamental for python as the language is designed for *readability*. The style guide is listed with the command `import this`:

```
import this

The Zen of Python³, by Tim Peters

Beautiful is better than ugly.
Explicit is better than implicit.
Simple is better than complex.
Complex is better than complicated.
Flat is better than nested.
Sparse is better than dense.
Readability counts.
Special cases aren't special enough to break the rules.
```

https://doi.org/10.1515/9783110629453-010

```
Although practicality beats purity.
Errors should never pass silently.
Unless explicitly silenced.
In the face of ambiguity, refuse the temptation to guess.
There should be one-- and preferably only one --obvious way to do it.
Although that way may not be obvious at first unless you're Dutch.
Now is better than never.
Although never is often better than *right* now.
If the implementation is hard to explain, it's a bad idea.
If the implementation is easy to explain, it may be a good idea.
Namespaces are one honking great idea -- let's do more of those!
```

The clear UI of Python can be seen when compared with the implementations of 45 languages for the Levenshtein algorithm.[4]

Levels of interaction, code types

Using Python in Chemistry means to access different types of integration with existing software. The following list gives the degree of implementation maturity, where level 5 is a pure python implementation with clean API design:

1. Low-level wrapper (Command line, Files) (GROMACS, NAMD, AMBER)
2. Library wrapper
3. Interoperability (e.g., C++ with Boost.Python) (RDKit)
4. Native implementation, no further API (called "single examples" in this book)
5. Native libraries and packages (pip)

Boost.Python[5] enables seamless interoperability between C++ and the Python programming language.

The ART of package design

The taxonomy of package management in programming languages[6] has always been under discussion. Package managers are programs that map relations between files and packages and between packages (dependencies), allowing the programmer to perform maintenance of systems.

Package management is instrumental for programming languages and operating systems, and yet it is neglected by both areas as an implementation detail. For this reason, it lacks the same kind of conceptual organization: we lack the terminology to classify them or to reason about their design trade-offs.

C++ integration example in Python

Using Xeus-Cling kernel[7,8] allows changing notebook cells to C++ language. Xeus: a C++ implementation of the Jupyter kernel protocol.

Polymorphism example

```
# C++ integration example in Jupyter notebook
# author: Jupyter Xeus team
# license: BSD 3-Clause License # code: github.com/...xeus-cling/xcpp.ipynb
# activity: active (2020)          # index: 10-1


// Class
class Foo
{ public:
   virtual ~Foo() {}
   virtual void print(double value) const
   { std::cout << "Foo value = " << value << std::endl; }
};
Foo bar; bar.print(1.2);


// Polymorphism of Foo
class Bar : public Foo
{ public:
   virtual ~Bar() {}
   virtual void print(double value) const
   { std::cout << "Bar value = " << 2 * value << std::endl; }
};


Foo* bar2 = new Bar; bar2->print(1.2);
delete bar2;
```

Python in R

The `reticulate package` lets you use Python and R together, seamlessly, within R code, in R Markdown documents, and in the RStudio™ IDE.

References

1. Python (programming language) (Wikipedia) https://en.wikipedia.org/wiki/Python_ (programming_language).
2. Mindfire Solutions. Advantages and Disadvantages of Python Programming Language https:// medium.com/@mindfiresolutions.usa/advantages-and-disadvantages-of-python-programming-language-fd0b394f2121.
3. Sweigart, A. The Zen of Python, Explained – The Invent with Python Blog https:// inventwithpython.com/blog/2018/08/17/the-zen-of-python-explained/.
4. 45 Levenshtein distance implementations (Wikibook) https://en.wikibooks.org/wiki/Algorithm_ Implementation/Strings/Levenshtein_distance.
5. Boost.Python https://www.boost.org/doc/libs/1_66_0/libs/python/doc/html/index.html.
6. Muhammad, H.; Real, L. C. V.; Homer, M. Taxonomy of Package Management in Programming Languages and Operating Systems. In *Proceedings of the 10th Workshop on Programming Languages and Operating Systems*; PLOS'19; Association for Computing Machinery: New York, NY, USA, 2019; pp 60–66. https://doi.org/10.1145/3365137.3365402.
7. xeus-cling (GitHub) https://github.com/QuantStack/xeus-cling.
8. Xeus-Cling: Run C++ code in Jupyter Notebook https://www.learnopencv.com/xeus-cling-run-c-code-in-jupyter-notebook/.

11 Python standard libraries and Conda

Python libraries and python packages play a vital role. Python packages are a set of python modules, while python libraries are a group of python functions aimed to carry out special tasks. The libraries are for very different things. Scikit-learn, for example, addresses more machine learning, whereas NumPy is used much more generally for scientific computing in Python. The libraries are at different levels of abstraction. Pandas is higher level than NumPy and, in fact, encapsulates its numeric features.

Conda, Anaconda, and Pydata ecosystem

Conda is a package manager and is both a command line tool and a python package. *Anaconda* is a set of packages.

Miniconda, is a smaller alternative to Anaconda that is just conda and its dependencies:

Miniconda installer = Python + conda
Anaconda installer = Python + conda + meta-package anaconda
meta-package anaconda = about 160 other Python packages for daily use in data science
Anaconda installer = Miniconda installer + conda install anaconda

```
# Working with Anaconda and environments

$ curl -O https://repo.anaconda.com/archive/Anaconda3-2019.03-Linux-x86_64.sh
  % Total   % Received % Xferd  Average Speed   Time    Time     Time  Current
                                 Dload  Upload   Total   Spent    Left  Speed
100 654M 100 654M    0       0  75.3M      0  0:00:08 0:00:08 --:--:-- 76.8M
$ sha256sum Anaconda3-2019.03-Linux-x86_64.sh
45c851b7497cc14d5ca060064394569f724b67d9b5f98a926ed49b834a6bb73a
Anaconda3-2019.03-Linux-x86_64.sh
$ bash Anaconda3-2019.03-Linux-x86_64.sh # or '-u' in case of reinstall


# How to install RDKit with Conda
$ conda create -c rdkit -n book
$ conda activate book
# Upgrade:
$ conda update --prefix /home/book/anaconda3 anaconda
$ conda update -c conda-forge jupyterlab
$ conda update --all
```

https://doi.org/10.1515/9783110629453-011

```
# Rename (with a trick)
$ source deactivate
$ conda create --name new_name --clone old_name
$ conda remove --name old_name --all
# it redownloads packages - (use --offline flag to disable it)
# time consumed and temporary double disk usage.
```

Anaconda

After installation of Anaconda, it includes 100+ of the most popular Python packages:[1]
- NumPy – support for large, multidimensional arrays and matrices
- NLTK – symbolic and statistical natural language processing (NLP)
- Numba – translates Python into fast C machine code with @jit
- Blaze – query data on different storage systems
- Matplotlib – popular plotting package
- h5py and PyTables can both access data stored in the HDF5 format, see Chapter 18

Here, as an excerpt, is the selected list[2] of "core" Python tools for data science:
- pandas – Data structures built on top of NumPy
- pandas_summary – Basic statistics using DataFrameSummary(df).summary()
- pandas_profiling – Descriptive statistics using ProfileReport
- sklearn_pandas – Helpful DataFrameMapper class

- seaborn – Python data visualization library based on matplotlib
- janitor – Clean messy column names
- missingno – Missing data visualization

Productivity and high-performance computing

- The Jupyter notebook provides IPython functionality and more in your web browser, allowing you to document your computation in an easily reproducible form.
- Cython extends Python syntax so that you can conveniently build C extensions, either to speed up critical code or to integrate with C/C++ libraries.
- Dask, Joblib, or IPyParallel for distributed processing with a focus on numeric data.

Quality assurance

- pytest, a framework for testing Python code
- numpydoc, a standard and library for documenting Scientific Python libraries

scikit-learn

The library features various classification, regression, and clustering algorithms including support vector machines and random forests. It is designed to interoperate with the Python numerical and scientific libraries, NumPy and SciPy.

scikit-image

This is a collection of algorithms for image processing and includes algorithms for segmentation, geometric transformations, color space manipulation, analysis, filtering, morphology, and feature detection.

SciPy

SciPy[3] (pronounced "Sigh Pie") is a Python-based ecosystem of open-source software for mathematics, science, and engineering. The SciPy ecosystem includes tools for data management and computation as well as productive experimentation and high-performance computing.

gensim

Gensim[4] is a FREE Python library with Scalable statistical semantics. It can analyze plain-text documents for semantic structure and retrieve semantically similar documents. Usage example is mol2vec (see Chapter 83.)
- Word2Vec model
- Doc2Vec model
- FastText model
- Similarity queries with annoy and Word2Vec
- LDA model

- Distance metrics
- Word movers' distance
- Text summarization
- Pivoted document length normalization

Inspection and "?"

The inspect[5] module provides several functions to help get information about live objects:
- modules
- classes
- methods
- functions

`inspect` can help examine the contents of a class and retrieve the source code of a method, extract, and format the argument list for a function including more complex tasks like listing the traceback.

? Example

```
mol?

> Type:        Mol
> String form: <rdkit.Chem.rdchem.Mol object at 0x000000000F70F490>
> File:        c:\users\book\.conda\envs\ahdira\lib\site-packages\rdkit\
                chem\rdchem.pyd

>
> Docstring:
> The Molecule class.
> In addition to the expected Atoms and Bonds, molecules contain:
>    - a collection of Atom and Bond bookmarks indexed with integers
>        that can be used to flag and retrieve particular Atoms or Bonds
>        using the {get|set}{Atom|Bond}Bookmark() methods.
```

inspect example

```
inspect.getmembers(mol)

> [('AddConformer', <bound method AddConformer of <rdkit.Chem.rdchem.Mol
object at 0x000000000F70F490>>),
> ('ClearComputedProps',
 <bound method ClearComputedProps of <rdkit.Chem.rdchem.Mol object at
0x000000000F70F490>>),
> ('ClearProp',
 <bound method ClearProp of <rdkit.Chem.rdchem.Mol object at
 0x000000000F70F490>>)
```

help example

```
help(mol.GetBondWithIdx(2))

> Help on Bond in module rdkit.Chem.rdchem object:
>
> class Bond(Boost.Python.instance)
>  |   The class to store Bonds.
>  |   Note: unlike Atoms, is it currently impossible to construct Bonds from
>  |   Python.
>  |
>  |   Method resolution order:
>  |       Bond
>  |       Boost.Python.instance
>  |       builtins.object
>  |
>  |   Static methods defined here:
>  |
>  |   ClearProp(...)
```

Dataframes

A dataframe is a multidimensional labeled data structure with columns of potentially different types.

pandas

Pandas is used for data manipulation and analysis. It offers data structures and operations for manipulating numerical tables and time series. Pandas allow importing data of various file formats such as CSV and Excel. Pandas allows various data manipulation operations.

Modin dataframe for 1TB+

pandas DataFrame is a lightweight DataFrame. Modin[6] transparently distributes the data and computation among processors. It uses DASK or ray as a backend.

```
# Using modin instead of pandas
# author: Gressling, T
# license: MIT License      # code: github.com/gressling/examples
# activity: single example # index: 11-1
!pip install modin[ray]     # or dask

import ray
ray.init(num_cpus=4) # limit cpus
import pandas as pd
pdf = pd.read_csv("molecules.csv") # classic pandas

import modin.pandas as pd
mdf = pd.read_csv("molecules.csv") # distributes among the 4 cpus
```

Excel comparison

With 1.2 billion Microsoft Office users[7], the Excel spreadsheet is the leading calculation tool for scientists too, so it is important to explain *in easy words*, the advantages of Jupyter notebooks.

Pros and cons[9]: Excel is suitable for light analytics when the task is straightforward and data size is small to moderate. Python and Pandas are much better for data cleaning and wrangling, higher level data analysis techniques, and a high amount of data, which is data science.

- Excel[10] starts at row 1, while Pandas starts at row ("index") 0
- Excel labels the columns with letters starting at A, while Pandas labels the columns with variable names

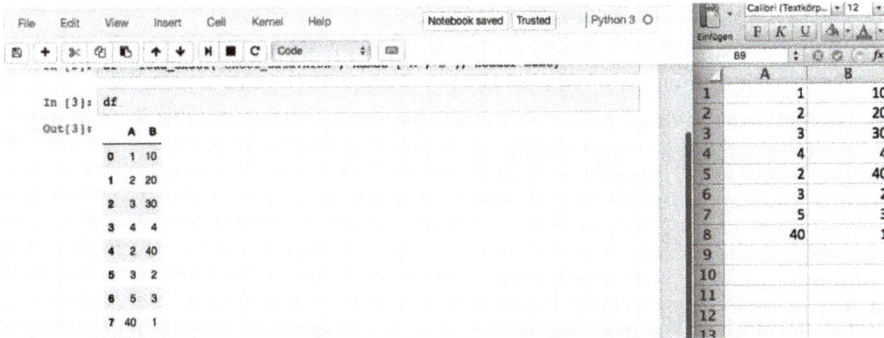

Figure 11.1: Pandas versus Excel layout[8] (Hamann).

Data operation comparison between Excel and dataframes

The way of thinking is different, so the following code compares the two approaches:

Filter Excel

```
cell[11] =IF( [@pH]>7, "Yes", "No" )
```

Filter Pandas

```
list_lab['alkaline'] = ['Yes' if x>7 else 'No' for x in list_lab['pH']]
```

Merge in Excel

VLOOKUP is an Excel function to look up and retrieve data from a specific column in a table:

```
cell: =VLOOKUP (value, table, col_index, [range_lookup])
```

Merge in Pandas

```
list_lab = pd.merge(pH, sample, how='left', on='sample_number')
```

– The first argument is the original dataframe,
– The second argument is the dataframe we are looking up values in,
– How specifies the type of join we want to make.
– On specifies the variable that we want to merge on (there's also left_on and right_on if the variables are called different things in each dataframe).

dask.dataframe

A Dask[12] DataFrame is a large parallel DataFrame composed of many smaller Pandas DataFrames, split along the index. These Pandas DataFrames may live on disk for larger-than-memory computing on a single machine or on many different machines in a cluster. The Dask dataFrame copies the Pandas API.

pyspark dataframe

In Apache Spark, a dataframe is a distributed *resilient* collection of rows under named columns. It is based on a special form of data storage (RDD). It is discussed in Chapter 19 (Jupyter and Spark).

cuDF/rapid.ai dataframe

cuDF[13] is a Python-based GPU dataframe library for working with data. It includes loading, joining, aggregating, and filtering data. It relies on NVIDIA® CUDA® primitives for low-level compute optimization and Apache Arrow, which is a cross-language development platform for in-memory data. The move to GPU allows for massive acceleration due to the many more cores GPUs have over CPUs. cuDF's API is a mirror of Pandas and, in most cases, can be used as a direct replacement.

```
# cuDF / rapid.ai dataframe
# author: Gressling, T
# license: MIT License      # code: github.com/gressling/examples
# activity: single example  # index: 11-2

import pandas as molCoord
import numpy as np
import cudf

pandas_df = molCoord.DataFrame(
        {'x': np.random.randint(0, 100000, size=100000),
         'y': np.random.randint(0, 100000, size=100000),
         'z': np.random.randint(0, 100000, size=100000)})

cudf_df = cudf.DataFrame.from_pandas(pandas_df)

# Timing Pandas
%timeit pandas_df.a.mean()
# Timing cuDF
%timeit cudf_df.a.mean() # 16x times faster with 1080 Ti GPU, CUDA 10
```

Lambda function with filter and map

Lambda functions in Python are defined inline and are limited to a single expression. It is a very helpful construct for data wrangling in the data science process.

```
# Lambda, filter and map example
# author: Gressling, T
# license: MIT License      # code: github.com/gressling/examples
# activity: single example  # index: 11-2

x = ['Python', 'programming', 'for', 'data science!']

print(sorted(x))
> ['Python', 'data science!', 'for', 'programming']

print(sorted(x, key=lambda arg: arg.lower()))
> ['data science!', 'for', 'programming', 'Python']

# filter
print(list(filter(lambda arg: len(arg) < 8, x)))
> ['Python', 'for']
```

map() is similar to filter() in that it applies a function to each item in an iterable, but it always produces a 1-to-1 mapping of the original items. The new iterable that map() returns will always have the same number of elements as the original iterable, which was not the case with filter():

```
print(list(map(lambda arg: arg.upper(), x)))
> ['PYTHON', 'PROGRAMMING', 'FOR', 'DATA SCIENCE!']
```

References

1. Packages with Python 3.6 / Anaconda documentation https://docs.anaconda.com/anaconda/packages/py3.6_win-64/.
2. Rohrer, F. Awesome Data Science with Python (GitHub): https://github.com/r0f1/datascience.
3. SciPy.org https://scipy.org/.
4. gensim: topic modelling for humans https://radimrehurek.com/gensim/apiref.html.
5. Inspect live objects (Python documentation) https://docs.python.org/2/library/inspect.html.
6. Seif, G. How to Speed up Pandas by 4x with one line of code (KDnuggets): https://www.kdnuggets.com/how-to-speed-up-pandas-by-4x-with-one-line-of-code.html.
7. Callaham, J. 1.2 billion Office users worldwide: https://www.windowscentral.com/there-are-now-12-billion-office-users-60-million-office-365-commercial-customers.
8. Hamann, D. data-tools Excel snippets (GitHub, MIT License): https://github.com/davidhamann/data-tools.
9. What are the pros and cons of using Python pandas library vs Excel VBA for data analysis? (Quora) https://www.quora.com/What-are-the-pros-and-cons-of-using-Python-pandas-library-vs-Excel-VBA-for-data-analysis.
10. Gandhi, A. Replacing Excel with Python (Medium): https://towardsdatascience.com/replacing-excel-with-python-30aa060d35e.
11. Piepenbreier, N. Learn How to do 3 Advanced Excel Tasks in Python) (Medium): https://towardsdatascience.com/learn-how-to-easily-do-3-advanced-excel-tasks-in-python-925a6b7dd081.
12. DataFrame – Dask documentation https://docs.dask.org/en/latest/dataframe.html.
13. cuDF / RAPIDS https://rapids.ai/.

12 IDE's and workflows

An integrated development environment (IDE) is a software application that provides comprehensive facilities to computer programmers for software development. An IDE normally consists of at least a source *code editor*, build *automation tools*, and a *debugger*.

The boundary between an IDE and other parts of the broader software development environment is not well-defined; sometimes, a version control system or various tools to simplify the construction of a graphical user interface (GUI) are integrated. Many modern IDEs also have a class browser, an object browser, and a class hierarchy diagram for use in object-oriented software development.[1]

Knime, Weka, and Rapidminer: data science software packages

Easy-to-use general-purpose software packages are critical to lowering the *barrier for entry* for domain scientists.[2] In comparison to raw Jupyter notebooks, here KNIME and JupyterLab are discussed.

KNIME example

The Konstanz Information Miner is a free and open-source data analytics, reporting, and integration platform. KNIME integrates various components for machine learning and data mining through its modular data pipelining concept.[3]

Through its modular data pipelining concept and with a GUI, as shown in figure 12.1, the IDE allows assembly of *nodes*, blending different data sources, including preprocessing (ETL: Extraction, Transformation, Loading), for modeling, data analysis, with only minimal programming. NodePit for KNIME allows searching and exploring nodes and workflows. Compared to Jupyter notebooks, these nodes can be combined in visually.

NodePit[5] KNIME node collection supporting versioning and node installation lists ~2,500 nodes. An example of the node description dialog is shown in figure 12.2.

https://doi.org/10.1515/9783110629453-012

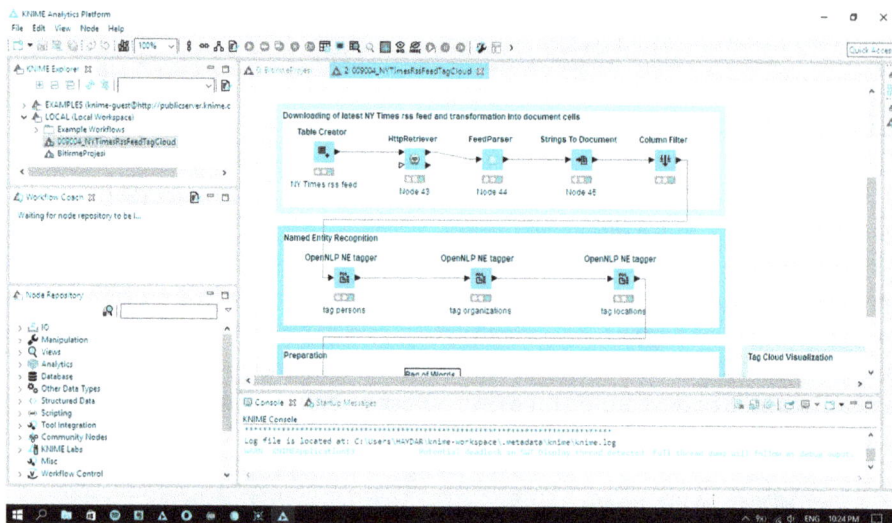

Figures 12.2: NodePit Screenshot *(Gressling)*.

Rich GUI development platforms versus programming in Jupyter notebooks

PRO IDE's:[6,7]

- **Code-free** creation of advanced analytics workflows
- More *casual* **users** who cannot possibly stay on top of the complexity of working in a programming environment
- GUI **ease of use:** visual representation intuitively explains which steps have been performed
- **Easy to track the different steps** and to isolate and fix specific workflow issues
- **Professional node support**, for example 150 Schrödinger nodes

CON IDE's:[8]
- Writing code **still is the most versatile way** to quickly and flexibly create a new analysis
- **Agility:** To get full value out of data, it is needed to be able to quickly try out a new routine. For an IDE, it is necessary to wait for the creation or adoption of *nodes* by the IT department
- But still, Python is required to **fully use** statistical analysis
- **Degree of freedom** and customization in default nodes
- A user interface can be **crowded** at times
- The IDE **tools for scriptwriting** and development are **not easy to use** or do not have many features
- Writing code makes it easier for **others to review**
- **Staging** to a **production**/real-time environment is easier with code

JupyterLab

JupyterLab is an interactive development environment for working with notebooks, code, and data. It has full support for Jupyter notebooks and enables you to use text editors, terminals, data file viewers, and other custom components side by side with notebooks in a tabbed work area.[9] JupyterLab follows the Jupyter Community Guides.[10]
- The new terminal is a tab view to use compared
- The ability to set out multiple windows easily, much like an IDE
- This will make working on a remote server so much nicer; just start Jupyter Lab and an ssh tunnel, and you have a terminal plus notebooks
- Remote server file editors in one browser window

Domain integrated workflow approaches

Besides native IDEs discussed in the earlier section, there are attempts for the creation of domain-specific solutions. Diagram 12.3 illustrates a generic data science workflow. Examples of an integration approach are the initiatives TeachOpenCADD for drug design and QwikMD for molecular dynamics (MD).

Example: QwikMD

Performing MD with NAMD and VMD[12] is a nonpython task. QwikMD ("QwikMD – Integrative Molecular Dynamics Toolkit for Novices and Experts") allows users to set up a molecular simulation fast, enabling the study of point mutations, partial dele-

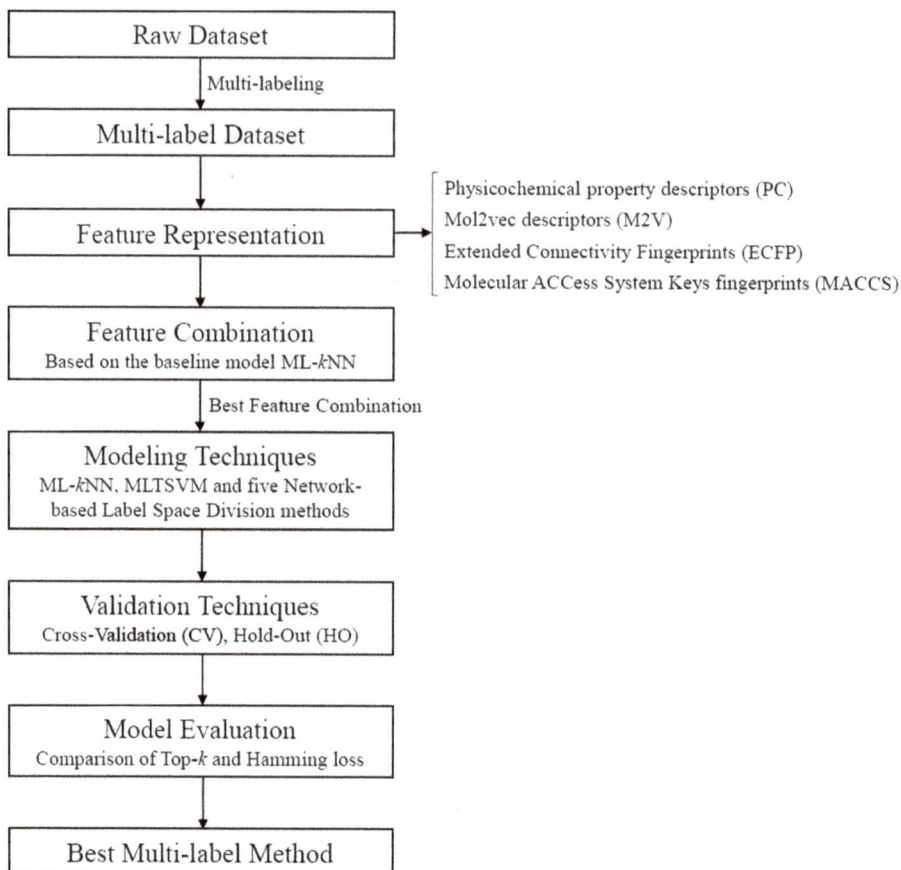

Figure 12.3: Flowchart of a multilabels modeling process *(Shan[11]).*

tions, or steering experiments. It provides a GUI for the standard MD workflow. NAMD is discussed in Chapter 53.

Example: TeachOpenCADD

A teaching platform for computer-aided drug design using open-source packages and data.[13,14] *The teaching material is offered in the following formats:*
- Coding-based Jupyter Notebooks on GitHub, so-called talktorials (talk + tutorial), that is, tutorials that can also be used in presentations
- GUI-based KNIME workflows on the KNIME Hub
- Open data resources are the ChEMBL and PDB databases for compound and protein structure data

- Open-source libraries utilized are RDKit, the ChEMBL web resource client and PyPDB, BioPandas (loading and manipulating molecular structures), and PyMOL for structural data visualization
- A conda yml file is provided to ensure an easy and quick setup of an environment containing all required packages.

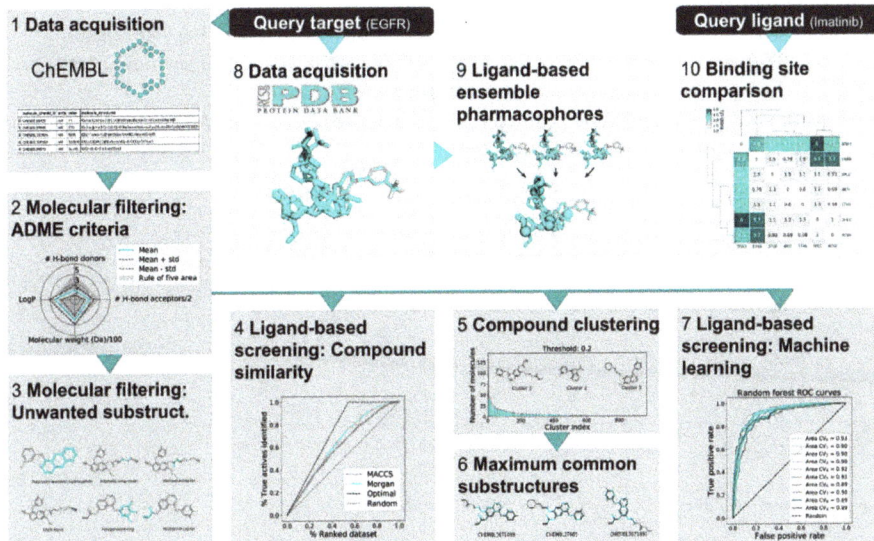

Figure 12.4: TeachOpenCADD workflow.[14]

References

1. Integrated development environment (Wikipedia) https://en.wikipedia.org/wiki/Integrated_development_environment.
2. Ferguson, A. What is data science? http://nas-sites.org/csr/files/2018/04/2.-Ferguson.pdf.
3. KNIME (Wikipedia) https://en.wikipedia.org/wiki/KNIME.
4. Görüntüsü, E. Knime screenshot (Wikimedia CC-BY-SA) https://commons.wikimedia.org/wiki/File:KNIMEEKRAN.png.
5. RDKit Salt Stripper (NodePit) https://nodepit.com/node/org.rdkit.knime.nodes.saltstripper.RDKitSaltStripperNodeFactory.
6. Berthold, M. To Code or Not to Code with KNIME (KDnuggets) https://www.kdnuggets.com/2015/07/knime-code-not-code.html.
7. What is the difference between Python and the data mining tools like Knime and Rapid Miner? – Quora https://www.quora.com/What-is-the-difference-between-Python-and-the-data-mining-tools-like-Knime-and-Rapid-Miner.
8. KNIME Analytics Platform Reviews & Ratings 2020 https://www.trustradius.com/products/knime-analytics-platform/reviews.

9. Wilhelm, F. Working efficiently with JupyterLab Notebooks https://florianwilhelm.info/2018/11/ working_efficiently_with_jupyter_lab/.
10. JupyterLab Documentation https://jupyterlab.readthedocs.io/en/stable/.
11. Shan, X.; Wang, X.; Li, C. -D.; Chu, Y.; Zhang, Y.; Xiong, Y.; Wei, D. -Q. Prediction of CYP450 Enzyme-Substrate Selectivity Based on the Network-Based Label Space Division Method http:// xbioinfo.sjtu.edu.cn/pdf/Shan-2019-Prediction%20of%20CYP450%20Enzyme-Substrat.pdf.
12. QwikMD – Integrative Molecular Dynamics Toolkit for Novices and Experts http://www.ks.uiuc. edu/Research/qwikmd/.
13. Sydow, D.; Morger, A.; Driller, M.; Volkamer, A. TeachOpenCADD: A Teaching Platform for Computer-Aided Drug Design Using Open Source Packages and Data. *J. Cheminform.* **2019**, *11* (1), 29. https://www.ncbi.nlm.nih.gov/pmc/articles/PMC6454689.
14. TeachOpenCADD (gitHub) https://github.com/volkamerlab/TeachOpenCADD.

13 Jupyter notebooks

A Jupyter notebook is an *interactive document* displayed in your browser, which contains source code, as well as rich text elements. This combination makes it extremely useful for explorative tasks where the source code, documentation, and even visualizations of your analysis are strongly intertwined. Due to this unique characteristic, Jupyter notebooks have achieved a strong adoption, particularly in the data science community.[1]

- *"Jupyter notebooks are a tool for exploration not for production"*[2]
- *"I've never seen any migration this fast. It's just amazing. IPython notebooks are really a killer app for teaching computing in science and engineering," (Lorena Barba)*
- *A "computational narrative – a document that allows researchers to supplement their code and data with analysis, hypotheses, and conjecture."*

  (Jupyter co-creator Brian Granger)

Why Jupyter notebooks?

Project Jupyter is a nonprofit organization created to "develop open-source software, open-standards, and services" for interactive computing across dozens of programming languages.[3]

- Analyzing workflows shows that the degree of spawning and merging is quite low, which means that most work within pipelines takes place in one main thread, supported by smaller branches that are merged.
- Providing data. Moreover, a huge GUI with a lot of empty or predefined values creates a confusing user experience.

Jupyter Notebooks are not an IDE – tests or code review should be moved out of a notebook into a .py file. The focus is still that Jupyter Notebooks are for exploration.[4]

Literate programming and the ART of computer programming

In 1983, Donald Knuth came up with a programming paradigm called literate programming. It is *"a methodology that combines a programming language with a documentation language, thereby making programs more robust, more portable, more easily maintained, and arguably more fun to write than programs that are written only in a high-level language. The main idea is to treat a program as a piece of literature, addressed to human beings rather than to a computer."*[5] This is introduced with Jupyter notebooks.

https://doi.org/10.1515/9783110629453-013

Perception

The format has exploded in popularity. Sharing site GitHub counted more than 4 million public Jupyter notebooks in September 2019; the prediction is 8M in 2021.

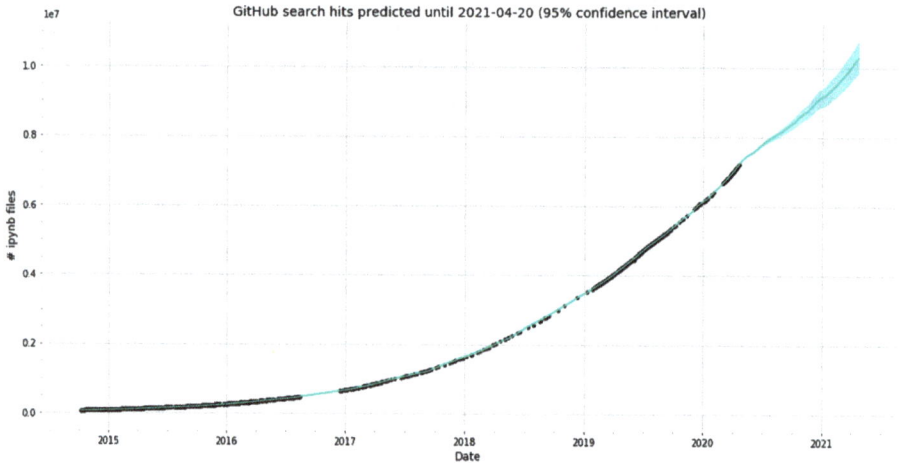

Figure 13.1: GitHub search hits 2015 to 2019 (2021 predicted) for ipynb[6] *(Parente)*.

Pro Jupyter

– Standardized layout
– A fusion between coding and documentation
– In GUIs like Rapidminer or Knime "Killed by GUI-Items"
– Millions of contributors, ready to use (no implementation of 'nodes')
– Follows the cloud paradigm, no local fat-client
– One GUI narrative: No differentiation between the Web Browser and Modelling Engine
– Broadly adopted

Con Jupyter

– Caveat[2]: It is Data Science, not a production environment!
– Encouraging poor coding practice, Data Scientists are not software engineers
– Makes it difficult to organize code logically, break it into reusable modules
– Hard to develop tests to ensure the code is working properly

- Lack of code versioning, Tools for software teams like git workflow cannot be applied
- Hard to test, a common meme is: *"Restart and run all ... or it didn't happen."*
- Nonlinear workflow
- Bad for running long asynchronous tasks

Big notebook infrastructures

Major companies like Netflix and Uber have big infrastructures.[7] They support Jupyter kernels locally on desktop systems, on remote JupyterHubs via Binder, and, in future, within Kubernetes runtimes by Kubeflow (see Chapters 6 and 31).

Runtimes and architecture

Jupyter implements a two-process model, with a kernel and a client. The client is the interface offering the user the ability to send code to the kernel. The kernel executes the code and returns the result to the client for display.

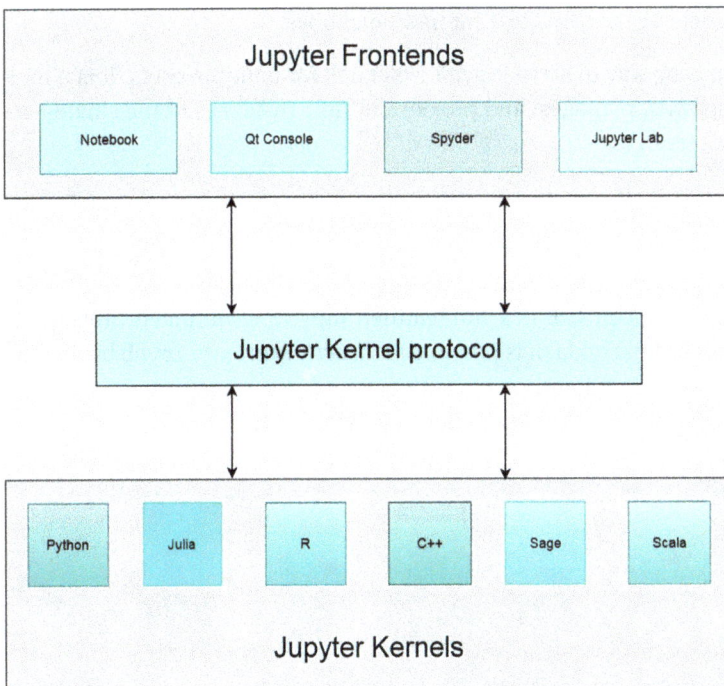

Figure 13.2: Jupyter architecture.[8]

Jupyter runtime environments

An application of a runtime environment (RTE) is within an operating system (OS) that allows that RTE to run, meaning from boot until power-down. In this case, it is the host of the Jupyter notebooks.

JupiterHub and Anaconda Jupyter

The classic approach to run a notebook is Anaconda, starting the server by $ jupyter notebook on the command prompt. **Important hint:** When using multiple environments within conda, a Jupyter notebook *can be attached to an environment only* by setting up a named kernel. It means that there are multiple IPython kernels for different virtualenvs (or conda) environments:

```
(base) $ conda install ipykernel # or pip install ipykernel
(base) $ source activate myenv
(myenv) $ python -m ipykernel install --user
                      --name myenv --display-name "Python (myenv)"
```

Then use this kernel Python (myenv) for new notebooks.

JupyterHub[9] is an easy way to serve Jupyter Notebook for multiple users. It is a multiuser hub that spawns, manages, and proxies multiple instances of the single-user Jupyter notebook server.

Visual Code (Microsoft)

Surprise: The free Microsoft IDE is a full-featured Jupyter environment. Instead of running notebooks in Anaconda, it is possible to ***directly start*** a *.ipynb by double-click in the Windows file explorer!

The direct inspection of variables including a data grid feels like JupyterLab:

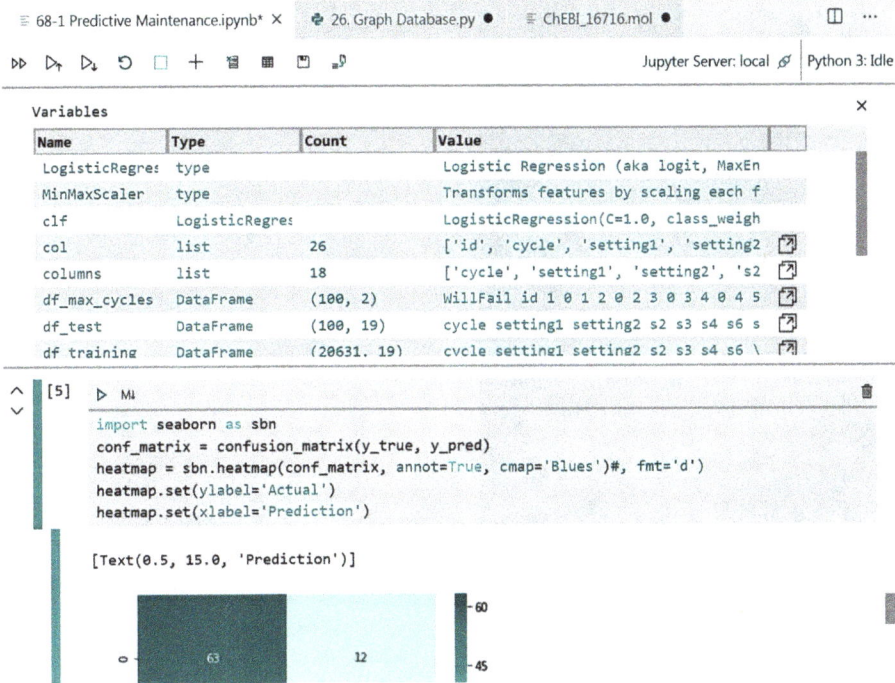

Figure 13.3: Microsoft Visual Code with Jupyter notebook *(Gressling)*.

Android

With *Pydroid* it is possible to run some notebooks on a mobile device. That works, at least for presentations. However, coding can also be performed, for example, during a 12-hour international flight.

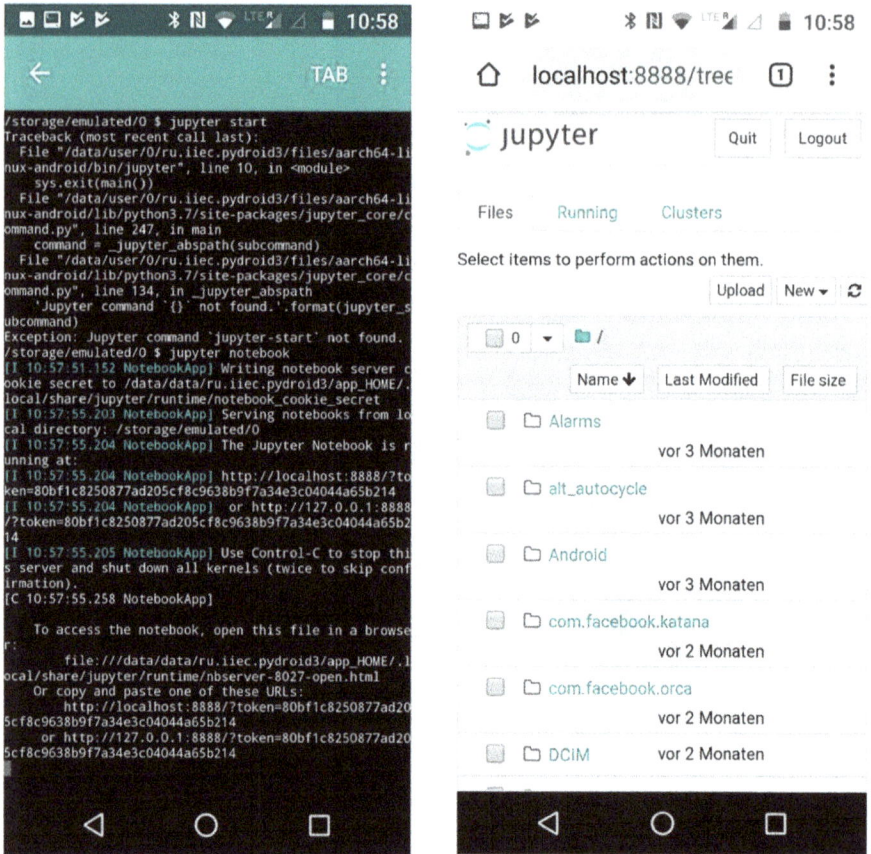

Figure 13.4: Jupyter on Android with Pydroid (Gressling).

Google Colab

If you are running a Chromebook or not having some of the above IDE's installed, it is fast and practical to use the online service of Google. There are limitations due to the tricky file handling and the IDE needs to be prepared before usage, but it is available by a browser and free.

```
# Runtime preparation Google Colab (in the first cell)
# author: Gressling, T
# License: MIT License      # code: github.com/gressling/examples
# activity: single example  # index: 13-1

!wget -c https://repo.continuum.io/miniconda/Miniconda3-latest-Linux-x86_64.sh
!chmod +x Miniconda3-latest-Linux-x86_64.sh
!time bash ./Miniconda3-latest-Linux-x86_64.sh -b -f -p /usr/local
!time conda install -q -y -c conda-forge rdkit

%matplotlib inline
import matplotlib.pyplot as plt
import sys
import os
sys.path.append('/usr/local/lib/python3.7/site-packages/')

from __future__ import print_function
from rdkit import rdBase
from rdkit import Chem
from rdkit.Chem import AllChem
from rdkit.Chem import Draw
from rdkit.Chem.Draw import IPythonConsole

m = Chem.MolFromSmiles('C1OC1')
m
> [figure]
```

References

1. Wilhelm, F. Working efficiently with JupyterLab Notebooks: https://florianwilhelm.info/2018/11/ working_efficiently_with_jupyter_lab/.
2. Mueller, A. 5 reasons why jupyter notebooks suck (towardsdatascience): https:// towardsdatascience.com/5-reasons-why-jupyter-notebooks-suck-4dc201e27086.
3. Project Jupyter (Wikipedia) https://en.wikipedia.org/wiki/Project_Jupyter.
4. Masnick, M. Pros and cons of Jupyter notebooks: https://masnick.blog/2018/08/28/ pros-and-cons.html.
5. Poulopoulos, D. Jupyter is now a full-fledged IDE (towardsdatascience): https:// towardsdatascience.com/jupyter-is-now-a-full-fledged-ide-c99218d33095.
6. Parente, P. nbestimate (MIT License): https://github.com/parente/nbestimate.
7. Netflix Technology Blog. Beyond Interactive: Notebook Innovation at Netflix https://medium.com/ netflix-techblog/notebook-innovation-591ee3221233.
8. xeus – Jupyter kernel protocol (BSD 3 License) https://github.com/jupyter-xeus/xeus.
9. JupyterHub documentation https://jupyterhub.readthedocs.io/en/latest/index.html.

14 Working with notebooks and extensions

JupyterLab extensions[1] can customize or enhance JupyterLab. They provide new themes, file viewers and editors, or renderers for rich outputs in notebooks. Extensions can provide an API for other extensions.

Kernels and extensions

Visual debugging with xpython kernel and JupyterLab

The JupyterLab debugger is a special Jupyter kernel. It is based on the native implementation of the Jupyter protocol xeus,[2] which is a library meant to facilitate the implementation of kernels for Jupyter. It takes the burden of implementing the Jupyter Kernel protocol.

– The standard Jupyter Kernel Gateway is a web server that provides headless access to Jupyter kernels. The application communicates with the kernels remotely through REST calls.[3] It is implemented with reference to the Swagger API standard.
– 100+ Kernels are listed in GitHub[3]

Launch a new Python notebook by selecting the xpython kernel:
– a variable explorer, a list of breakpoints and a source preview
– the possibility to navigate the call stack (next line, step in, step out, etc.)
– the ability to set breakpoints intuitively, next to the line of interest
– flags to indicate where the current execution has stopped

```
$ conda install xeus-python -c conda-forge
```

Then, run Jupyter Lab and on the sidebar, search for the Extension Manager and enable it. Search for the debugger extension. Create a new xpython notebook and compose a simple function.
– Support for rich mime type rendering in the variable explorer.
– Support for conditional breakpoints in the UI.
– Enable the debugging of Voilà dashboards from the JupyterLab Voilà preview extension.
– Enable debugging with as many kernels as possible.

https://doi.org/10.1515/9783110629453-014

Start JupyterLab using:

```
$ jupyter lab
```

nbdev

Complement Jupyter by adding support for **literate programming** with nbdev[4] (see Chapter 12):
- automatic creation of python modules from notebooks, following best practices
- synchronization of any changes back into the notebooks
- automatic creation of searchable, hyperlinked documentation from the code
- pip installers readily uploaded to PyPI
- testing
- continuous integration
- version control conflict handling

nbdev enables software developers and data scientists to develop well-documented python libraries, following best practices without leaving the Jupyter environment.

```
pip install nbdev
```

Jupyter_contrib_nbextensions

J. Bercher[5] has compiled a collection of about 60 community-contributed unofficial extensions that add functionality to the Jupyter notebook within JupyterLab. These extensions are mostly written in Javascript and will be loaded locally in the browser. Some examples are:[6,7]
- 2to3 Python migrator
- Hide input/hide input All
- Table of contents
- ExecuteTime
- Freeze

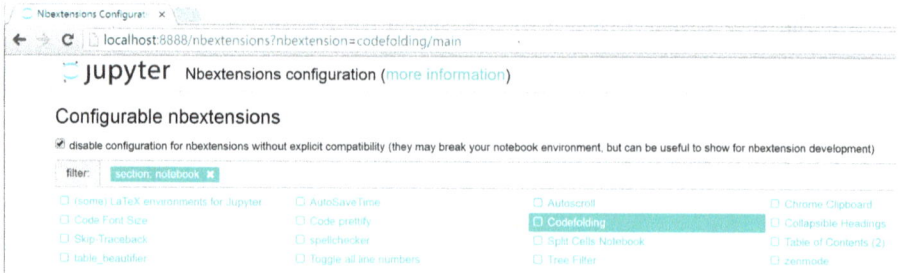

Figure 14.1: nbextensions.[6]

Handling notebooks

Viewing ipynb's online

Jupyter `nbviewer.jupyter.org` is a web application behind The Jupyter Notebook Viewer.[8] The source code is available on Git.[9] To view local files online, not integrated with GitHub,[8] provide the ipynb files by a trick: load to gdrive and paste the public link to the viewer.

Another possibility for real online working is to use Colab. Colaboratory is a free Jupyter notebook environment that runs in the cloud and stores its notebooks on Google Drive. Colab was originally an internal Google project.

Export static ipynb's

Reading offline is just an emergency solution. PDF export is not recommended due to page breaks. Use the default format, HTML, instead. Sometimes, there is no active environment. So, a static output has to be used.[10] Use nbconvert:

```
$ jupyter nbconvert <input notebook> --to <output format>
```

- You can also export notebooks directly from the Jupyter environment
- by a web service[11]

Voilà and Jupyter widgets

Voilà[12] turns Jupyter notebooks into interactive standalone web applications.

```
$ conda install -c conda-forge voila
or pip install voila

$ voila <NOTEBOOK>.ipynb
```

Papermill

is a tool for parameterizing, executing, and analyzing Jupyter notebooks. Papermill can spawn multiple notebooks with different parameter sets and execute them concurrently:
- parameterize notebooks
- add a new cell tagged with injected-parameters with input parameters in order to overwrite the values in parameters
- execute notebooks
- looks for the parameters cell and treats this cell as defaults for the parameters passed in at execution time
- If no cell is tagged with parameters, the injected cell will be inserted at the top of the notebook

```
# papermill example
# author: Gressling, T
# License: BSD 3 License       # code: github.com/gressling/examples
# activity: single example  # index: 14-1

!pip install papermill

import papermill as pm

pm.execute_notebook(
   'path/to/input.ipynb',
   'path/to/output.ipynb',
   parameters = dict(mol='COc(c1)cccc1C#N', bonds=[1, 3, 4, 8, 11])
)


Papermill supports the following name handlers for input and output:

Local file system: local
HTTP, HTTPS protocol: http://, https://
Amazon Web Services: AWS S3 s3://
Azure: Azure DataLake Store, Azure Blob Store adl://, abs://
Google Cloud: Google Cloud Storage gs://
```

Rerun notebooks through papermill will reuse the injected-parameters cell from the prior run. In this case, Papermill will replace the old injected-parameters cell with the new run's inputs.

Papermill can also help collect and summarize metrics from a collection of notebooks.[13]

References

1. JupyterLab 2.1.0 documentation https://jupyterlab.readthedocs.io/en/stable/user/extensions. html.
2. xeus-python (GitHub) https://github.com/jupyter-xeus/xeus-python.
3. Jupyter Kernel Gateway https://jupyter-kernel-gateway.readthedocs.io/en/latest/.
4. nbdev: use Jupyter Notebooks for everything https://www.fast.ai/2019/12/02/nbdev/.
5. jupyter_contrib_nbextensions (GitHub) https://github.com/ipython-contrib/jupyter_contrib_ nbextensions.
6. Bercher, J.-F. jupyter_contrib_nbextensions (BSD 3 License) https://github.com/jfbercher/ jupyter_contrib_nbextensions.
7. contrib_nbextensions 0.5.0 documentation https://jupyter-contrib-nbextensions.readthedocs. io/en/latest/.
8. nbviewer – A simple way to share Jupyter Notebooks https://nbviewer.jupyter.org/.
9. nbviewer (GitHub) https://github.com/jupyter/nbviewer.
10. How to Export Jupyter Notebooks into Other Formats https://www.blog.pythonlibrary. org/2018/10/09/how-to-export-jupyter-notebooks-into-other-formats/.
11. Jupyter Notebooks (ipynb) Viewer and Converter https://htmtopdf.herokuapp.com/ipynbviewer/.
12. voila (GitHub) https://github.com/voila-dashboards/voila.
13. papermill (GitHub) (BSD 3 License) https://github.com/nteract/papermill.

15 Notebooks and Python

pip[1] is a de facto standard package-management system used to install and manage software packages written in Python. Many packages can be found in the default source for packages and their dependencies. The Python Package Index, abbreviated as PyPI is the official third-party software repository for Python. This ecosystem is the base of coding in Jupyter.

Calling notebooks from notebooks

Store your high-level APIs in a separate notebook. Via magic switch, %run[2] call the notebook with the functions. Keep the function *calls* visible in the master notebook:

```
# Notebook interaction examples
# author: Gressling, T
# license: MIT License       # code: github.com/gressling/examples
# activity: single example   # index: 15-1

%run ./called_notebook.ipynb
callThis()
```

via nbrun

automating[3] the execution of Jupyter notebooks as well as passing arguments for parameterization can be done with nbrun. It offers a simple, yet effective way to remove code duplication and manual steps, allowing to build automated and self-documented analysis pipelines. It contains a single function run_notebook() that allows the program to call/execute other notebooks.

```
nb = 'analyseSubStructure.ipynb'
args = {'mol': 'OCCc1c(C)[n+](cs1)Cc2cnc(C)nc2N'}

run_notebook(nb, out=f'convertSMILES/executed_notebook_{mol}', nb_kwargs=args)
```

https://doi.org/10.1515/9783110629453-015

via nbparameterise

nbparameterise[4] is a tool to run notebooks with input values. When you write the notebook, these are defined in the first code cell, with regular assignments:

```
# Example: analyseSubStructure.ipynb
mol = 'OCCc1c(C)[n+](cs1)Cc2cnc(C)nc2N'
```

nbparameterise handles finding and extracting these parameters and replacing them with input values. Then run the notebook with the new values.

```
import nbformat
from nbparameterise import (extract_parameters, replace_definitions, values)

with open("analyseSubStructure.ipynb") as f:
    nb = nbformat.read(f, as_version=4)

# Get a list of Parameter objects
orig = extract_parameters(nb)

# Update one or more parameters,
# replaces 'OCCc1c(C)[n+](cs1)Cc2cnc(C)nc2N' (Thiamin)
params = values(orig, mol='CC(=O)OCCC(/C)=C\C[C@H](C(C)=C)CCC=C')

# Make a notebook object with these definitions and execute it.
new_nb = replace_definitions(nb, params)
```

Extract functionality from the notebook into a *.py file

Extensive notebook of statements and expressions without any structure leads to name collisions and confusion. Since notebooks allow the execution of cells in a different order, this can be extremely harmful.

For these reasons, do some cleanup (refactoring) and create a function in a *.py file instead:

```
def get_mol():
    mol='CC(=O)OCCC(/C)=C\C[C@H](C(C)=C)CCC=C'
    return mol
```

Then move it into a module analyseSubStructure.py that can be created within our directory. Now, inside the notebook, just import and use it:

```
from analyseSubStructure import get_mol

print(get_mol())
> CC(=O)OCCC(/C)=C\C[C@H](C(C)=C)CCC=C
```

This leads to another best practice – Integrating classic programming.

intake

Intake[5] is a Python library for accessing data in a simple and uniform way. It consists of three parts:
1. A lightweight plugin system for adding data loader drivers for new file formats and servers (like databases, REST endpoints or other cataloging services)
2. A cataloging system for specifying these sources in simple YAML syntax, or with plugins that read source specs from some external data service
3. A server–client architecture that can share data catalog metadata over the network, or even stream the data directly to clients if needed

A graphical data browser is available in the Jupyter notebook environment. It will show the contents of any installed catalogs, plus allow for selecting local and remote catalogs to browse and select entries from these.

```
# intake example
# author: Gressling, T
# license: MIT License      # code: github.com/gressling/examples
# activity: single example  # index: 15-2

! intake example
Creating an example catalog...
 Writing us_states.yml
 Writing states_1.csv
 Writing states_2.csv

import intake
catalog = intake.open_catalog('us_states.yml')
dataset = intake.open_csv('states_*.csv')
print(ds)
> <intake.source.csv.CSVSource object at 0x1163882e8>
```

The YAML Catalog:

```
sources:
  states:
    description: US state information from [CivilServices](https://civil.
                 services/)
    driver: csv
    args:
      urlpath: '{{ CATALOG_DIR }}/states_*.csv'
    metadata:
      origin_url:
      'https://github.com/CivilServiceUSA/us-states/blob/v1.0.0/data/states.csv'
```

debugging with ipdb

ipdb[6] (pdb for Jupyter) exports functions to access the IPython debugger, which features tab completion, syntax highlighting, better tracebacks, and better introspection with the same interface as the pdb module.

Using Jupyter notebook, begin the cell with magic command **%%debug**. Then an ipdb line will be shown at the bottom of the cell, which will help navigate through the debugging session:
- n- execute the current line and go to the next line.
- c- continue execution until the next breakpoint.

Make sure to restart the kernel.

```
# debugging with ipdb
# author: Gressling, T
# license: MIT License        # code: github.com/gressling/examples
# activity: single example  # index: 15-3

import ipdb
ipdb.set_trace()
ipdb.set_trace(context=5) # will show five lines of code

analyseSubStructure.get_mol(mol)

OR
%pdb

> ipdb> n, or print(mol) or c
```

hvplot

hvPlot[7] offers an alternative for the static plotting API provided by Pandas and other libraries, with an interactive Bokeh-based plotting API that supports panning, zooming, hovering, and clickable/selectable legends with the benefits of modern, interactive plotting libraries for the web, like Bokeh and HoloViews.

```python
# Example hvPlot
# author: holoviz.org
# license: BSD 3   # code: github.com/holoviz/hvplot
# activity: single example   # index: 15-4

import pandas as pd, numpy as np
idx = pd.date_range('1/1/2000', periods=1000)
df = pd.DataFrame(np.random.randn(1000, 4), index=idx,
columns=list('ABCD')).cumsum()

import hvplot.pandas # noqa
df.hvplot()
```

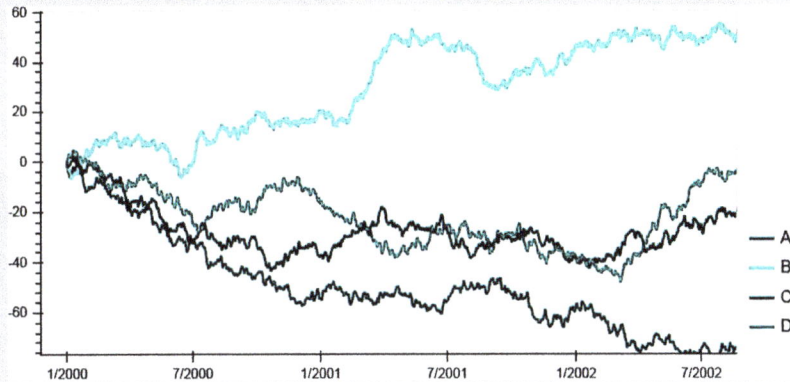

Figure 15.1: hvplot example.

UML

PlantUML is an open-source tool allowing users to create UML diagrams from plain text language.

```
# UML example
# author: Gressling, T
# license: MIT License      # code: github.com/gressling/examples
# activity: single example  # index: 15-5


!pip install iplantuml


%%plantuml
@startuml
Device -> M2: Temperature 23.4
M2->M3: situation.update('23.4');
M3->M3: checkTemperature();
M3->M2: situation.update(OVER_MAX)
M2->Device: HEATER_OFF
@enduml
```

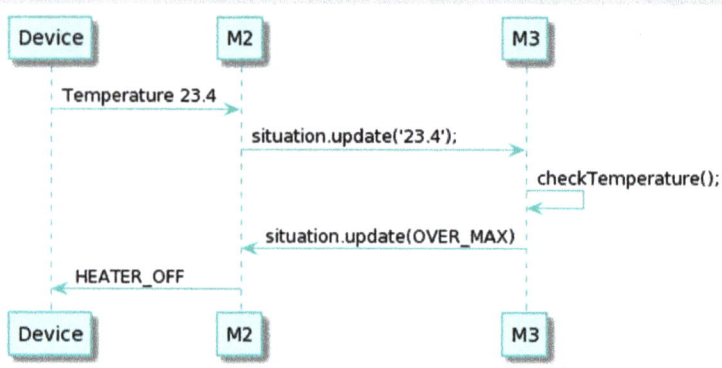

Figure 15.2: PlantUML Example (*Gressling*).

tqdm

For long-running calculations, it is recommended to show a progress meter. To do
that, wrap any iterable with **tqdm(iterable)** and in the next cell, persistent meter-
ing is generated.[8]

```
# tqdm progress bar example
# author: Gressling, T
# license: MIT License       # code: github.com/gressling/examples
# activity: single example   # index: 15-6

from tqdm import tqdm
max_atoms=10000
for i in tqdm(range(max_atoms)):
    ...
76%|██████████████████████|        | 7568/10000 [00:33<00:10, 229.00it/s]
```

References

1. pip (package manager) (Wikipedia) https://en.wikipedia.org/wiki/Pip_(package_manager).
2. Running a Jupyter notebook from another notebook https://stackoverflow.com/
   questions/49817409/running-a-jupyter-notebook-from-another-notebook.
3. :Ingargiola, A. nbrun (GitHub): https://github.com/tritemio/nbrun.
4. Kluyver, T. nbparameterise (GitHub) https://github.com/takluyver/nbparameterise.
5. intake 0.5 documentation https://intake.readthedocs.io/en/latest/quickstart.html.
6. ipdb https://pypi.org/project/ipdb/.
7. Developers, H. hvPlot documentation https://hvplot.holoviz.org/.
8. tqdm – A Fast, Extensible Progress Bar for Python (GitHub) https://github.com/tqdm/tqdm.

16 Versioning code and Jupyter notebooks

A component of software configuration management,[1] version control, also known as revision control or source control, is the management of changes to documents, computer programs, large web sites, and other collections of information.

GitHub[2] offers the distributed version control and source code management (SCM) functionality of Git, plus its own features. It provides access control and several collaboration features such as bug tracking, feature requests, task management, and wikis for every project.

Jupyter and other GUI versioning and difference analysis

As GUI tools like KNIME do not have a text-based approach, a difference between two versions can only be in a *descriptive* way, which means that the visual difference is written as a text log.

As Jupyter is a mixture of text and GUI, a real text-by-text difference is also not easy. But program code is available, so at least these parts can be compared.

Code versioning core concepts

1. **Branching** is a core concept, GitHub flow is based upon it. There is only one rule: anything in the master branch is always deployable. A Branch is to create an environment to try out new ideas. Changes that are made do not affect the master branch. The branch would not be merged until it is ready to be reviewed.
2. Add **commits**. Whenever adding, editing, or deleting a file, a commit happens in the branch. This process keeps track of the work progress. Commits also create a transparent history that others can follow to understand. By writing clear commit messages, it is easier to follow along and provide feedback.
3. Open a **Pull Request** initiates discussion about commits. Anyone can see exactly what changes would be merged if they accept the pull request.
4. **Deploy** from a branch for final testing in production before merging to master.
5. **Merge**. If the changes have been verified in production, merge the code into the master branch.

Fork a repository means creating a local copy (clone) of the entire project. Forking is typically done once when beginning work on a project. A branch is a lightweight thing that is often temporary and may be deleted. A fork is a new project that is based on a previous project. Forking is a concept while cloning is a process. Forking is just

https://doi.org/10.1515/9783110629453-016

containing a separate copy of the repository and there is no command involved. Cloning is done through the command 'git clone' and it is a process of receiving all the code files to the local machine.

To understand the relationships within and across repositories, *navigating code directly* in GitHub is possible.

GitHub actions/CIDC

With CIDC, it is possible to automate, customize, and execute software development workflows with GitHub actions. The programmer can discover, create, and share actions to perform any job and combine the actions in a customized workflow.

In software engineering, CI/CD or CICD generally refers to the combined practices of continuous integration and continuous delivery (continuous deployment).

Difference notebooks with nbdime in JupyterLab

After extension installation, nbdime's buttons[3,4] in the notebook appear. Clicking the git button will open a new tab, showing the difference between the last commit and the currently saved version of the notebook:

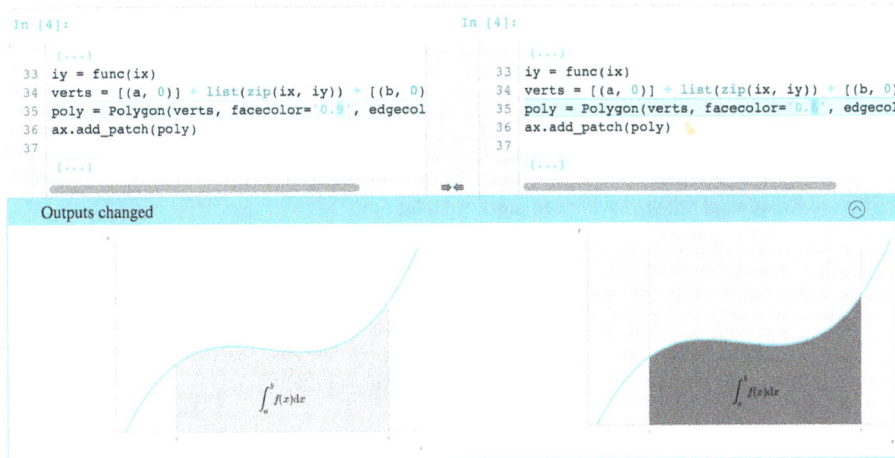

Figure 16.1: nbdime[3] difference.

Verdant

In JupyterLab, the Verdant[5,6] plugin also provides a *difference analysis*. It is an extension that automatically records the history of all experiments that run in a Jupyter notebook and stores them in a .ipyhistory JSON file.

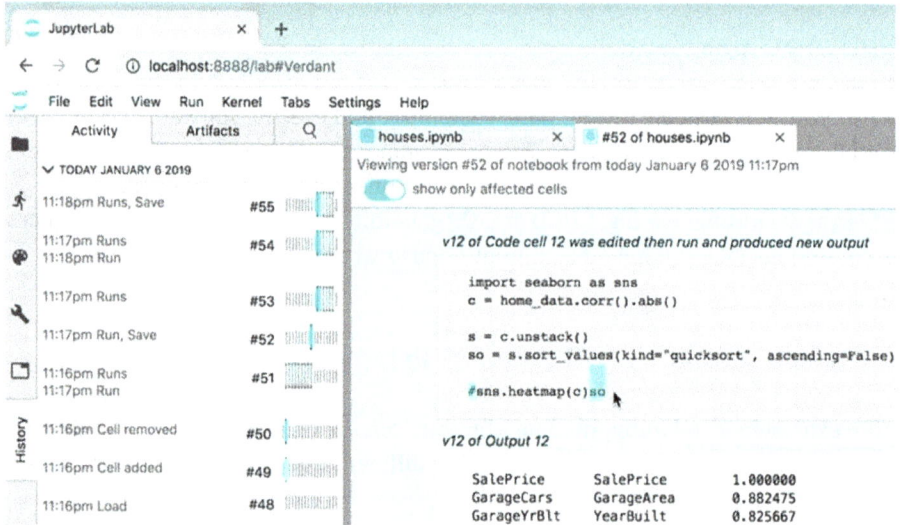

Figure 16.2: Verdant extension *(Kery)*.

References

1. Version control (Wikipedia) https://en.wikipedia.org/wiki/Version_control.
2. GitHub (Wikipedia) https://en.wikipedia.org/wiki/GitHub.
3. nbdime – diffing and merging of Jupyter Notebooks (BSD 3 License); https://nbdime. readthedocs.io/en/stable/.
4. nbdime (GitHub) https://github.com/jupyter/nbdime.
5. Kery, Mary Beth. Verdant (MIT License) https://github.com/mkery/Verdant.
6. Kery, M. B. Towards Effective Foraging by Data Scientists to Find Past Analysis Choices (CHI 2019) http://www.cs.cmu.edu/~NatProg/papers/paper092-Kery-CHI2019.pdf.

17 Integration of Knime and Excel

Before the Python era, there were at least two software ecosystems that were used extensively. One class was GUI workflow tools like Knime and Rapidminer, and the other was Microsoft Excel; so integration from Jupyter to both must be feasible.

Knime

Started in 2004 by a team of software engineers at the University of Konstanz, the first version of KNIME was released 2006. KNIME has over 15,000 actual users in industry and also research groups (2012).[1]

Using Python *in* KNIME

A number of nodes[2] are available that use Python from KNIME:
- Python Scripting nodes provide the data from KNIME as a *Pandas Dataframe* (in and out)
- Complex data transformations, where for them aren't nodes available
- Prototyping and experimenting with new computational methods
- Taking advantage of the machine-learning tools available in scikit-learn (complement KNIME's built-in capabilities)
- Save the executed workflow preserves input data, code, and results all together

It is recommended to use Anaconda and then create a new Python environment for use inside of KNIME:[3]

```
$ conda create -y -n py35_knime python=3.5 pandas jedi
```

Using Jupyter notebooks *in* Knime

The knime_jupyter package, which is available in the Python Script nodes, loads the code present in a notebook. Indicate load_notebook to only import code from notebook cells that have been *tagged,* for example, *"knime".* Create tags for a cell by using the View->Cell Toolbar menu.

https://doi.org/10.1515/9783110629453-017

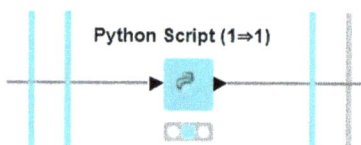

Then, call `load_notebook` with the path and the name of the notebook. The notebook object that is returned provides access to all the functions and variables defined in the notebook itself.

Call KNIME *from* Python

It works by constructing a *Python dictionary object* that has the parameters to send to the web service. The results return as a Python object.

```python
# Call KNIME node by WebService
# author: Landrum, Greg
# license: CC BY 4.0          # code: knime.com/blog/blending-knime-and-python
# activity: active (2020)     # index: 17-1

import requests, json
from requests import auth

def call_KNIME_WebService(serviceName, values, user, pw,
                baseURL="http://<<SERVER>>:8080/webportal/rest/v4/
                repository/"):
    url = "%s%s:job-pool"%(baseURL,serviceName)
    authd = auth.HTTPBasicAuth(user,pw)
    reqbody = json.dumps({"input-data":{"jsonvalue":qry}})
    resp = requests.post(url,auth=authd,data=reqbody,
        headers={"Content-Type":"application/json"})

    if resp.ok and resp.status_code==200:
        outvs = resp.json()['outputValues']
        return tuple(outvs.values())
    raise RuntimeError(resp.status_code)

qry=dict(smiles['c1ccccc1C(=O)O'])
node='Lifesciences/demo/WebServices/1_BasicDescriptors'
result=call_KNIME_WebService(node, qry, user, pw)
```

Use Jupyter to access workflow via KNIME.exe

Install[4] the KNIME python package from PyPI by doing pip install knime.[4]

```
# Work with a workflow via KNIME.exe
# author: Gressling, T
# license: MIT License      # code: github.com/gressling/examples
# activity: single example  # index: 17-2

import knime
knime.executable_path = "KNIME_PATH/knime.exe"
workspace = "KNIME_PATH/KNIME workspaces/book"
workflow = "test_workflow"
knime.Workflow(workflow_path=workflow, workspace_path=workspace)
# shows the workflow in the notebook!

in = pd.DataFrame({'smiles':['COCc1cc(C=O)ccc10']})

with knime.Workflow(workflow_path=workflow,
                    workspace_path=workspace) as KWorkflow:
    KWorkflow.data_table_inputs[0] = in
    KWorkflow.execute()

result=KWorkflow.data_table_outputs[0]
```

Microsoft Excel

It has been more than 35 years since Microsoft Excel was released; it has become one of the most popular tools in workplaces, setting the *.csv standard for data files (see Chapter 27). It is a program that helps *sketch and calculate*. Excel has add-ins that enhance its built-in features. It is not a core professional statistics tool, still, it has lots of advantages. Excel (2016) has 484 functions.

Excel *.CSV read and write

```
import pandas as pd
# load csv
df = pd.read_csv('some_data.csv', low_memory=True, header=None)
# or pd.read_excel(...)

# Save Results
df.to_csv('some_data-out.csv', sep=',', encoding='utf-8', index=False)
```

Comparison between Excel and Data Frames, see Chapter 10.

Excel in Python (with openpyx)

openpyxl[5] is a Python library to read and write Excel 2010. The `excel.exe` program need not be running.

```
# Excel in Python with openpyx
# author: Gressling, T
# license: MIT License      # code: github.com/gressling/examples
# activity: single example  # index: 17-3

from openpyxl import Workbook
wb = Workbook()

# active worksheet
ws = wb.active
# Data can be assigned directly to cells
ws['A1'] = 'Ethanol'
ws['B1'] = 'CCO'
# Rows can also be appended
ws.append([1, 2, 3])
# Calculations and formula
ws["A3"] = "=LEN(B1)"

# Save the file
wb.save("smiles.xlsx")
```

Python in Excel (VBA with xlwings)

The python library `xlwings` allow the program to call Python scripts through VBA and pass data between the two.[6]

```
(book) c:\datascience>pip install xlwings
Collecting xlwings
   Downloading xlwings-0.18.0.tar.gz (595 kB)
     |███████████████████████████| 595 kB 192 kB/s
Collecting comtypes
   Downloading comtypes-1.1.7.zip (180 kB)
     |███████████████████████████| 180 kB 187 kB/s

c:\datascience>xlwings addin install
> xlwings 0.18.0
> Successfully installed the xlwings add-in! Please restart Excel.
```

Hint: Use $ `xlwings quickstart <ProjectName>` for experiments. Use $ `where python` to find the PYTHONPATH in Anaconda. With ALT+F11, when you enter the VBA editor, you find:

```
# Use Python in Excel with xlwings
# author: Gressling, T
# license: MIT License      # code: github.com/gressling/examples
# activity: single example  # index: 17-4
```

```vba
' EXCEL VBA Part
' Look for a Python Script in the same location as the spreadsheet
' Look for a Python Script with the same name as the spreadsheet
' (but with a .py extension)
' From the Python Script, call the function "main()"
Sub SampleCall()
    python = Left(ThisWorkbook.Name,
                    (InStrRev(ThisWorkbook.Name,".", -1, vbTextCompare) - 1))
    RunPython("import " & python & ";" & python & ".main()")
End Sub
```

```python
# Python part
import xlwings as xw

def main():
    wb = xw.Book.caller()
    wb.sheets[0].range("A1").value = "Data science in chemistry!"

if __name__ == "__main__":
    xw.books.active.set_mock_caller()
    main()
```

It is also possible to work with the Excel engine via COM/OLE manipulation on Windows OS level out of python.

References

1. KNIME (Wikipedia) https://en.wikipedia.org/wiki/KNIME.
2. Landrum, G. Setting up the KNIME Python extension: https://www.knime.com/blog/setting-up-the-knime-python-extension-revisited-for-python-30-and-20.
3. Landrum, G. You don't have to choose! Blending KNIME and Python:https://www.knime.com/blog/blending-knime-and-python.
4. Landrum, G. KNIME and Jupyter: https://www.knime.com/blog/knime-and-jupyter.
5. openpyxl – A Python library to read/write Excel https://openpyxl.readthedocs.io/en/stable/.
6. Andreou, C. How to Supercharge Excel With Python: https://towardsdatascience.com/how-to-supercharge-excel-with-python-726b0f8e22c2.

Data engineering

18 Big data

**Big data is a field that treats ways to analyze, systematically extracted informa-
tion from, or otherwise, deal with data sets that are too large or complex to be
dealt with by traditional data-processing application software.[1,2] It is any collection of data sets *so large and complex* that it becomes difficult to process using
on-hand data management tools or traditional data processing applications.**

The challenges[3] include capturing big amounts of information (IoT), data storage
techniques (HDFS, RDD), data analysis (Spark), and transfer (Flume, Kafka). Apache
Hadoop software collection is *one example* of managing this ecosystem.

The first documented use of the term "big data" appeared in a 1997 paper by scientists at NASA, describing the problem they had with visualization.[4]

HDFS and RDDs

The core of Apache Hadoop consists of a storage part, known as *Hadoop Distributed
File System*[5] (HDFS) and a processing part, which is a *MapReduce*[6] programming
model. Hadoop splits files into large blocks and distributes them across nodes in a
cluster. The blocks are called Resilient Datasets, RDD.

An RDD (Resilient Distributed Dataset) is the fundamental data structure of
Apache Spark. It is an *immutable collection of objects* which computes on the different
nodes of the cluster. Resilient means: they can only be *created, deleted, or copied.*
Not updated ... and that gives the speed of big data processing.

*RDDs hide all the complexity of transforming and distributing data automatically
across multiple nodes.* This is implemented by a scheduler running on a cluster.

Examples see Python implementation in Chapter 19.

https://doi.org/10.1515/9783110629453-018

Python example for HDFS

```
# Access HDFS file system
# author: Gressling, T
# license: MIT License       # code: github.com/gressling/examples
# activity: single example  # index: 18-1

from hdfs3 import HDFileSystem
hdfs = HDFileSystem(host='localhost', port=8020)
hdfs.ls('/user/data')
hdfs.put('local-file.txt', '/user/data/remote-file.txt')
hdfs.cp('/user/data/file.txt', '/user2/data')
with hdfs.open('/user/data/file.txt') as f:
    data = f.read(10000000) # huge amount!

with hdfs.open('/user/data/file.csv.gz') as f:
    df = pandas.read_csv(f, compression='gzip', nrows=100000)
```

Flume

Apache Flume is a distributed software for *collecting, aggregating, and moving large amounts of log data*. It has an architecture based on streaming data flows. It is fault-tolerant with failovers and recovery mechanisms. Using this data ingestion pattern is common with predictive maintenance in the chemical industry, see Chapter 68.

Python example

```
# Capture production logs with FLUME
# author: Gressling, T
# license: MIT License       # code: github.com/gressling/examples
# activity: single example  # index: 18-2

# stream.py
from pyspark import SparkContext
from pyspark.streaming import StreamingContext
from pyspark.streaming.flume import FlumeUtils

if __name__ == "__main__":
    sc = SparkContext(appName="data_production_lot1");
    ssc = StreamingContext(sc, 30)
    stream = FlumeUtils.createStream(ssc, "127.0.0.1", 55555)
    stream.pp(rint()
```

```
# capture.py
import requests
import random

class log_capture():
  def __init__(self):
      self.url = "https://<<YOUR SERVER>>/api/production/"
      self.rand = str(random.randint(0, 99)) # random data
      self.r = requests.get(self.url + self.rand)

  def get_r(self):
      return(self.r.text)

if __name__ == "__main__":
    import os
    with open("log_capture.txt", "w") as file_in:
        file_in.write(log_capture().get_r())
    os.system("cat log_capture.txt")
```

```
# flume.conf
agent.sources = tail-file
agent.channels = c1
agent.sinks=avro-sink
# define source and sink
agent.sources.tail-file.type = exec # important!
agent.sources.tail-file.command = python /home/book/capture.py
agent.sinks.avro-sink.type = avro
agent.sinks.avro-sink.hostname = 127.0.0.1
agent.sinks.avro-sink.port = 55555
```

Kafka

Kafka is a distributed publish-subscribe *messaging system* that maintains feeds of messages in partitioned (RDD) and replicated topics. There are three players in the Kafka ecosystem: producers, topics (run by brokers), and consumers.

Messaging brokers: difference between Kafka and MQTT

Kafka is a messaging broker with a *transient store*, which consumers can subscribe and listen to. It's an *append-only log*, which consumers can pull from. It is designed for big amounts of data.

MQTT is a messaging broker for a machine to machine *communication*. The purpose is to hold a communication channel alive and have reliable messaging. It is designed for robustness.

Kafka python example

```
# Massive data broker KAFKA
# author: Gressling, T
# license: MIT License      # code: github.com/gressling/examples
# activity: single example  # index: 18-3

from kafka import KafkaProducer
producer = KafkaProducer(bootstrap_servers=['localhost:9092'])
# Asynchronous example
future = producer.send('chemprocess-data', b'raw_bytes')
producer.flush()

from kafka import KafkaConsumer
# To consume latest messages and auto-commit offsets
consumer = KafkaConsumer('chemprocess-data',
                    group_id='chemprocess-group',
                    bootstrap_servers=['localhost:9092'])
for message in consumer:
    print ("%s:%d:%d: key=%s value=%s" % (message.topic,
            message.partition, message.offset,
            message.key, message.value))
```

Role in chemistry

Also in chemistry, we have extremely large data sets that must be analyzed computationally to reveal patterns and associations. In the past few years, a new field of research has emerged:
– **Big data approach to analytical chemistry**[7]
– **Data-centric science for materials innovation**[8]
– **Data-driven materials science: status, challenges, and perspectives**[9]
– **Framing the role of big data and modern data science in chemistry**[10]
– **Big data of materials science: critical role of the descriptor**[11]
– **Big data analytics in chemical engineering**[12]

References

1. Grüning, B. A.; Lampa, S.; Vaudel, M.; Blankenberg, D. Software engineering for scientific big data analysis https://www.researchgate.net/publication/333326758_Software_engineering_ for_scientific_big_data_analysis.
2. Big data (Wikipedia) https://en.wikipedia.org/wiki/Big_data.
3. Press, Gil. 12 Big Data Definitions: What's Yours? https://www.forbes.com/sites/ gilpress/2014/09/03/12-big-data-definitions-whats-yours/.
4. Cox, M.; Ellsworth, D. Application-controlled demand paging for out-of-core visualization https://dl.acm.org/doi/10.5555/266989.267068.
5. Filesystem Interface – Apache Arrow https://arrow.apache.org/docs/python/filesystems.html.
6. MapReduce (Wikipedia) https://en.wikipedia.org/wiki/MapReduce.
7. Dubrovkin, J. Big Data Approach to Analytical Chemistry https://www.researchgate.net/ publication/260553457_Big_Data_approach_to_analytical_chemistry.
8. Tanaka, I.; Rajan, K.; Wolverton, C. Data-centric science for materials innovation https://www.cambridge.org/core/services/aop-cambridge-core/content/view/ FCF9EEAB5DF45813DD48793BB231BD8D/S0883769418002051a.pdf.
9. Himanen, L.; Geurts, A.; Foster, A. S.; Rinke, P. Data-Driven Materials Science: Status, Challenges, and Perspectives https://arxiv.org/pdf/1907.05644.pdf.
10. Che-Workshop. Framing the Role of Big Data and Modern Data Science in Chemistry https://www. nsf.gov/mps/che/workshops/data_chemistry_workshop_report_03262018.pdf.
11. Ghiringhelli, L. M.; Vybiral, J.; Levchenko, S. V.; Draxl, C.; Scheffler, M. Big data of materials science: critical role of the descriptor http://arxiv.org/abs/1411.7437v2.
12. Chiang, L.; Lu, B.; Castillo, I. Big Data Analytics in Chemical Engineering: http://arjournals. annualreviews.org/doi/full/10.1146/annurev-chembioeng-060816-101555.

19 Jupyter and Spark

Apache Spark[1] is an open-source distributed general-purpose cluster-computing framework. Spark provides an interface for programming entire clusters with implicit data parallelism and fault tolerance.

It is used[2] for processing and analyzing a large amount of data. Just like Hadoop MapReduce, it also works with the system to distribute data across the cluster and process the data in parallel.

PySpark

Spark is implemented in Scala. PySparks is a Python-based wrapper on top of the Scala API. PySpark communicates via the Py4J library. Py4J then allowo talking to JVM (JavaVirtualMachine)-based code, then accessing Hadoop data storage:

```
(query) -> PySpark -> Spark -> Scala -> JVM -> Hadoop (raw-data)
```

The reason is that Scala is function-based and is much easier to parallelize than python. PySpark has a way of handling parallel processing, without the need for the threading or multiprocessing modules. All of the complicated communication and synchronization between threads, processes, and even different CPUs is handled by Spark.

Setup

How to create test environments:
- **Docker** There are a few distributions for PySpark environments
  ```
  $ docker run -it --rm -p 8888:8888 jupyter/pyspark-notebook
  ```
- **Databrics**
- **local setup**[4] (not recommended)

Tool 'findspark'

Sometimes, PySpark it not on sys.path, by default. You can address this by adding PySpark to sys.path at runtime. The package findspark[5] does that for you.

https://doi.org/10.1515/9783110629453-019

```
$ pip install findspark

import findspark
findspark.init()
import pyspark
sc = pyspark.SparkContext(appName="predictiveMaintenance_plant")
```

Code example

```
# PySpark Example
# author: Gressling, T
# license: MIT License       # code: github.com/gressling/examples
# activity: single example   # index: 19-1

import pyspark
sc = pyspark.SparkContext('local[*]')

txt = sc.textFile('file:////tmp/SMILES.txt')
print(txt.count())

python_lines = txt.filter(lambda line: 'CO' in line.lower())
print(CO_lines.count())
```

The entry point is the SparkContext object. This object allows you to connect to a Spark cluster and create RDDs (see Chapter 18). The local[*] string indicates using a local cluster (single-machine mode).

RDD example

The code creates an iterator of 10,000,000 elements and then uses parallelize() to distribute it into 50 partitions:

```
# RDD Example on Hadoop
# author: Gressling, T
# license: MIT License       # code: github.com/gressling/examples
# activity: single example   # index: 19-2

manyElements = range(10000000)
rdd = sc.parallelize(manyElements, 50)

# work on RDD
odds = rdd.filter(lambda x: x % 2 != 0) # filter to odd numbers
odds.take(10)
> [1, 3, 5, 7, 9, 11, 13, 15, 17, 19]
```

`Parallelize()` turns that iterator into a distributed set of numbers and so offers the powerful capability of Spark's infrastructure. This example uses the RDD's `filter()` method instead of Python's built-in `filter()`. With the same result, the RDD-based method operation is *distributed across several* CPUs or nodes, instead of running locally in the Python interpreter.

Alternative: Apache Toree

For ajJupyter connection to a local spark cluster, you can also use Apache Toree.[6] Apache Toree (previously Spark Kernel) acts as the middleman between the application and a Spark cluster.[7] It is a kernel for the Jupyter Notebook platform that provides interactive access to Apache Spark and exposes the Spark programming model in Python.

Applications wanting to work with Spark can be located remotely from a Spark cluster and an Apache Toree Client or Jupyter Client can be used to communicate with an Apache Toree Server.

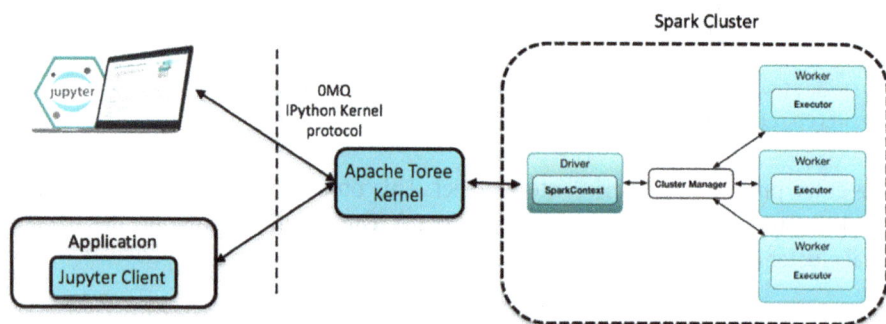

Figure 19.1. Apache ToreeAarchitecture.

- Notebooks send snippets of code, which are then executed by Spar ; and the results are returned directly to the application.
- Apache Toree supports visualizations that integrate directly with Spark Data Frames.
- Apache Toree provides magis that enhances the user experience, manipulating data coming from Spark tables or data. An example is: `%%sql  select * from molecules or %%dataframe molecules`.

SQLContext and SparkSession

Spark Context: Prior to Spark 2.0.0, sparkContext was used as a channel to access all spark functionality. After SPARK 2.0.came,

SparkSession provides a single point of entry to interact with underlying Spark functionality and allows programming Spark with DataFrame and Dataset APIs. Depreciated: Spark Session.[8]

All the functionality available in `sparkContext` are also available in `sparkSession`. In order to use APIs of SQL, HIVE, and Streaming, there is no need to create separate contexts, as `sparkSession` includes all the APIs.

```
# Spark configuration example
# author: Gressling, T
# license: MIT License        # code: github.com/gressling/examples
# activity: single example    # index: 19-3

conf = (SparkConf()
    .setAppName('COVID-19-Dataset')
    .setMaster('spark://head_node:56887')
    .setAll([('spark.executor.memory', '80g'),
    ('spark.executor.cores', '12'),
    ('spark.cores.max', '12'),
    ('spark.driver.memory', '80g')]))
conf.set('spark.authenticate', True)
conf.set('spark.authenticate.secret', 'secret-key')

sc = SparkContext(conf=conf)
sql_context = SQLContext(sc)
```

Spark code example

There are several examples of how to query data from a Big Data source, using RDKit. An implementation example is described by *Lovrić* for PySpark and RDKit.[9]

```
# PySpark and RDKit
# author: Lovrić, Mario Graz/Zagreb (Et al.)
# license: MIT License
# code: onlinelibrary.wiley.com/.../10.1002/minf.201800082
# activity: single example    # index: 19-4

from pyspark import SparkConf, SparkContext
from pyspark.sql import SQLContext
from rdkit import Chem
from rdkit import DataStructs
```

```
def clean(y):
    # parameters: y: RDD SMILES; Return: cleaned isomeric SMILES
    rdd2smile = y[0].encode("ascii", "ignore") # RDD to string
    chx = Chem.MolFromSmiles(rdd2smile)
    cleanedIsomericSMILES = Chem.MolToSmiles(chx, isomericSmiles=True)
    return cleanedIsomericSMILES


# list of descriptors from the RDKit Descriptors modules
descriptors = list(np.array(Descriptors._descList)[:, 0])
# the calculator module from RDKit
calculator = MoleculeDescriptors.MolecularDescriptorCalculator(descriptors)
df = sql_context.read.parquet(filepath)
sqlContext.read.parquet


# Example: diclofenac
l1 = FingerprintMols.FingerprintMol(Chem.MolFromSmiles('C1=CC=C(C...C2Cl)Cl'))
smiles_rdd = df.select('can_smiles').limit(num_lines).rdd.repartition(1000)
cleaned_rdd = smiles_rdd.map(clean)
cleaned_df = cleaned_rdd.map(lambda x: (x,)).toDF(['smiles'])
                .dropna(thresh=1, subset=('smiles'))
smiles_clean_rdd = cleaned_df.rdd.repartition(1000).persist()
mol_rdd = smiles_clean_rdd.map(lambda x: Chem.MolFromSmiles(rdd2smile(x)))


# assign results of the query for substructure 'CO' to sub_rdd
sub_rdd = mol_rdd.map(lambda x: x.HasSubstructMatch(Chem.MolFromS-
miles('CO')))
count = sub_rdd.collect().count(True)


mol_rdd = smiles_clean_rdd.map(lambda x:
            Chem.MolFromSmiles(rdd2smile(x)))
fng_rdd = mol_rdd.map(lambda x: _fng_mol(x))
sim_sol = fng_rdd.map(lambda x: _sim(l1, x))
            .filter(lambda x: x == 1.0).countByValue()


desc_data_rdd = smiles_clean_rdd.map(
    lambda x: comp_desc(rdd2smile(x), calculator))
descriptors_df = desc_data_rdd.map(
    lambda x: desc_dict(x)).toDF(descriptors)
descriptors_df.write.parquet('descriptor_data.parquet')
```

Spark applications in chemistry

- Efficient iterative virtual screening with Apache Spark and conformal pre-diction[10]
- Large-scale virtual screening on public cloud resources with Apache Spark[11]

References

1. Apache Spark (Wikipedia) https://en.wikipedia.org/wiki/Apache_Spark.
2. What are some good uses for Apache Spark? https://www.quora.com/What-are-some-good-uses-for-Apache-Spark.
3. Lee, L. First Steps With PySpark and Big Data Processing: https://realpython.com/pyspark-intro/.
4. Vázquez, F. How to use PySpark on your computer https://towardsdatascience.com/how-to-use-pyspark-on-your-computer-9c7180075617.
5. Benjamin, R. K. findspark (GitHub): https://github.com/minrk/findspark.
6. Toree (Apache) https://toree.apache.org/.
7. Apache Toree (IBM) https://developer.ibm.com/open/projects/apache-toree/.
8. SparkSession vs SparkContext in Apache Spark https://data-flair.training/forums/topic/sparksession-vs-sparkcontext-in-apache-spark/.
9. Lovrić, M.; Molero, J. M.; Kern, R. PySpark and RDKit: Moving towards Big Data in Cheminformatics: https://onlinelibrary.wiley.com/doi/full/10.1002/minf.201800082.
10. Ahmed, L.; Georgiev, V.; Capuccini, M.; Toor, S.; Schaal, W.; Laure, E.; Spjuth, O. Efficient iterative virtual screening with Apache Spark and conformal prediction: https://link.springer.com/content/pdf/10.1186%2Fs13321-018-0265-z.pdf.
11. Capuccini, M.; Ahmed, L.; Schaal, W.; Laure, E.; Spjuth, O. Large-scale virtual screening on public cloud resources with Apache Spark: https://jcheminf.biomedcentral.com/articles/10.1186/s13321-017-0204-4.

20 Files: structure representations

Molecular descriptors[1] play a fundamental role in computational chemistry. The way molecules, thought of as real bodies, are transformed into numbers, they allow the mathematical treatment of chemical information contained in the molecule.

Notations and languages

Since the last thirty years, a lot of notations and languages were defined to persist structures into symbols:
- ChEMBL IDs (via ChEMBL look-up
- SMILES, Canonical Smiles and the arbitrary target specification (SMARTS), an extension of SMILES for the specification of substructural queries
- Nomenclature (via OPSIN), Images (via OSRA)
- SYBYL Line Notation
- Molecular Query Language
- Chemistry Development Kit
- International Chemical Identifier (InChI), InChI Strings (via RDKit), InChI Keys (via ChEMBL look-up)

SMILES

The simplified molecular-input line-entry system (SMILES) is a specification in the form of a line notation for describing the structure of chemical species using short ASCII strings.

SMILES strings can be imported by most molecule editors for conversion back into two-dimensional drawings or three-dimensional models of the molecules.

```
# Copper(II) sulfate: [Cu+2].[O-]S(=O)(=O)[O-]
# Vanillin: O=Cc1ccc(O)c(OC)c1 and COCc1cc(C=O)ccc1O
```

SMARTS

SMILES arbitrary target specification (SMARTS) is a language for specifying substructural patterns in molecules. The SMARTS line notation is expressive and allows extremely precise and transparent substructural specification and atom typing.[2]

https://doi.org/10.1515/9783110629453-020

```
# combine acid oxygen and tetrazole nitrogen in a definition of
# oxygen atoms that are likely to be anionic
# under physiological conditions: [$([OH][C,S,P]=O),$([nH]1nnnc1)]
```

Canonical Smiles

A SMILES string is a way to represent a 2D molecular graph as a 1D string. In most cases, there are many possible SMILES strings for the same structure. Canonicalization is a way to determine which of all possible SMILES will be used as the reference SMILES for a molecular graph.[3]

Implementation example in different chemical toolkits

Parse two SMILES strings and convert them into canonical form. Check that the results give the same string (Python ***assert*** ... tests if a condition returns True)

```
# Convert SMILES to canonical SMILES comparison
# author: (various)
# license: CC-BY-SA (no changes)
# code: Convert_a_SMILES_string_to_canonical_SMILES
# activity: collection of examples # index: 20-1
```

```
# OpenBabel/Pybel
import pybel
smiles = ["CN2C(=O)N(C)C(=O)C1=C2N=CN1C", "CN1C=NC2=C1C(=O)N(C)C(=O)N2C"]
cans = [pybel.readstring("smi", smile).write("can") for smile in smiles]
assert cans[0] == cans[1]
```

```
# Indigo
from indigo import *
indigo = Indigo()
mol1 = indigo.loadMolecule("CN2C(=O)N(C)C(=O)C1=C2N=CN1C")
mol2 = indigo.loadMolecule("CN1C=NC2=C1C(=O)N(C)C(=O)N2C")
```

```
mol1.aromatize()
mol2.aromatize()
assert mol1.canonicalSmiles() == mol2.canonicalSmiles())
```

```
# OpenEye
from openeye.oechem import *
def canonicalize(smiles):
  mol = OEGraphMol()
  OEParseSmiles(mol, smiles)
  return OECreateCanSmiString(mol)

assert (canonicalize("CN2C(=O)N(C)C(=O)C1=C2N=CN1C") ==
        canonicalize("CN1C=NC2=C1C(=O)N(C)C(=O)N2C"))
```

```
# RDKit
from rdkit import Chem

smis = ["CN2C(=O)N(C)C(=O)C1=C2N=CN1C", "CN1C=NC2=C1C(=O)N(C)C(=O)N2C"]
cans = [Chem.MolToSmiles(Chem.MolFromSmiles(smi),True) for smi in smis]
assert cans[0] == cans[1]
```

```
# Cactvs
Prop.Setparam('E_SMILES',{'unique':True})
s1=Ens('CN2C(=O)N(C)C(=O)C1=C2N=CN1C').new('E_SMILES')
s2=Ens('CN1C=NC2=C1C(=O)N(C)C(=O)N2C').new('E_SMILES')
if (s1!=s2): raise RuntimeError('SMILES not equal')
```

File structure converter

OpenBabel

OpenBabel is a C++ toolkit. It has capabilities for reading and writing over 80 molecular file formats as well as capabilities for manipulating molecular data.[4] Also, the OpenBabel project provides converting the attributes of atoms and molecules through the Pybel[5] module.

OPSIN and CIRpy

Another interesting converter is OPSIN,[6] the Open Parser for Systematic IUPAC nomenclature by the University of Cambridge. It currently supports 33 nomenclatures. This is available as JAVA *.jar packages. A python alternative is CIRpy.[7] It is the Python interface for the Chemical Identifier Resolver (CIR).

References

1. Molecular descriptor (Wikipedia) https://en.wikipedia.org/wiki/Molecular_descriptor.
2. SMILES arbitrary target specification (Wikipedia) https://en.wikipedia.org/wiki/SMILES_arbitrary_target_specification.
3. Convert a SMILES string to canonical SMILES (Chemistry Toolkit) https://ctr.fandom.com/wiki/Convert_a_SMILES_string_to_canonical_SMILES.
4. O'Boyle, N. M.; Morley, C.; Hutchison, G. R. Pybel: a Python wrapper for the OpenBabel cheminformatics toolki https://www.ncbi.nlm.nih.gov/pmc/articles/PMC2270842. https://doi.org/10.1186/1752-153X-2-5.
5. OpenBabel. Pybel https://open-babel.readthedocs.io/en/latest/UseTheLibrary/Python_Pybel.html.
6. OPSIN. Open Parser for Systematic IUPAC Nomenclature https://opsin.ch.cam.ac.uk/.
7. Swain, M. CIRpy https://cirpy.readthedocs.io/en/latest/.

21 Files: other formats

A file format is a way that information is encoded for storage designed for a very particular type of data. The file should have a published specification describing the encoding method. The type can be identified by a filename extension (*.mol) or self-describing internal metadata (header, *.cif, or *.cml) or extern metadata (chemical/MIME). They can be unstructured (raw memory dumps) or structured (XML, JSON).

Chemical Markup Language (CML)

Chemical Markup Language (CML) is an XML-based open standard for representing molecular and other chemical data. CompChem[1] adds computational chemistry semantics on top of the CML schema.

- The XML definition contains a set of hundreds of chemical name tags:
  - CMLReactl for reactions
  - CMLSpec for spectral data
  - Crystallography
  - polymers (PML)
- Supported by OpenBabel, PyBel, Jmol, Avogadro

For use with standard XML / XSLT mechanisms[2] in Python. Example file "Spectral_DataSet.xml":

```
<!-- reference: Peter Murray-Rust[3] --->
<!DOCTYPE CML PUBLIC "-//CML//DTD CML//EN">
<CML>
 <MOL>
 <XLIST TITLE="Spectral DataSet" BUILTIN="INFRARED">
  <XLIST BUILTIN="SPECTRUM" CONTENT="GRAPH" DISPLAY="CONTINUOUS">
  <ARRAY SIZE="3735" DICTNAME="XVALUES" TYPE="FLOAT"
         CONVENTION="JCAMP" TITLE="X-Axis">
 399.195 400.159 401.123 402.088 403.052 404.016 404.98 405.945 406.909
 408.837 409.802 410.766 411.73 412.694 413.659 414.623 415.587 416.551
 ...
  </ARRAY>
  </XLIST>
 </XLIST>
 </MOL>
</CML>
```

https://doi.org/10.1515/9783110629453-021

Example code for reading the data

```python
# Parsing data from file Chemical Markup Language (CML)
# author: Gressling, T
# license: MIT License        # code: github.com/gressling/examples
# activity: single example  # index: 21-1
import xml.etree.ElementTree as ET
tree = ET.parse('Spectral_DataSet.xml')
root = tree.getroot()

# query elements
print(root.tag)
> 'CML'

# query elements
for xlist in root.findall('XLIST'):
   dictName = xlist.get('DICTNAME')
   print(name)
> HEADERINFORMATION
> MISCELLANEOUS
> REQUIREDSPECTRALPARAMETERS
> ...
```

MDL Molfile CTAB

Chemical table[4] file (CT file) is a family of text-based chemical file formats that describe molecules and chemical reactions. One format, for example, lists each atom in a molecule, the x-y-z coordinates of that atom, and the bonds among the atoms. The molfile consists of some header information, the Connection Table containing atom info, then bond connections and types, followed by sections for more complex information.

```python
# file extension: *.mol
# MIME-TYPE: chemical/x-mdl-molfile
# https://www.ebi.ac.uk/chebi/searchId.do?chebiId=CHEBI:30746
# Benzoic Acid

molfile= open('ChEBI_30746.mol', 'r')
print(molfile.read())
```

```
> Marvin 06110722242D
> 9  9  0  0  0  0           999 V2000
>   -0.7145    0.4125   0.0000 C   0  0  0  0  0  0  0  0  0  0  0  0
>   -0.7145   -0.4125   0.0000 C   0  0  0  0  0  0  0  0  0  0  0  0
...
>    1.4289    1.6499   0.0000 O   0  0  0  0  0  0  0  0  0  0  0  0
> 1  2  2  0  0  0  0
> 8  3  1  0  0  0  0
> 4  1  1  0  0  0  0
...
> 7  8  1  0  0  0  0
> 8  9  2  0  0  0  0
>M END
```

SDF file format

SDF[4] is one of a family of chemical-data file formats developed by MDL; it is intended especially for structural information. "SDF" stands for structure-data file, and SDF files actually wrap the molfile (MDL Molfile) format. A feature of the SDF format is its ability to include associated data.

```
# fileextension: *.sdf      # MIME-TYPE: chemical/x-mdl-molfile
# SDF File Format
# example: github.com/NicolasCARPi/example-files/example.sdf

molecules = Chem.SDMolSupplier('example.sdf')
molecules
> <rdkit.Chem.rdmolfiles.SDMolSupplier at 0x1819c070>

mol = [m for m in molecules]
len(mol)
> 321
```

File formats in MD, MM, and QM

Many topology and coordinate file formats have been created since the last 50 years.

Structure files

- Amber (*.prmtop)
- CHARMM/NAMD (*.pdb)
- CHARMM/NAMD (*.psf)
- CHARMM (*.crd)
- Gromacs (*.gro)
- Tinker (*.xyz)

Trajectory files

- Amber (*.nc, *.netcdf), velocities
- Amber (*.crd, *.mdcrd), velocities (if NetCDF)
- Amber (*.inpcrd, *.rst, *.rst7)
- CHARMM/NAMD (*.dcd)
- CHARMM/NAMD (*.pdb), concatenated PDBs or sets of PDBs
- Gromacs (*.xtc)
- Gromacs (*.trr), velocities, pressure, virial, forces, only velocities available with trajectory class
- Tinker (*.arc)

GROMACS

The GROMACS file format[5] family was created for use with the molecular simulation software package GROMACS. It closely resembles the PDB format but was designed for storing the output from molecular dynamics simulations, so it allows for additional numerical precision and optionally retains information about particle velocity as well as a position at a given point in the simulation trajectory.

CHARMM

The CHARMM molecular dynamics package can read and write[5] a number of stand-ard chemical and biochemical file formats; however, the CARD (coordinate) and PSF (protein structure file) are largely unique to CHARMM. The CARD format is fixed-column-width, resembles the PDB format, and is used exclusively for storing atomic coordinates. The PSF file contains atomic connectivity information (which describes atomic bonds) and is required before beginning a simulation.

References

1. Phadungsukanan, W.; Kraft, M.; Townsend, J. A.; Murray-Rust, P. The semantics of Chemical Markup Language (CML) for computational chemistry : CompChem: https://www.ncbi.nlm.nih.gov/pmc/articles/PMC3434037.
2. The ElementTree XML API – Python documentation https://docs.python.org/2/library/xml.etree.elementtree.html.
3. Murray-Rust, P. CML Examples: http://www.ch.ic.ac.uk/omf/cml/doc/examples/.
4. Chemical table file (Wikipedia) https://en.wikipedia.org/wiki/Chemical_table_file.
5. Chemical file format (Wikipedia) https://en.wikipedia.org/wiki/Chemical_file_format.

22 Data retrieval and processing: ETL

Extract, transform, load (ETL) is the general procedure of copying data from one or more sources into a destination system that represents the data differently from the sources or in a different context than the sources.[1]

In the data science process, it is a time-consuming step before data-wrangling starts. The process involves extracting data from homogeneous or heterogeneous sources as well as transformation processes like data cleansing and death them into a proper storage format or structure. The design goal is preparing for querying and analysis. The last step is the insertion of data into the final target database.

The ETL cycle consists of the following execution steps:
1. Initiation
2. Build reference data
3. Extract (from sources)
4. Validate
5. Transform (clean, apply business rules, check for data integrity, create aggregates or disaggregates)
6. Stage (load into staging tables, if used)
7. Audit reports (for example, on compliance with business rules. Also, in case of failure, helps to diagnose/repair)
8. Publish (to target tables)
9. Archive

Data cleansing example: load, filter, save

```
# Simple ETL example
# author: Gressling, T
# license: MIT License        # code: github.com/gressling/examples
# activity: single example     # index: 22-1


import pandas as pd

# STEP 1 - EXTRACT
df = pd.read_csv('logP_dataset.csv', low_memory=True, header=None)


# STEP 2 - TRANSFORM (here: Filter)
# filter example row for 1-bromo-1-methylsulfanylpropane
# https://pubchem.ncbi.nlm.nih.gov/compound/88089970
df = df.loc[df[0] == 'CCC(SC)Br', :]
```

https://doi.org/10.1515/9783110629453-022

```
STEP 3 - LOAD
df.to_csv('logP_dataset-filter.csv', sep=',', encoding='utf-8', index=False)
```

Bonobo

Bonobo[2,3] is a lightweight extract-transform-load (ETL) framework for Python. It provides tools for building data transformation pipelines using plain python primitives and executing them in parallel. In Chapter 26 more data pipeline framworks are shown.

Here is an example of a bonobo ETL framework with graphical output:

```
# ETL with bonobo
# author: Gressling, T
# license: MIT License          # code: github.com/gressling/examples
# activity: single example      # index: 22-1

!pip install bonobo
import bonobo

def generate_data():
    yield 'Data'
    yield 'science'
    yield 'in'
    yield 'chemistry'

def uppercase(x: str):
    return x.upper()

def output(x: str):
    print(x)

graph = bonobo.Graph(
    generate_data,
    uppercase,
    output,
)
graph.add_chain(output, _input=generate_data)
graph
```

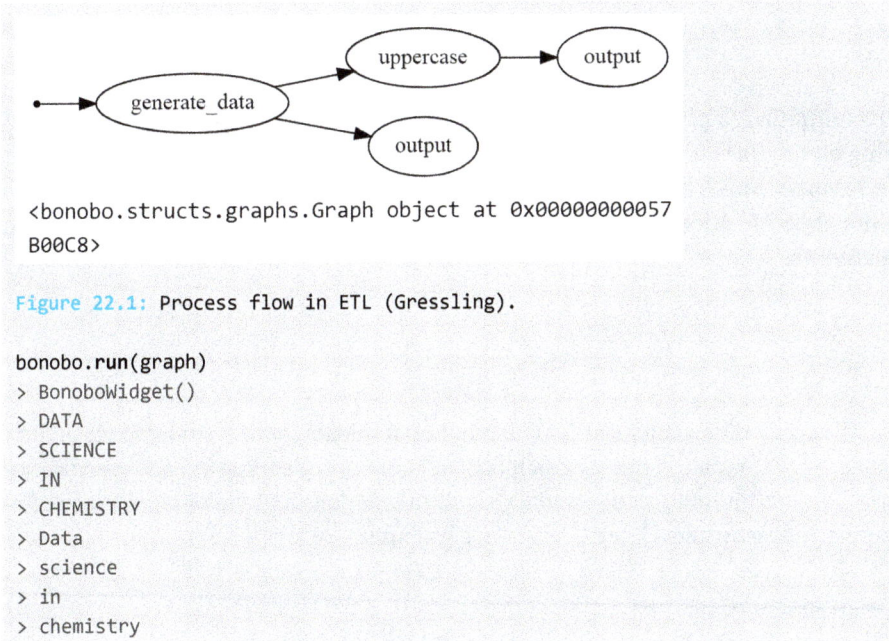

```
<bonobo.structs.graphs.Graph object at 0x00000000057
B00C8>
```

Figure 22.1: Process flow in ETL (Gressling).

```
bonobo.run(graph)
> BonoboWidget()
> DATA
> SCIENCE
> IN
> CHEMISTRY
> Data
> science
> in
> chemistry
```

References

1. Extract, transform, load (Wikipedia) https://en.wikipedia.org/wiki/Extract,_transform,_load.
2. bonobo https://github.com/python-bonobo/bonobo.
3. Bonobo · Data-processing for humans · Python ETL https://www.bonobo-project.org/.

23 Data pipelines

In computing, a pipeline,[1] also known as a data pipeline, is a set of data processing elements connected in series, where the output of one element is the input of the next one. The elements of a pipeline are often executed in parallel or in time-sliced fashion. Some amount of buffer storage is often inserted between elements.

aiiDA

aiiDA[2] is a workflow manager for computational science, with a strong focus on provenance, performance, and extensibility. aiiDA supports workflows, data provenance (track inputs, outputs, and metadata) of all calculations and high-performance computing with schedulers like SLURM. Code example:

```
# aiida workflow
# author: Aiida Team http://www.aiida.net
# author: THEOS/MARVEL, EPFL (Switzerland); Bosch RTC in Cambridge MA
# license: MIT License              # code: github.com/aiidateam/aiida_demos
# activity: active (2019)           # index: 23-1

from aiida import load_dbenv, Node, Code, Calculation, CalculationFactory,
...

PwCalculation = CalculationFactory('quantumespresso.pw')

qb = QueryBuilder()
qb.append(Code, filters={'label':'pw-SVN-piz-daint'}, tag='code')
qb.append(PwCalculation, output_of='code', tag='calculation',
          filters={'attributes.jobresource_params.num_machines':1}
    )
qb.append(ParameterData, input_of='calculation',
          filters={'attributes.CONTROL.calculation':'vc-relax'})
qb.append(ParameterData, output_of='calculation',
project=('attributes.wall_time_seconds'))

all_walltimes = qb.all()
print "Number of calculations:", len(all_walltimes)

generate_query_graph(qb.get_json_compatible_queryhelp(), '1.png')
Image(filename='1.png')
```

AiDA is available as a docker image. It also supports quantum espresso.

https://doi.org/10.1515/9783110629453-023

Build pipelines with Pandas using pdpipe

pdpipe[3] is a simple framework for serializable, chainable, and verbose Pandas pipelines. The library processes pipelines that can be broken down or composed together and adhere to scikit-learn's transformer API.

```python
# Build pandas pipelines using pdpipe
# author: Gressling, T
# license: MIT License        # code: github.com/gressling/examples
# activity: single example   # index: 23-2

import pdpipe as pdp
df = pd.DataFrame(
    data=[[smiles, 0.7633, 'HIGH'], [smiles, 0.5664, 'MED'], [smiles, 0.8977,
'LOW']],
    index=['mol1', 'mol2', 'mol3'],
    columns=['smiles', 'fingerprint', 'tc'] # tc is 'toxicity class'
    )

pipeline = pdp.ColDrop('smiles')
pipeline+= pdp.OneHotEncode('mass')
# pipeline+= ...
# pipeline+= ...

df_out = pipeline(df)
df_out
>           fingerprint  tc_HIGH    tc_MED
>    mol1      0.7633        1          0
>    mol2      0.5664        0          1
>    mol3      0.8977        0          0
```

ChEMBL structure pipeline

ChEMBL protocols can also model pipelines, for example,[4] to standardize and salt strip molecules. This is shown in the following example:

```python
# Salt strip in ChEMBL structure pipeline
# author: Gressling, T
# license: MIT License        # code: github.com/gressling/examples
# activity: single example   # index: 23-3

o_molblock = """
  Marvin  10310613082D
```

```
 6  6  0  0  0  0            999 V2000
   0.7145   -0.4125    0.0000 C   0  0  0  0  0  0  0  0  0  0  0  0
   0.0000   -0.8250    0.0000 C   0  0  0  0  0  0  0  0  0  0  0  0
   0.7145    0.4125    0.0000 C   0  0  0  0  0  0  0  0  0  0  0  0
   0.0000    0.8250    0.0000 C   0  0  0  0  0  0  0  0  0  0  0  0
  -0.7145   -0.4125    0.0000 C   0  0  0  0  0  0  0  0  0  0  0  0
  -0.7145    0.4125    0.0000 C   0  0  0  0  0  0  0  0  0  0  0  0
 2  1  2  0  0  0  0
 3  1  1  0  0  0  0
 4  3  2  0  0  0  0
 5  2  1  0  0  0  0
 6  4  1  0  0  0  0
 5  6  2  0  0  0  0
M  END"""
from chembl_structure_pipeline import standardizer
std_molblock = standardizer.standardize_molblock(o_molblock)

# Get the parent molecule
parent_molblock, _ = standardizer.get_parent_molblock(o_molblock)

# Check the molecule
from chembl_structure_pipeline import checker
issues = checker.check_molblock(o_molblock)
```

References

1. Pipeline (computing) (Wikipedia) https://en.wikipedia.org/wiki/Pipeline_(computing).
2. aiida-core (GitHub) https://github.com/aiidateam/aiida-core.
3. pdpipe – Easy pipelines for pandas dataframes (GitHub) https://pdpipe.github.io/pdpipe/.
4. ChEMBL Structure Pipeline (GitHub) https://github.com/chembl/ChEMBL_Structure_Pipeline.

24 Data ingestion: online data sources

Software as a service (SaaS[1]) is a software licensing and delivery model in which software is licensed on a subscription basis and is centrally hosted. It is sometimes referred to as "on-demand software."

Due to the huge amount of information accumulated in chemistry, using online resources has become a standard way of data ingestion. This chapter gives examples of different implementation patterns.

- Pubchem
- ChemSpider
- Wikipedia and Wikidata
- CHEMBL/ChEBI
- ZINC
- datasetsearch.research.google.com
- Other online resources

PubChem

PubChem is a database of chemical molecules and their activities against biological assays. The system[2,3] is maintained by the National Center for Biotechnology Information (NCBI) in the U.S. Millions of compound structures and descriptive datasets can be freely downloaded. PubChem contains substance descriptions and small molecules with fewer than 1,000 atoms and 1,000 bonds. More than 80 database vendors contribute to the growing PubChem database.

Python example

The database is accessible from python via pubchempy[4]:

```
# Query Pubchem online
# author: Swain, Matt  (Et al.)
# license: MIT License      # code: github.com/.../PubChemPy
# activity: on hold (2017)  # index: 24-1


from pubchempy import get_compounds, Compound

comp = Compound.from_cid(1423)
print(comp.isomeric_smiles)
> CCCCCCCNC1CCCC1CCCCCC(=O)O
```

https://doi.org/10.1515/9783110629453-024

```
comps = get_compounds('Aspirin', 'name')
print(comps[0].xlogp)
> 1.2

# A full list of options is available at the PUG REST specification5
```

ChemSpider

ChemSpider is owned by the Royal Society of Chemistry. The database[6] contains information on more than 77 million molecules from over 270 data sources.

ChemSpiPy

ChemSpiPy[7] is a Python wrapper that allows simple access to the web APIs offered by ChemSpider. The aim is to provide an interface for users to access and query the database. After registration, an API-KEY is required:

```
# Query ChemSpider online
# author: Gressling, T
# license: MIT License        # code: github.com/gressling/examples
# activity: single example    # index: 24-2

from chemspipy import ChemSpider
cs = ChemSpider('<YOUR-API-KEY>')
c1 = cs.get_compound(236)  # Specify compound by ChemSpider ID
c2 = cs.search('benzene')  # Search using name, SMILES, InChI, InChIKey, etc.
```

Wikipedia and Wikidata

Wikidata[8] is a free and collaborative Linked Open Data (LOD) knowledge base that can be edited by humans and machines. The project was started by the Wikimedia Foundation as an effort to centralize links, information, and enable queries. All of the data there is free (under the CC0).

For many daily laboratory questions, chemists just use Wikipedia via the browser. Using python, Wikidata can be queried by using its SPARQL[9] endpoint that enables the chemist to run advanced queries or by using its REST API. Also, Wikipedia is open to python.

Python example

```
# Query Wikidata and Wikipedia
# author: Gressling, T
# license: MIT License         # code: github.com/gressling/examples
# activity: single example     # index: 24-3


import wikipedia

# query wikipedia to get a list
print(wikipedia.search("benzoic"))
> ['Benzoic acid', 'Benzoic anhydride', 'Sodium benzoate', 'Benzaldehyde',
> 'Saccharin', 'Benzyl benzoate', 'Benzoyl peroxide', 'Work-up (chemistry)' ...

# get one single article https://en.wikipedia.org/wiki/Perovskite_(structure)
wiki = wikipedia.page("Benzoic acid")
wiki.summary
> 'Benzoic acid is a white (or colorless)  solid with the formula C6H5CO2H.
> It is the simplest aromatic carboxylic acid. The name is derived ...

#!pip install qwikidata
from qwikidata.entity import WikidataItem, WikidataLexeme, WikidataProperty
from qwikidata.linked_data_interface import get_entity_dict_from_api

# create an item representing "benzoic acid"
entity = get_entity_dict_from_api("Q191700")
item = WikidataItem(entity)
item
> WikidataItem(label=benzoic acid, id=Q191700, description=chemical compound,
> aliases=['Retardex', 'E210', 'phenylcarboxylic acid'

# create a property representing "subclass of"
subclassItem = get_entity_dict_from_api("P3117")
subclass = WikidataProperty(subclassItem)
subclass
> WikidataProperty(label=DSSTox substance ID, id=P3117, description=DSSTox
> substance identifier (DTXSID) used in the
> Environmental Protection Agency CompTox Dashboard, aliases=['DTXSID'])
```

CHEMBL/ChEBI

Chemical Entities of Biological Interest, also known as ChEBI, is a manually curated database and ontology of molecular entities focused on "small" chemical compounds.

Currently, it contains over 1.9 million of compounds extracted from over 72,000 documents. Altogether within the ChEMBL, there are over 65 million of bioactivity measurements gathered from over 1.1 million bioassays.[10]

To have independent services with separated responsibilities, it is possible to docker chembl:

```
$ docker pull chembl/unichem_rhel7
```

Python via libChEBIpy

is a Python API[11–13] for accessing the ChEBI database. Ontologies are discussed further in Chapter 88.

```
# Query ChEBI ontology via Python
# author: Gressling, T; based on libChEBIpy documentation
# license: MIT License        # code: github.com/gressling/examples
# activity: single example    # index: 24-4

chebi_entity = ChebiEntity('CHEBI:15903')
print chebi_entity.get_name()

for outgoing in chebi_entity.get_outgoings():
  target_chebi_entity = ChebiEntity(outgoing.get_target_chebi_id())
  print outgoing.get_type() + '\t' + target_chebi_entity.get_name()

> beta-D-glucoseis_enantiomer_of beta-L-glucose
> has_role epitope
> is_a D-glucopyranose
```

Query via REST

The CheMBL Web Services[10,14] provides simple reliable programmatic access to the data stored in the ChEMBL database. RESTful API approaches are quite easy to master in most languages but still require writing a few lines of code.

```
# Query ChEBI data via REST
# author: Gressling, T; based on documentation by Kott
# license: MIT License        # code: github.com/gressling/examples
# activity: single example    # index: 24-5

#!pip install chembl_webresource_client

from chembl_webresource_client import *
```

```
compounds = CompoundResource()

c = compounds.get('CHEMBL1')
print c
> {u'smiles': u'COc1ccc2[C@@H]3[C@H](COc2c...C(C)(C)[C@@H]6COc7cc(OC)
ccc7[C@H]56',
>   u'chemblId': u'CHEMBL1', u'passesRuleOfThree':
>   u'No', u'molecularWeight': 544.59, u'molecularFormula': u'C32H32O8'...}

print compounds.drug_mechnisms('CHEMBL1642')
> [{u'chemblId': u'CHEMBL1862', ..., u'mechanismOfAction':
> u'Stem cell growth factor receptor inhibitor'}]
```

ZINC

The ZINC database (recursive acronym: ZINC is not commercial) is a curated collection of commercially available chemical compounds. ZINC is different from other chemical databases because it aims to represent the biologically relevant, three-dimensional form of the molecule. ZINC is updated regularly. It may be downloaded and used free of charge.[15]

Python access

smilite[16] is a Python module to download and analyze SMILES strings from ZINC:

```
# Query ZINC database
# author: Raschka[16], Sebastian
# License: GNU General Public License v3.0 # code: github.com/rasbt/smilite
# activity: active (2020)      # index: 24-6

import smilite
import sys

smile_str = ''
simple_smile_str = ''

zinc_id = ZINC01234567
backend = 'zinc15'

smile_str = smilite.get_zinc_smile(zinc_id, backend=backend)
if smile_str:
        simple_smile_str = smilite.simplify_smile(smile_str)
print('{}\n{}\n{}'.format(zinc_id, smile_str, simple_smile_str))

> C[C@H]1CCCC[NH+]1CC#CC(c2ccccc2)(c3ccccc3)O
> CC1CCCCN1CCCC(C2CCCCC2)(C3CCCCC3)O
```

datasetsearch.research.google.com

Google Dataset Search is a search engine from Google that helps researchers locate online data that is freely available for use. The company launched the service on September 5, 2018, and stated that the product was targeted at scientists and data journalists. Google Dataset Search complements Google Scholar, the company's search engine for academic studies, and reports.[17, 18]

This API is not to be confused with the google patent search API that is discussed in Chapter 91.

Other online resources

- cactus.nci.nih.gov (OSRA chemical OCR server, structure lookup service)[19]
- rcsb.org (PDB code access); REST API[20]
- webbook.nist.gov/chemistry[21]
- crystallography.net[22]
- List of chemical databases in Wikipedia[23]
- NMR[24]

Security concerns

By contacting Internet services for specific property data retrievals, the API may expose your structure data to third parties. This should be avoided for confidential information.

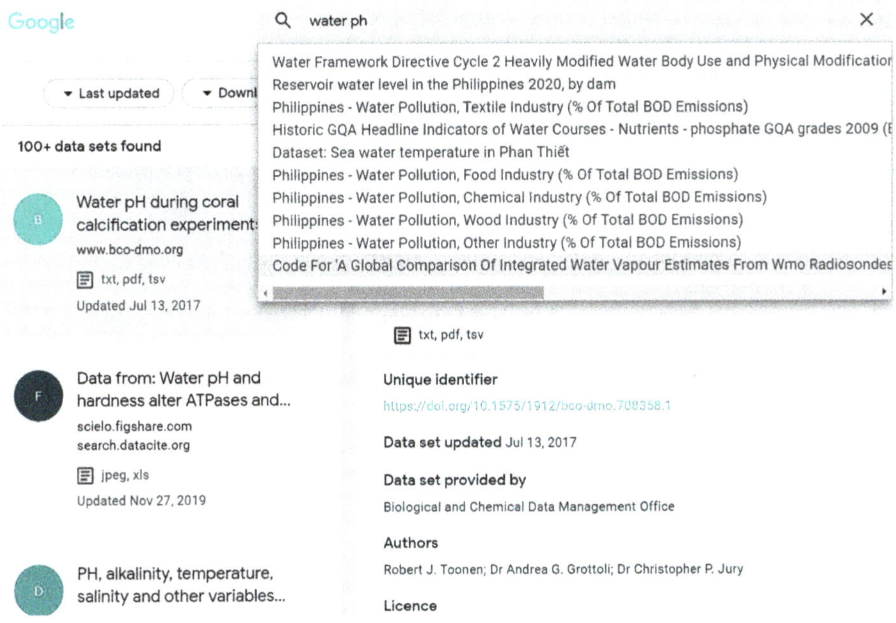

Figure 24.1: Google Dataset Search *(Google)*.

References

1. Software as a service (Wikipedia) https://en.wikipedia.org/wiki/Software_as_a_service.
2. PubChem (Wikipedia) https://en.wikipedia.org/wiki/PubChem.
3. PubChem – NCBI https://pubchem.ncbi.nlm.nih.gov/.
4. PubChemPy (PyPI) https://pypi.org/project/PubChemPy/1.0/.
5. PubChem REST interface: PUG https://pubchemdocs.ncbi.nlm.nih.gov/pug-rest.
6. ChemSpider (Wikipedia) https://en.wikipedia.org/wiki/ChemSpider.
7. ChemSpiPy documentation https://chemspipy.readthedocs.io/en/latest/guide/intro.html.
8. Wikidata (Wikidata) https://www.wikidata.org/wiki/Wikidata:Main_Page.
9. Janakiev, N. Where do Mayors Come From: Querying Wikidata with Python and SPARQL http://janakiev.com/blog/wikidata-mayors/.
10. Bachorz, R. A. The ChEMBL database of molecules in a Docker environment https://medium.com/@rbachorz/the-chembl-database-of-molecules-in-a-docker-environment-6cdc64d4f0e.
11. libChEBIpy (GitHub) https://github.com/libChEBI/libChEBIpy.
12. Swainston, N.; Hastings, J.; Dekker, A.; Muthukrishnan, V.; May, J.; Steinbeck, C.; Mendes, P. libChEBI: an API for accessing the ChEBI database https://www.ncbi.nlm.nih.gov/pmc/articles/PMC4772646/.
13. libChEBI API Documentation http://libchebi.github.io/libChEBI%20API.pdf.
14. Kott. A python client for accessing ChEMBL web services http://chembl.blogspot.com/2014/05/a-python-client-for-accessing-chembl.html.
15. ZINC database (Wikipedia) https://en.wikipedia.org/wiki/ZINC_database.

16. Raschka, S. smilite – Python module to download and analyze SMILES from ZINC (GitHub) https://github.com/rasbt/smilite.
17. Google Dataset Search (Wikipedia) https://en.wikipedia.org/wiki/Google_Dataset_Search.
18. Dataset Search: https://datasetsearch.research.google.com/search?query=water%20ph&docid=U1Q%2FRCO4g7IpcJb6AAAAAA%3D%3D.
19. CACTUS – Chemical Structure Lookup Service https://cactus.nci.nih.gov/cgi-bin/lookup/search.
20. RCSB PDB – REST Web Service http://www.rcsb.org/pdb/software/rest.do.
21. NIST Chemistry WebBook – National Institute of Standards and Technology https://webbook.nist.gov/chemistry/.
22. Crystallography Open Database: Search results http://www.crystallography.net/cod/result.php?spacegroup=P%20-1.
23. Category:Chemical databases (Wikipedia) https://en.wikipedia.org/wiki/Category:Chemical_databases.
24. Simulate and predict NMR spectra http://www.nmrdb.org/.

25 Designing databases

A database is an organized collection of data, generally stored and accessed electronically from a computer system. Where databases are more complex they are often developed using formal design and modeling techniques.[1]

Relational databases and SQL

The Structured Query Language (SQL) was standardized in the mid-nineties. Together with DBMS (Database Management Systems), it became the fundament of the first wave of digitalization. Use cases are transaction-correct data (constant) as well as storage systems with specific response times. Examples of ANSI SQL statements:

```
Create table crystal_structure (pkey integer asc, name varchar(20), atom
varchar(2), x integer, y integer, z integer, spacegroup varchar(20))

// Select only the info we need
Select distinct atom, x, y, z from crystal_structure a where
a.name='perovskite';
```

Entity-relationship diagrams are used to document and design a relational database system; this example is a tiny Laboratory Management Information System (LIMS):

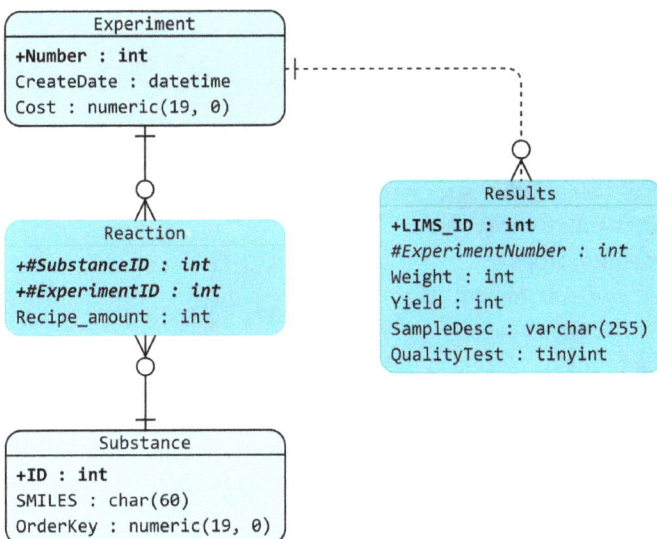

Figure 25.1: Entity-relationship (ER) diagram *(Gressling)*.

https://doi.org/10.1515/9783110629453-025

DB example

psycopg2[2] is the most popular PostgreSQL database adapter for the Python programming language. Its main features are the complete implementation of the Python DB API 2.0 specification and the thread safety (several threads can share the same connection). It was designed for heavily multithreaded applications that create and destroy lots of cursors and make a large number of concurrent "INSERT"s or "UPDATE"s.

```
# SQL with a relational database
# author: Gressling, T
# License: MIT License          # code: github.com/gressling/examples
# activity: single example    # index: 25-1

import psycopg2
conn = psycopg2.connect("dbname='example_db' user='dbuser'
                         host='localhost' password='password'")
cur = conn.cursor()

cur.execute("""SELECT * from results where yield>80""")
rows = cur.fetchall()

for row in rows:
    print("   ", row[0])
```

ipython-sq[3] introduces `%sql` magic. Connect to a database and issue SQL commands within the notebook.

Non-SQL databases

Different types of data, as well as other structures, led to the development of NoSQL databases. Examples of the four basic types:
- **Wide column:** Cassandra, HBase
- **Document:** Apache CouchDB, MongoDB
- **Key-value:** Berkeley DB, Couchbase
- **Graph:** Apache Giraph, Neo4J

Neo4J

Graph-DB's are of special interest: Chemistry is based on structure representations; and in artificial intelligence, the network topologies are also structures. And finally a relevant field of application are Knowledge Graphs.

The main difference to relational databases is that the data model of relational databases is defined outside the data, whereas, in graph databases the model is defined by the data. No data, no model.

Please refer to Chapters 88 and 89 for details on graphs. Here, the technical aspects are discussed. Neo4J is also involved in ontology modeling discussed in Chapter 88.

Cypher

To query a graph database, the standard language is CYPHER:

```
MATCH (person:labworker)-[:PERFORMS]->(experiment:name)
RETURN person.labworker AS worker, experiment.name AS experiment
```

Here is an example of graph databases that work with python, see also Chapter 89 where networks are discussed. Use py2neo[4,5] library for connecting the Jupyter notebook to the Neo4j Server. It is a very simple library that connects a python application with the Neo4j database. Run "`pip install py2neo`" from the shell to install this library. Also, a neo4j database is easy to install[6] and set up.

```
# Neo4J graph example
# author: Gressling, T
# license: MIT License        # code: github.com/gressling/examples
# activity: single example    # index: 25-2


# from neo4j import GraphDatabase
# !pip install py2neo
from py2neo import Graph
graph = Graph("bolt://localhost:7687", auth=("neo4j", "password"))
graph.delete_all()

# define the entities of the graph (nodes)
from py2neo import Node

laboratory = Node("Laboratory", name="Laboratory 1")
lab1 = Node("Person", name="Peter", employee_ID=2)
lab2 = Node("Person", name="Susan", employee_ID=4)
sample1 = Node("Sample", name="A-12213", weight=45.7)
```

```
sample2 = Node("Sample", name="B-33443", weight=48.0)
# shared sample between two experiments
sample3 = Node("Sample", name="AB-33443", weight=24.3)
experiment1 = Node("Experiment", name="Screening-45")
experiment2 = Node("Experiment", name="Screening/w/Sol")

graph.create(laboratory | lab1 | lab2 | sample1 | sample2 | experiment1 |
experiment2)

# Define the relationships of the graph (edges)
from py2neo import Relationship

graph.create(Relationship(lab1, "works in", laboratory))
graph.create(Relationship(lab2, "works in", laboratory))
graph.create(Relationship(lab1, "performs", sample1))
graph.create(Relationship(lab2, "performs", sample2))
graph.create(Relationship(lab2, "performs", sample3))
graph.create(Relationship(sample1, "partof", experiment1))
graph.create(Relationship(sample2, "partof", experiment2))
graph.create(Relationship(sample3, "partof", experiment2))
graph.create(Relationship(sample3, "partof", experiment1))

import neo4jupyter
neo4jupyter.init_notebook_mode()
neo4jupyter.draw(graph)
```

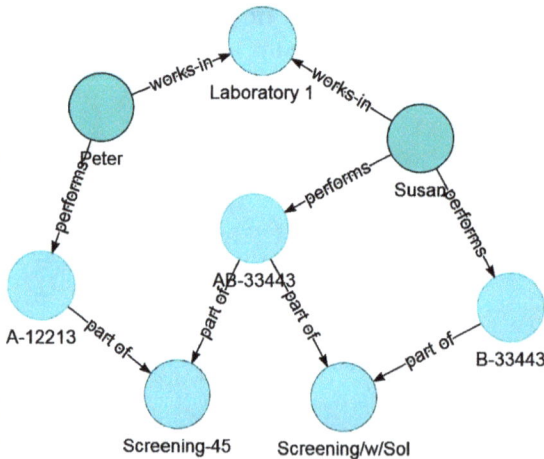

Figure 25.2: Graph of neo4j database content.

Big data

As the focus of data science is pattern detection, one of the major building blocks is a huge amount of data. *'Big Data'* can be structured or unstructured. The term and examples of using Hadoop, Big data, and Spark have been discussed in Chapters 18 and 19.

References

1. Database (Wikipedia) https://en.wikipedia.org/wiki/Database.
2. psycopg2 – Python-PostgreSQL Database Adapter (PyPi) https://pypi.org/project/psycopg2/.
3. ipython-sql – RDBMS access (PyPI) https://pypi.org/project/ipython-sql/.
4. hello-world – neo4j with jupyter (GitHub) https://nicolewhite.github.io/neo4j-jupyter/hello-world.html.
5. Maeztu, G. neo4jupyter (GitHub): https://github.com/merqurio/neo4jupyter.
6. Neo4j Desktop Setup https://neo4j.com/download-thanks-desktop.

26 Data science workflow and chemical descriptors

The cross-industry standard process for data mining, known as CRISP-DM[1], is an open standard process model that describes common approaches used by data mining experts. It is the most widely-used analytics model. Feature engineering is the process of using domain knowledge to extract features from raw data via data mining techniques.

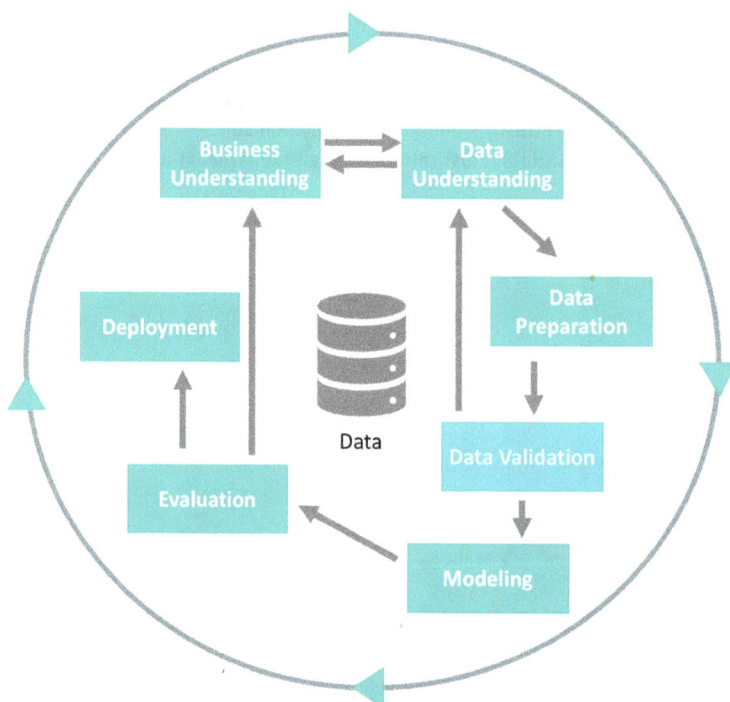

Figure 26.1: Data science process CRISP-DM[1] *(Majid)*.

Exploratory data analysis aka feature engineering is the most important part of machine learning – not model fitting! Data science newcomers move to complex models too soon.

That means that simple algorithms are often the winning algorithms for such datasets. Feature engineering is what differentiates a winning solution from others stuck in the design of complex artificial neural networks.

https://doi.org/10.1515/9783110629453-026

Roles

Especially in bigger companies,[2] the data science process can be divided into different
roles:

- **data exploration** – occurs early in a project; includes viewing sample data,
  running queries for statistical profiling
- **data preparation** – iterative task; includes cleaning, standardizing, transforming, denormalizing, and aggregating data; the most time-intensive task of a
  project!
- **data validation** – recurring task; occurs as part of data exploration, data preparation, development, pre-deployment, and post-deployment phases
- **productionalization** – deploying code to production, backfilling datasets, training models, validating data, and scheduling

1 Data governance

- Question underlying *assumptions* about the data
- Identify how to resolve *discrepancies* in data sources
- Evaluating if *new data sources* are valuable

2 Understand the data

Summary statistics and visualizations
- Percentiles identify the *range* for most of the data
- Averages and medians describe the central *tendency*
- Correlations can indicate *relationships*

Visualize
- Box plots identify *outliers*
- Density plots and histograms show the *spread*
- Scatter plots can describe *relationships*

3 Data cleansing

- *Missing data* affects some models more than others
- *Outliers* removal
- Perhaps data needs to be *aggregated*
- *Rescale variables* (e.g. standardizing or normalizing)
- *Calculation* of surrogate data

4 Data wrangling

A *feature* is an attribute or property shared by all of the independent units on which analysis or prediction is to be done. It is a characteristic that might help when solving the problem:
- *strongly relevant* (i.e., the feature has information that does not exist in any other feature),
- *relevant*
- *weakly relevant* (some information that other features include) or
- *irrelevant* (even if some features are irrelevant, having too many is better than missing those that are important)

5 Find descriptors (feature engineering)

Features are derived from descriptors. Calculations obtaining the descriptors by using classic cheminformatics play a vital role in data science.

The transformation of data with the help of cheminformatic into descriptors and features introduces implicit domain knowledge to data science in chemistry. The molecular descriptor[3] is the final result of a logic and a mathematical procedure that transforms chemical information encoded within a symbolic representation of a molecule into a useful number or the result of some standardized experiment.

The molecular descriptors can be divided into two categories:
- **experimental measurements** such as dipole moment or polarizability, and physicochemical properties
- **theoretical molecular descriptors** derived from a symbolic representation. The main classes of theoretical molecular descriptors are:
    - **0D**-descriptors (constitutional descriptors, count descriptors)
    - **1D**-descriptors (list of structural fragments, fingerprints)
    - **2D**-descriptors (graph invariants)
    - **3D**-descriptors (3D-MoRSE, WHIM, GETAWAY, quantum-chemical descriptors, size, steric, surface and volume descriptors)
    - **4D**-descriptors (derived from GRID or CoMFA methods, Volsurf)

Classic descriptors in chemistry

This list contains an excerpt from classic molecular descriptors[4] (of ~2,000). Calculation examples are spread in part B of this book. Mostly, they are the result of calculations within MM, MD, and QC:

1) Boiling point (**BP**)
2) Melting point (**MP**)
3) Heat capacity at T
   constant (**CT**)
4) Heat capacity at P
   constant (**CP**)
5) Entropy (**S**)
6) Density (**DENS**)
7) Enthalpy of
   vaporization (**HVAP**)
8) Standard enthalpy
   of vaporization (**DHVAP**)
...

9) Enthalpy of formation (**HFORM**)
10) Standard enthalpy
    of formation (**DHFORM**)
11) Motor octane number (**MON**)
12) Molar refraction (**MR**)
13) Acentric factor (**AcenFac**)
14) Total surface area (**TSA**)
15) Octanol-water
    partition coefficient (**LogP**)
16) Molar volume (**MV**)
17) Log water solubility (**logSw**)
18) Total surface area (**TSA**)
...

Structural descriptors

Another class of descriptors plays an important role in organic chemistry. Examples are `ring counts, aromaticity,` or `graph diameter`. These examples are discussed in Chapter 58 and code is implemented by RDKit. Another example of feature engineering is the use of Morgan fingerprints (e.g., as an input layer in neural nets), discussed in Chapter 60.

6 Categorize the problem

For more details, see also Chapter 33. To go further to model selection the problem, category has do be determined:

By *input*:
- `Labeled data -> supervised learning`
- `Unlabelled data + find structure ->  unsupervised learning`
- `Interacting with an environment -> reinforcement learning`

By *output*:
- `Number -> regression`
- `Class -> classification`

```
-   Assign to input labels -> clustering
-   Anomaly detection
```

Understand constraints
- Storage capacity?
- Does prediction have to be fast, real-time?
- Does learning have to be fast?

7 Choosing a data science algorithm (model)

Different algorithms have different feature engineering requirements (step 5). Feature engineering is the process of going from *raw data to data that is ready for modeling*:
- Makes the models *easier to interpret*
- *Captures* more complex *relationships*
- *Reduces data redundancy* and dimensionality

Details are discussed in Chapters 33 and 38. Factors affecting the choice of an algorithm are:
- Goals (scientific or business)
- Cost of pre-processing work
- Accuracy
- Explainability
- Scalability
- Model complexity

Example of an accuracy test

Test the accuracy of a prediction using the following metrics: *recall, precision, f1-score, and accuracy*. In order to calculate these metrics, a confusion matrix[5] for the cross-validation dataset predictions is required:

```
# Confusion matrix example
# author: Gressling, T
# license: MIT License        # code: github.com/gressling/examples
# activity: single example    # index: 26-1

from sklearn.metrics import confusion_matrix

def plot_confusion_matrix(cm, classes, title, cmap=plt.cm.Blues):
```

```
...
    thresh = cm.max() / 2.
    for i, j in itertools.product(range(cm.shape[0]), range(cm.shape[1])):
        plt.text(j, i, cm[i, j],
                    horizontalalignment="center",
                    color="white" if cm[i, j] > thresh else "black")
matrix = confusion_matrix(y_CV, y_pred)
plot_confusion_matrix(matrix, classes=[0, 1], title='Confusion matrix')
plt.show()
```

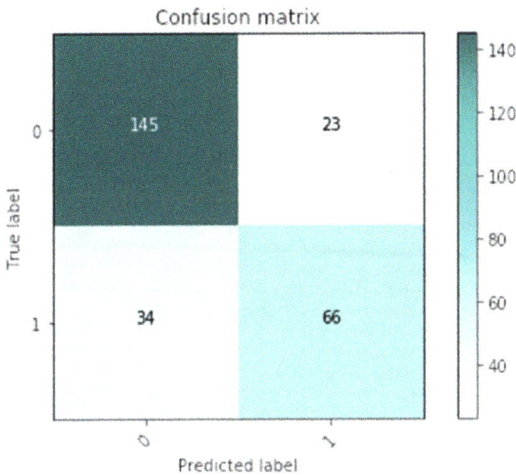

Figure 26.2: Confusion matrix example for cross-validation of dataset predictions *(Gressling)*.

The metrics were calculated as described below:
- P = precision = true positives / predicted positives
- R = recall = true positives / actual positives
- f1 = 2 * ((RP)/(R+P))
- accuracy = (true positives + true negatives)/total

8 Business aspects

Investment considerations: *caveats* to consider before setting up a data science project:
- Be aware of the complexity
- Manage the continuous change

References

1. Cross-industry standard process for data mining (Wikipedia) https://en.wikipedia.org/wiki/Cross-industry_standard_process_for_data_mining.
2. Beyond Interactive: Notebook Innovation at Netflix https://medium.com/netflix-techblog/notebook-innovation-591ee3221233.
3. Molecular Descriptors http://www.moleculardescriptors.eu/books/handbook.htm.
4. Descriptors Dataset http://www.moleculardescriptors.eu/dataset/dataset.htm.
5. sklearn.metrics.confusion_matrix documentation (BSD License) https://scikit-learn.org/stable/modules/generated/sklearn.metrics.confusion_matrix.html.

Data science as field of activity

27 Community and competitions

Comparing students that live on-campus in learning communities to those that live off-campus shows that students that participate in learning communities are more likely to persist in graduation.[1] Learning communities offer an advanced kind of educational or "pedagogical" design.

Communities

Community and competitions in slack or others sponsored by private institutions are a good starting point[2] for learning data science[3]. Some entry addresses (Q1/2020) are:
- Data Quest: https://www.dataquest.io/chat
- Data Science Salon: https://datasciencesalon.slack.com/
- IBM Data Science Community: https://community.ibm.com/.../datascience
- Deep Learning on Udacity: https://slackin.udacity.com/deep-learning
- AI Researchers (slack): https://ai-researchers-invite.herokuapp.com/

Obtaining help and reading forums

Depending on the topic, there are data science-centered forums to join and ask Questions[4]:

Model training
- Kaggle forum
- Data Science Reddit

Coding Python or R
- Stack overflow

Blogs
- Medium[5]
- Data Science Central
- KDnuggets
- KNIME Blog[6]

https://doi.org/10.1515/9783110629453-027

Kaggle

Kaggle, for example, is a complete project-based learning environment including data science hosting:

– **Competitions**
– **Datasets** (not only related to competitions) with preview

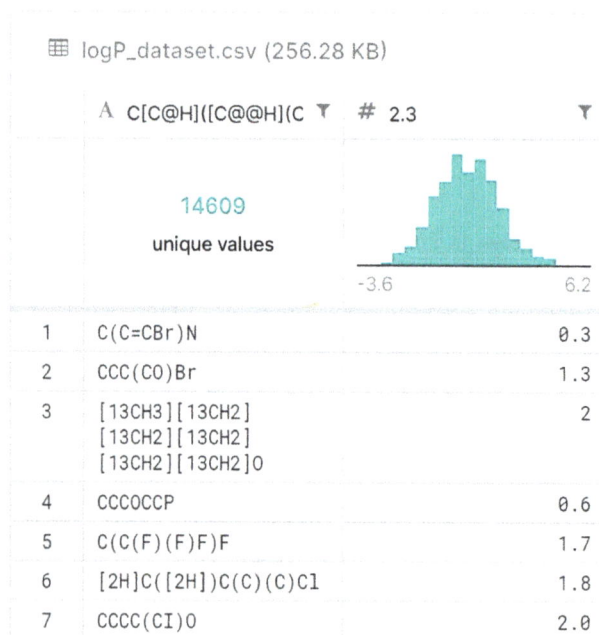

⊞ logP_dataset.csv (256.28 KB)

A C[C@H]([C@@H](C ▼	# 2.3 ▼
14609	
unique values	
	-3.6 6.2
1 C(C=CBr)N	0.3
2 CCC(CO)Br	1.3
3 [13CH3][13CH2] [13CH2][13CH2] [13CH2][13CH2]O	2
4 CCCOCCP	0.6
5 C(C(F)(F)F)F	1.7
6 [2H]C([2H])C(C)(C)Cl	1.8
7 CCCC(CI)O	2.0

Figure 27.1: Kaggle dataset preview[7] *(Kisin)*.

– **Kernels** are Kaggle's version of Jupyter Notebooks to share code along with a lot of visualizations, outputs, and explanations. *The "Kernels" tab takes you to a list of public kernels, which people use to showcase some new tool or share their expertise or insights about some particular dataset*
– **Learn**. Kaggle contains free, practical, hands-on courses that cover the minimum prerequisites needed to quickly get started in the field. Interact with no more passive reading!

On Kaggle you can observe – as discussed in Chapter 26 – that data wrangling aka exploratory data analysis and feature engineering are the most important parts of machine learning – not model fitting. The comments and explanations of the solution paths from the authors and winners of the competitions are worth reading[8] and provide a lot of inspiration.

Kaggle example: predicting molecular properties

For the *Chemistry and Mathematics in Phase Space* (CHAMPS)[9] challenge,[10,11] the goal was to predict magnetic interactions between atoms.

"In this competition,[9] you will be predicting the scalar_coupling_constant between atom pairs in molecules, given the two atom types (e.g., C and H), the coupling type (e.g., 2JHC), and any features you are able to create from the molecule structure (xyz) files."

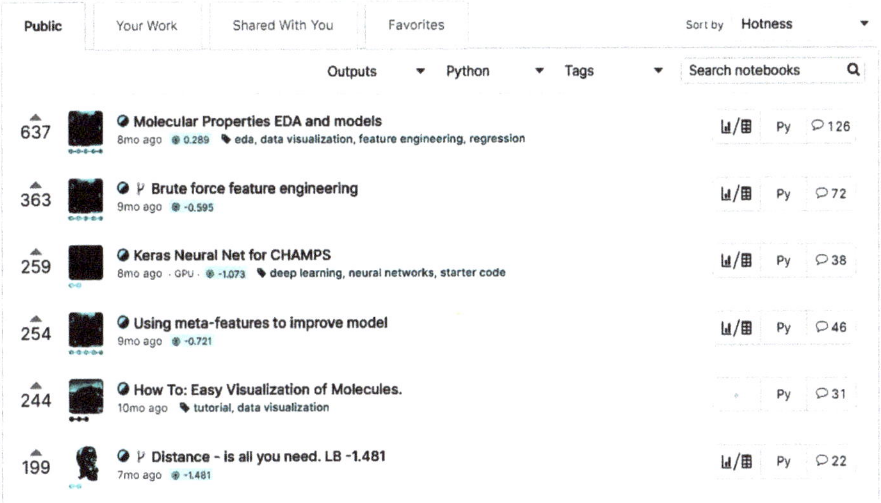

Figure 27.2: Interesting notebooks in a Kaggle competition (*Screenshot by Gressling*).

References

1. Learning community (Wikipedia) https://en.wikipedia.org/wiki/Learning_community.
2. Formulatedby. 15 Data Science Slack Communities to Join (Medium) https://towardsdatascience.com/15-data-science-slack-communities-to-join-8fac301bd6ce.
3. Agarwal, N. Use Kaggle to start your ML/ Data Science journey (Medium): https://towardsdatascience.com/use-kaggle-to-start-and-guide-your-ml-data-science-journey-f09154baba35.
4. The Importance of Community in Data Science https://www.datasciencecentral.com/profiles/blogs/the-importance-of-community-in-data-science.
5. Search and find – Medium https://medium.com/search?q=data%20science.
6. KNIME Community Forum https://forum.knime.com/.
7. Kisin, V. Tutorial ML In Chemistry Research. RDkit & mol2vec (License Apache 2.0): https://www.kaggle.com/vladislavkisin/tutorial-ml-in-chemistry-research-rdkit-mol2vec.
8. Torrubia, A. Kaggle Solution Path: Predicting Molecular Properties: https://www.kaggle.com/c/champs-scalar-coupling/discussion/106468.

9. Bristol University, Cardiff University, Imperial College and Leeds University. CHAMPS https://champsproject.com/.
10. Exploring Molecular Properties Data https://www.kaggle.com/robikscube/exploring-molecular-properties-data.
11. Predicting Molecular Properties https://www.kaggle.com/c/champs-scalar-coupling.

28 Data science libraries

There is no alternative that comes in one single package. While scikit-learn is not exceptionally good at anything, it can do a lot of things at once.

mlpack.org

mlpack[1,2] is a fast, flexible machine learning library written in C++ that aims to provide fast, extensible implementations of cutting-edge machine learning algorithms. mlpack provides these algorithms as Python bindings. It is built on the Armadillo linear algebra library, the ensmallen numerical optimization library, and parts of Boost. mlpack is licensed under the 3-clause BSD license.

dlib.net

Dlib[3,4] contains a wide range of machine learning algorithms; all modularly designed and simple to use. Dlib is used in a wide range of applications including robotics, embedded devices, mobile phones, and large high-performance computing environments.

shogun-toolbox.org

The Shogun[5] ML toolbox promotes this mission through its features (accessible, open-source) with a focus on ML education and developing its contributors. Shogun is one of the oldest and largest open source ML platforms.

scikit-learn.org

This library[6] is extensively used and discussed in this book. It is possible to replace parts of Scikit-Learn with packages like NLTK, Gensim, SciPy, and Matplotlib but they do not come in one whole package like the Scikit-Learn. Their specific implementations though are better in detail.

Non-python alternatives are generally found in non-python environments like R, Matlab, SPSS, SAS, and Stata.

https://doi.org/10.1515/9783110629453-028

References

1. mlpack 3: a fast, flexible machine learning library (10.21105/joss.00726) https://www.mlpack.org/files/mlpack3.pdf.
2. mlpack.org https://www.mlpack.org/.
3. King, D. E. Dlib-ml: A Machine Learning Toolkit http://www.jmlr.org/papers/volume10/king09a/king09a.pdf.
4. dlib C++ Library – Machine Learning http://dlib.net/ml.html.
5. Shogun Machine Learning – Home https://www.shogun-toolbox.org/.
6. scikit-learn: machine learning in Python, Documentation https://scikit-learn.org/stable/.

29 Deep learning libraries

By all measures, TensorFlow is the undisputed leader.[1,2]

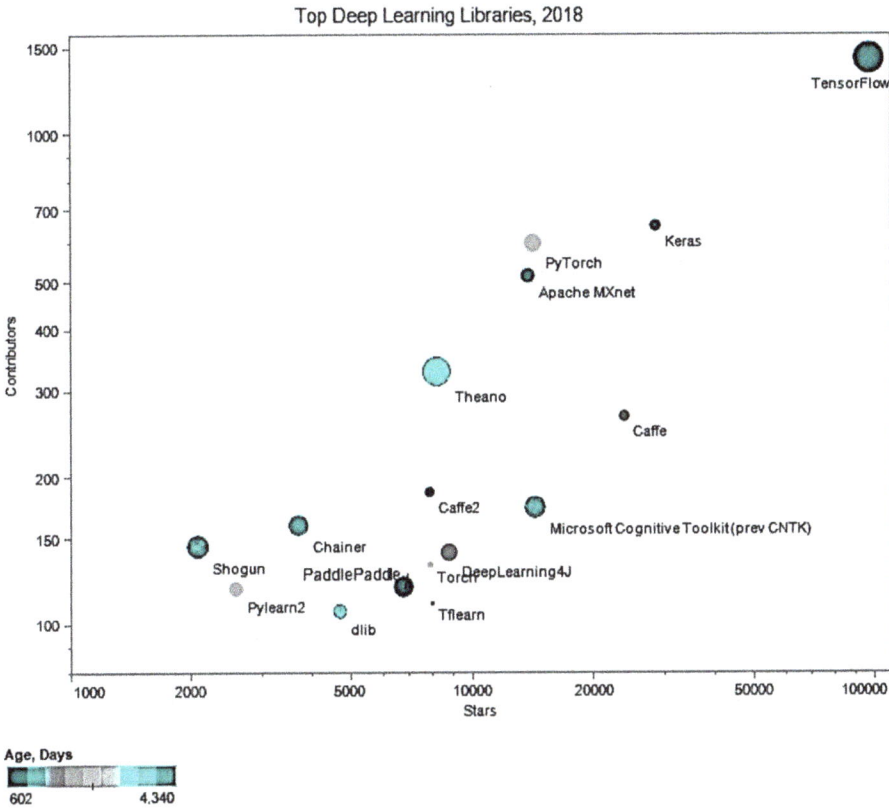

Figure 29.1: Top 16 open-source deep learning libraries[1] by Github stars and contributors; log scale on both axes and circle size represents the number of commits *(Gregory Piatetsky)*.

For comparison of the programming style and the API design, it is recommended to read the reference implementations on MNIST,[3,4] which every framework provides as *"Hello world of the neural network."* Here is a small excerpt with shortened code:

```
# MNIST comparison in major NN frameworks
# author: Gressling, T
# license: MIT License        # code: github.com/gressling/examples
# activity: single example    # index: 29-1
```

https://doi.org/10.1515/9783110629453-029

```
# Keras
model = Sequential()
  model.add(Conv2D(32, kernel_size=(3, 3), activation='relu', input_
  shape=input_shape))
  model.add(Conv2D(64, (3, 3), activation='relu'))
  model.add(MaxPooling2D(pool_size=(2, 2)))
  model.add(Dropout(0.25))
  model.add(Flatten())
  model.add(Dense(128, activation='relu'))
  model.add(Dropout(0.5))
  model.add(Dense(num_classes, activation='softmax'))

# pytorch
super(Net, self).__init__()
  self.conv1 = nn.Conv2d(1, 32, 3, 1)
  self.conv2 = nn.Conv2d(32, 64, 3, 1)
  self.dropout1 = nn.Dropout2d(0.25)
  self.dropout2 = nn.Dropout2d(0.5)
  self.fc1 = nn.Linear(9216, 128)
  self.fc2 = nn.Linear(128, 10)

# Theano
def net(batch_size, width, height, channels):
  l_input = L.InputLayer(shape=(batch_size, width, height, channels),
name='input')
  l_conv_1 = L.Conv2DLayer(l_input, num_filters=20, filter_size=(5,5),
name='conv1')
  l_pool_1 = L.MaxPool2DLayer(l_conv_1,  pool_size=(2,2),  name='pool1')
  l_conv_2 = L.Conv2DLayer(l_pool_1, num_filters=50, filter_size=(5,5),
name='conv2')
  l_pool_2 = L.MaxPool2DLayer(l_conv_2,  pool_size=(2,2), name='pool2')
  l_flatten = L.FlattenLayer(l_pool_2, name='flatten')
  l_dense_1 = L.DenseLayer(l_flatten, num_units=500, name='dense1')
  l_dense_2 = L.DenseLayer(l_dense_1, num_units=10, name='dense2',
nonlinearity=softmax)
  return l_dense_2
```

Model interoperability is discussed in Chapter 30.

TensorFlow

TensorFlow[5] is a free and open-source software library for dataflow and differentiable programming across a range of tasks. It is a symbolic math library and is also used for machine learning applications such as neural networks.

It is used for both research and production at Google. TensorFlow was developed by the Google Brain team.

Theano

Theano[6] is a Python library and optimizing compiler for manipulating and evaluating mathematical expressions, especially matrix-valued ones.

In Theano, computations are expressed using a NumPy-Esque syntax and compiled to run efficiently on either CPU or GPU architectures.

Keras

Keras is an open-source neural network library written in Python. It is capable of running on top of TensorFlow, Microsoft Cognitive Toolkit, R, Theano or PlaidML.

Designed to enable fast experimentation with deep neural networks, it focuses on being user-friendly, modular, and extensible.[7,8]

Pytorch

PyTorch[9] is an open-source machine learning library based on the Torch library, used for applications such as computer vision and natural language processing.

It is primarily developed by Facebook's AI Research lab.

OpenNMT-py: Open-Source Neural Machine Translation is a Pytorch port for an open-source (MIT) neural machine translation system. It is designed to try out new ideas in translation, summary, image-to-text, and morphology. It is used in IBM RXN, a tool for synthesis prediction.

References

1. Clark, D. Top 16 Open Source Deep Learning Libraries and Platforms (KDnuggets) https://www.kdnuggets.com/2018/04/top-16-open-source-deep-learning-libraries.html.
2. Comparison of deep-learning software (Wikipedia) https://en.wikipedia.org/wiki/Comparison_of_deep-learning_software.
3. Mnist cnn – Keras Documentation https://keras.io/examples/mnist_cnn/.
4. Ertam, F.; Aydın, G. Data classification with deep learning using Tensorflow https://ieeexplore.ieee.org/document/8093521/.
5. Wikipedia contributors. TensorFlow (Wikipedia) https://en.wikipedia.org/wiki/TensorFlow.
6. Theano (software) (Wikipedia) https://en.wikipedia.org/wiki/Theano_(software).
7. Keras (Wikipedia) https://en.wikipedia.org/wiki/Keras.
8. Keras https://keras.io/.
9. PyTorch (Wikipedia) https://en.wikipedia.org/wiki/PyTorch.

30 ML model sources and marketplaces

Trade involves the transfer of ideas or services from one entity to another in exchange for money. An ML market is a *service* where entities regularly gather for the purchase and sale of algorithms. The pricing of ML models can also be fixed by bidding in auctions. ML model performance is the indicator of quality (i.e image recognition benchmark).

Where to obtain models

Tensorflow Hub

TensorFlow Hub[1] is a library to foster the publication, discovery, and consumption of reusable parts of machine learning models. Figure 30.1 shows the user interface. There is also a Python API:

```
# Running a model in Tensorflow Hub
# author: Google
# License: Apache 2.0      # code: tensorflow.org/hub
# activity: active (2020)  # index: 30-2

#!pip install "tensorflow_hub>=0.6.0"
#!pip install "tensorflow>=2.0.0"
import tensorflow as tf
import tensorflow_hub as hub

# universal sentence embedding model, nnlm-en-dim128
# token-based text embedding-trained model
module_url = "https://tfhub.dev/google/nnlm-en-dim128/2"
embed = hub.KerasLayer(module_url)
embeddings = embed(["A long sentence.", "single-word", "http://example.com"])
print(embeddings.shape)
> (3,128)
```

ModelDepot.io

Is a platform[2] for discovering, sharing, and discussing easy to use and pretrained machine learning models.

https://doi.org/10.1515/9783110629453-030

Algorithms and model packages on AWS marketplace

"With AWS Marketplace,[3,4] you can browse and search for hundreds of machine learning algorithms and models in a broad range of categories, such as computer vision, natural language processing, speech recognition, text, data, voice, image, video analysis, fraud detection, predictive analysis, and more."

IBM Model Asset eXchange (MAX)

Free, deployable, and trainable code – a place[5] for developers to find and use free and open-source deep learning models

Apple Core ML

Models can be used with Core ML,[6] Create ML, and Xcode. They are available in a number of sizes and architecture formats. Refer to the model's associated Xcode project for guidance on how best to use the model.

Model interoperability

The Open Neural Network Exchange Format (ONNX[7]) is a standard for exchanging deep learning models.[8] It promises to make deep learning models portable, thus preventing vendor lock-in. ONNX is designed to allow framework interoperability, the fundament of exchanging models. ONNX was accepted as a graduate project in Linux Foundation AI.

Each computation dataflow graph is a list of nodes that forms a graph. Nodes have inputs and outputs. Each node is a call to an operator. Metadata documents the graph.

```
# Keras -> ONNX -> CNTK (Microsoft)
# author: ONNX Team
# license: MIT License
# code: github.com/onnx/onnx-ecosystem/.../keras_onnx.ipynb
# activity: active (2020) # index: 30-1


import onnxmltools

# Export model
from keras.models import load_model
```

```
input_keras_model = 'model.h5'
output_onnx_model = 'model.onnx'
keras_model = load_model(input_keras_model)
onnx_model = onnxmltools.convert_keras(keras_model)
onnxmltools.utils.save_model(onnx_model, output_onnx_model)

# Import and run
import cntk as runtime # CNTK is microsoft
# Import the Chainer model into CNTK via the CNTK import API
z = runtime.Function.load('model.onnx',
        device=runtime.device.cpu(), format=runtime.ModelFormat.ONNX)
# run
predictions = z.eval({z.arguments[...]}))
```

It is possible to use direct converters.[9]

ONNX runtime

ONNX runtime is a performance-focused engine for ONNX models, which inferences efficiently across multiple platforms and hardware (Windows, Linux, and Mac and on both CPUs and GPUs). ONNX runtime has proved to considerably increase performance over multiple models.

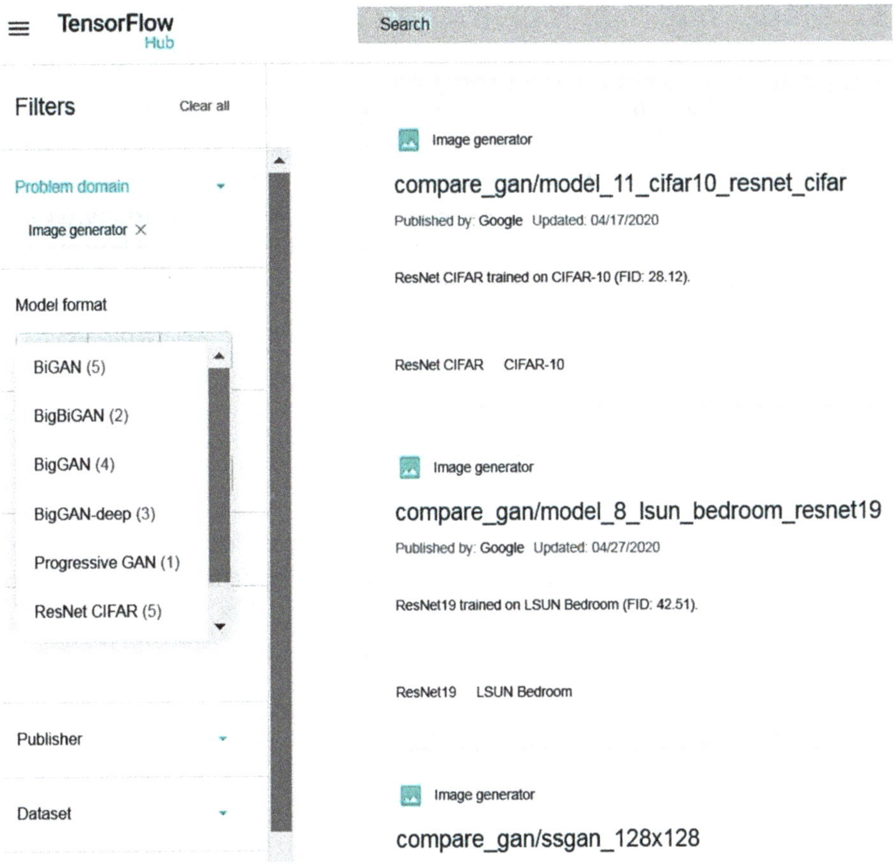

☰ **TensorFlow**
    Hub

Search

Filters   Clear all

Problem domain   ▼

  Image generator ✕

Model format

  BiGAN (5)

  BigBiGAN (2)

  BigGAN (4)

  BigGAN-deep (3)

  Progressive GAN (1)

  ResNet CIFAR (5)

Publisher   ▼

Dataset   ▼

🖼 Image generator

compare_gan/model_11_cifar10_resnet_cifar

Published by: **Google** Updated: 04/17/2020

ResNet CIFAR trained on CIFAR-10 (FID: 28.12).

ResNet CIFAR   CIFAR-10

🖼 Image generator

compare_gan/model_8_lsun_bedroom_resnet19

Published by: **Google** Updated: 04/27/2020

ResNet19 trained on LSUN Bedroom (FID: 42.51).

ResNet19   LSUN Bedroom

🖼 Image generator

compare_gan/ssgan_128x128

Figure 30.1: Tensorflow Hub user interface *(Google)*.

References

1. TensorFlow Hub https://www.tensorflow.org/hub.
2. ModelDepot https://modeldepot.io/.
3. Find and Subscribe to Algorithms and Model Packages on AWS Marketplace https://docs.aws.amazon.com/sagemaker/latest/dg/sagemaker-mkt-find-subscribe.html.
4. Machine learning solutions in AWS Marketplace https://aws.amazon.com/marketplace/solutions/machine-learning/#algorithms.
5. Models https://developer.ibm.com/exchanges/models/.
6. Apple Inc. Machine Learning – Models – Apple Developer https://developer.apple.com/machine-learning/models/.
7. ONNX | Home https://onnx.ai/.
8. Dillon. What every ML/AI developer should know about ONNX https://blog.paperspace.com/what-every-ml-ai-developer-should-know-about-onnx/.
9. Shuai, Y. deep-learning-model-convertor (GitHub) https://github.com/ysh329/deep-learning-model-convertor.

31 Model metrics: MLFlow and Ludwig

In software engineering, a software development process is a process of dividing software development work into distinct phases to improve design, product management, and project management. It is also known as a software development life cycle (SDLC). The methodology may include the predefinition of specific deliverables and artifacts that are created and completed by a project team to develop or maintain an application.[1]

This chapter contains the ML aspect of MLOps. In Chapter 6, the MLOps process from IT perspective was discussed.

Track metrics

Building many different models requires different codebases for the ML models and many graphs or notebooks and metrics to keep track of for optimizing *code and hyperparameters* of each model. Usually, this work is done with Excel and a few tracking and model optimization services (examples):
- Neptune: neptune.ml
- Google Tensorboard: tensorflow.org/tensorboard
- Uber Michelangelo: uber.com/michelangelo
- Weights and Biases: wandb.com
- Comet: comet.ml

These optimization services are still necessary but must be seen in the broader context of the ML lifecycle.

MLFlow

MLFlow already has the ability to track metrics, parameters, and artifacts as part of experiments, package models, and reproducible ML projects, and to deploy models to batch or real-time serving platforms.[2]

The MLflow community (Github[3]) has over 120 contributors from over 40 companies that have contributed code to the project, and over 200 companies are using MLflow.[4] Install the server from PyPI via pip `install mlflow` (requires conda).

https://doi.org/10.1515/9783110629453-031

```
Commands:

  artifacts     Upload, list, and download
                artifacts from an MLflow artifact...
  azureml       Serve models on Azure ML.
  db            Commands for managing an MLflow tracking database.
  experiments   Manage experiments.
  models        Deploy MLflow models locally.
  run           Run an MLflow project from the given URI.
  runs          Manage runs.
  sagemaker     Serve models on SageMaker.
  server        Run the MLflow tracking server.
  ui            Launch the MLflow tracking UI for local viewing of run...
```

To call the UI, use the statement `MLFlow ui` in your working directory and access it
by open `http://localhost:5000` then choose an experiment.

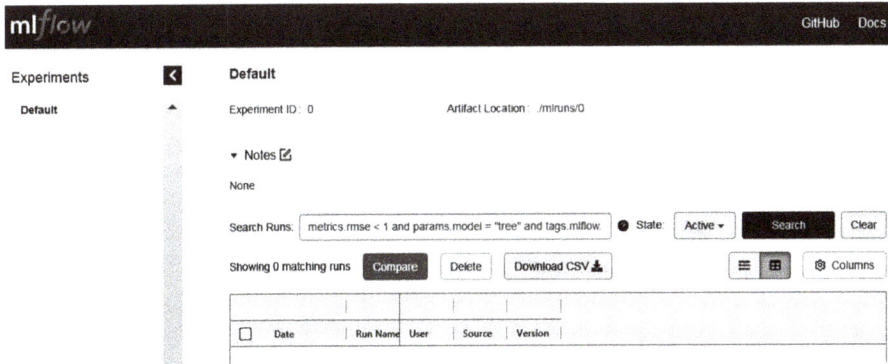

Figure 31.1: User Interface of MLFlow.

MLFlow organizes data science artifacts into **two categories**:
- Experiments – The problem to be solved, created around a dataset
- Runs – Each individual "attempt" at feature engineering and model training

MLFlow has three **main components**:
- Experiment Tracking to record and query experiments: code, data, config, and results (tracking server)
- Reproducible Projects with a packaging format to reproduce runs on any platform
- Models and Model Management based on models general format

Example experiment tracking

Use the MLflow Tracking API to start a run and log parameters, metrics, and artifacts (files) from the data science code. Once MLFlow is configured to point to the experiment ID, each execution will begin to log and capture any metrics. Focus on accuracy metrics like the RMSE, R-squared, or the mean absolute error. Focus on the time of execution of the `model.fit` and the `model.predict` – there is *more than just accuracy* to choose a good model!

```python
# MLflow Tracking API
# author: Databricks[5]; Teichmann[6]; MLFlow[7]
# license: various / not stated
# code: gitlab.com/jan-teichmann/ml-flow-ds-project
# activity: single example          # index: 31-1

import mlflow
import mlflow.sklearn
experiment = 'iris_classification'
mlflow.set_tracking_uri(os.environ['MLFLOW_TRACKING_URI'])
mlflow.set_experiment(experiment)

 with mlflow.start_run() as run:
    model = LogisticRegression(
        C=0.1, solver='saga', multi_class='multinomial',
        penalty='l1', max_iter=2000
    )
    # log metric
    mlflow.log_param('C', 0.1)
    mlflow.log_param('penalty', 'l1')


    # Log artifact (output file)
    with open("output.txt", "w") as f:
        f.write("Classification checkpoint message")
    mlflow.log_artifact("output.txt")
    ...

print(run.info)
# <RunInfo: artifact_uri='s3://ar

client = mlflow.tracking.MlflowClient()
r = client.get_run(run_id)
print(r)
# <Run: data=<RunData: metrics={'f1_score': 0.9458874458874459, 'n_iter_': 1401.0},
# params={...}, tags={'mlflow.source.name': ... 'mlflow.source.type': 'LOCAL'}>,
# experiment_id='1', ...'>>

experiments = client.list_experiments()
# returns a list of mlflow.entities.Experiment
```

Example projects

The file created in YAML syntax specifies a name and a Conda or Docker environment. It contains more detailed information about each entry point and defines a command to run the entry points.

```
# MLFlow environment definition (YAML)
# author: Gressling, T
# license: MIT License        # code: github.com/gressling/examples
# activity: single example    # index: 31-2

name: structure_prediction
conda_env: structure_prediction.yaml
# Can have a docker_env instead of a conda_env, e.g.
# docker_env:
#     image:  mlflow-docker-structure_prediction
entry_points:
  main:
    parameters:
      data_file: path
      regularization: {type: float, default: 0.1}
    command: "python train.py -r {regularization} {data_file}"
  validate:
    parameters:
      data_file: path
    command: "python validate.py {data_file}"
```

Example model: Each model is a directory containing files together with an MLmodel file in the root of the directory. Flavors are the key concept in MLFlow models– they are a convention that deployment tools can use to *understand the model.* MLflow provides several standard flavors that might be useful in applications like Keras (keras), PyTorch (pytorch), and Scikit-learn (sklearn).

```
# Directory written by mlflow.sklearn.save_model(model, "my_model")
my_model/
├── MLmodel
└── model.pkl
```

Serving. With MLflow, you can quickly deploy a local model with the following command:

```
mlflow serve -m path_to_the_model -p 1234 # port
```

Uber Ludwig

Deep learning pipeline. It is very time-consuming and code-intensive to find the right model architecture and hyperparameters. It also requires knowledge of algorithms used and state-of-the-art techniques. Ludwig provides a toolbox that allows training and testing a deep learning model without writing code.[8]

A declarative approach to train a deep learning model by providing a file containing the meta-information in a configuration file (YAML). Specify some information about the features contained, for example, dependent or independent variables.

Example project

```
# Install Ludwig
pip install git+https://github.com/uber/ludwig
# additional libs, i.e. python -m spacy download en
```

```
# model_definition.yaml
input_features:
  -
  name: text
  type: SMILES[9]
output_features:
  -

  name: REAC_toxicity
  type: category
```

Ludwig performs a random data split into training and validation. Then, Ludwig creates the test set, preprocesses them, and then builds a model with the specified encoders and decoders, also displaying the training process inside a console and a TensorBoard-style capability.

```
$ ludwig train -data_smiles molecules.dat -model_definition_file
model_definition.yaml
$ ludwig visualize -visualization learning_curves -training_stats
results/training_stats.json
$ ludwig predict -data_csv path/to/molecules.dat -model_path /path/to/model
```

Also, with the Ludwig Python API, an equivalent workflow can be implemented:

```
# UBER Ludwig model training
# author: Gressling, T
# license: MIT License        # code: github.com/gressling/examples
# activity: single example    # index: 31-3

from ludwig import LudwigModel
import pandas as pd

df = pd.read_csv('molecules.dat')
model_definition = {
    'input_features':[
        {'name':'text', 'type':'SMILES'},
    ],
    'output_features': [
        {'name': 'REAC_toxicity', 'type': 'category'}
    ]
}

model = LudwigModel(model_definition)
train_stats = model.train(df)
model.close()
```

References

1. Software development process (Wikipedia) https://en.wikipedia.org/wiki/Software_development_process.
2. Mewald, C. Introducing the MLflow Model Registry https://www.wtwjasa.com/introducing-the-mlflow-model-registry-machine-learning-model-hub/.
3. mlflow (Github) https://github.com/mlflow/mlflow.
4. Thunder Shiviah, M. S. Managing the Complete Machine Learning Lifecycle with MLflow https://www.slideshare.net/databricks/managing-the-complete-machine-learning-lifecycle-with-mlflow.
5. MLflow Quick Start (Python) – Databricks https://docs.databricks.com/_static/notebooks/mlflow/mlflow-quick-start-python.html.
6. Teichmann, J. Iris Batch Scoring – ml-flow-ds-project (GitLab) https://gitlab.com/jan-teichmann/ml-flow-ds-project.
7. MLflow documentation https://www.mlflow.org/docs.
8. Tanner, G. Introduction to Uber's Ludwig https://towardsdatascience.com/introduction-to-ubers-ludwig-cdaa67245cfa.
9. Parsing SMILES http://www.dalkescientific.com/writings/diary/archive/2007/06/25/smiles_states.html.

Introduction to ML and AI

32 First generation (logic and symbols)

Many *early* AI programs used the same basic algorithm. To achieve some goal (like winning a game or proving a theorem), they proceeded *step by step* towards it (by making a move or a deduction) as if searching through a maze, backtracking whenever they reached a dead end. This paradigm was called "*reasoning as a search.*"

The principal difficulty was that, for many problems, the number of possible paths through the "maze" was simply astronomical (a situation known as a "combinatorial explosion"). Researchers would reduce the search space by using heuristics or "rules of thumb" that would eliminate those paths that were unlikely to lead to a solution.[1]

Symbols are one possible way of transferring reward signals learned which are the base for transfer learning. Every abstract category asserts an analogy between all the disparate objects to another via symbols. Like the carboxyl group is, in symbols, (C(=O)OH) and we transfer our knowledge about carboxyl groups to another context with the help of the term (C(=O)OH). The symbols may be part of different molecules but share common properties.

Sympy

SymPy is a Python library for symbolic mathematics. It aims to become a full-featured computer algebra system (CAS) while keeping the code as simple as possible in order to be comprehensible and easily extensible. SymPy is written entirely in Python.[2]

```
# Symbolic mathematics with Sympy
# author: Gressling, T
# license: MIT License       # code: github.com/gressling/examples
# activity: single example   # index: 32-1

from sympy import symbols

x, y = symbols('x y')
expr = x + 2*y
expr
> x + 2*y

solve(x**2 - 2, x)
> [-√2, √2]
```

https://doi.org/10.1515/9783110629453-032

Logic theorist

1956 – The first demonstration of the logic theorist (LT) was written by Allen Newell, J.C. Shaw, and Herbert A. Simon (Carnegie Institute of Technology, now Carnegie Mellon University or CMU). This is often called the first AI program.[1] 1959 – The General Problem Solver (GPS) was created by Newell, Shaw, and Simon while at CMU.

A program called General Problem Solver[3] (GPS) is another example of the earliest AI programs in existence, succeeding the Logic Theorist Program.

```python
# General problem solver implementation
# authors: Belotti, Jonathon; Connelly, Daniel
# license: not stated
# code: github.com/thundergolfer/the-general-problem-solver
# activity: single example  # index: 32-2

def gps(initial_states, goal_states, operators):
    # Find a sequence of operators that will achieve all of the goal states.
    # Returns a list of actions that will achieve all of the goal states
    return [state for state in final_states if state.startswith(prefix)]

# Achieving subgoals
def achieve_all(states, ops, goals, goal_stack):
    # Achieve each state in goals and make sure they still hold at the end.
    # The goal stack keeps track of our recursion: which preconditions are we
    # trying to satisfy by achieving the specified goals?
    return states

def achieve(states, operators, goal, goal_stack):
    # Achieve the goal state using means-ends analysis.
    # Identifies an appropriate and applicable operator--one that contains the
    # goal state in its add-list and has all its preconditions satisfied.
    for op in operators:
        result = apply_operator(op, states, operators, goal, goal_stack)

def apply_operator(operator, states, ops, goal, goal_stack):
    # Applies operator and returns the resulting states.
    return [state for state in result if state not in delete_list] + add_list

start = example_data[choice]['start']
finish = example_data[choice]['finish']
ops = copy.deepcopy(example_data[choice]['ops'])

for action in gps(start, finish, ops):
    print(action)
```

The original program code in machine language is available.[4]

References

1. History of artificial intelligence (Wikipedia) https://en.wikipedia.org/wiki/History_of_artificial_intelligence.
2. SymPy documentation https://docs.sympy.org/latest/tutorial/intro.html.
3. Belotti, J. the-general-problem-solver (GitHub) https://github.com/thundergolfer/the-general-problem-solver.
4. Newell, A.; Simon, H. A. The logic theory machine – a complex information processing system http://shelf1.library.cmu.edu/IMLS/MindModels/logictheorymachine.pdf.

33 Second generation (shallow models)

Judea Pearl's influential 1988 book *Probabilistic Reasoning in Intelligent Systems* brought probability and decision theory into AI[1]. Among the many new tools in use were Bayesian networks, hidden Markov models, information theory, stochastic modeling, and classical optimization. Precise mathematical descriptions were also developed for "computational intelligence" paradigms like neural networks and evolutionary algorithms.

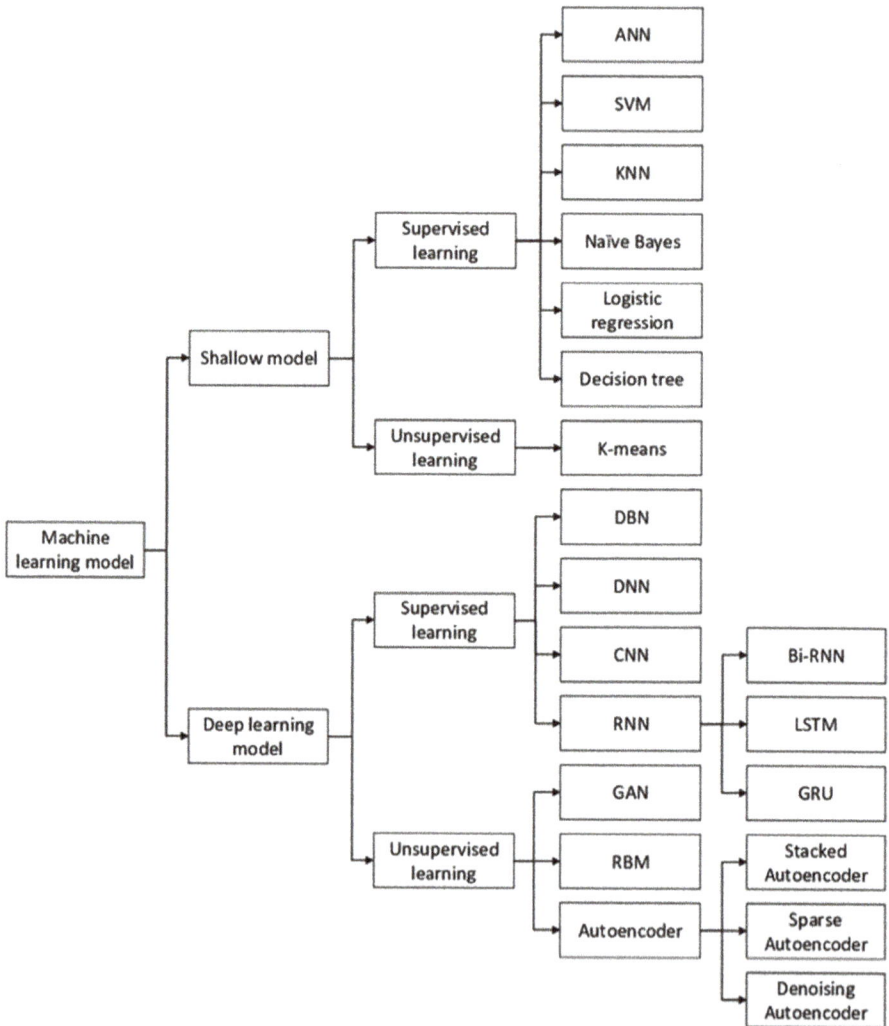

Figure 33.1: Taxonomy of machine learning algorithms[2] *(Liu, CC BY)*.

https://doi.org/10.1515/9783110629453-033

Taxonomy of supervised and unsupervised learning

- Supervised learning – labeled data; learning an input–output **relation**
- Unsupervised learning – unlabeled data; learning data **structure**

ML is *Software 2.0* – probabilities and weights

As there are fundamental differences between coding **traditional code** – **Software 1.0** and train **ML models** – **Software 2.0**[3]:

- Data Science and ML code handle the *inherent non-deterministic characteristics* of statistical modeling. ML models are not guaranteed to behave the same way, when trained twice. Traditional code that can be easily unit tested.
- *ML models are the result of a training process* where the result can be large (GB, TB), and that is served differently from *ML code* itself.
- Collaboration in ML between experimentation, testing, and production deployments are not identical to approval processes in software engineering, but sometimes similar.

It is very crucial for machine learning enthusiasts to know and understand the basic and important machine learning algorithms in order to keep up with the current trend. In this article, 10 basic algorithms are listed that play very important roles in the machine learning era.

Four basic *types* of statistical machine learning (shallow)

- Regression
- Classification (two-class, multiclass)
- Clustering
- Anomaly detection

A comparison plot[4] of classifiers on synthetic datasets is shown in Figure 33.3. The 30 diagrams illustrate the nature of decision boundaries. As real datasets are not used, the intuition conveyed by these examples does not necessarily carry over to reality.

What algorithm to use?

Figure 33.2 contains a path of how an algorithm can be chosen. Particularly in high-dimensional spaces, data can more easily be separated linearly and the simplicity of classifiers such as naive Bayes and linear SVMs might lead to better generalization than is achieved by other classifiers.

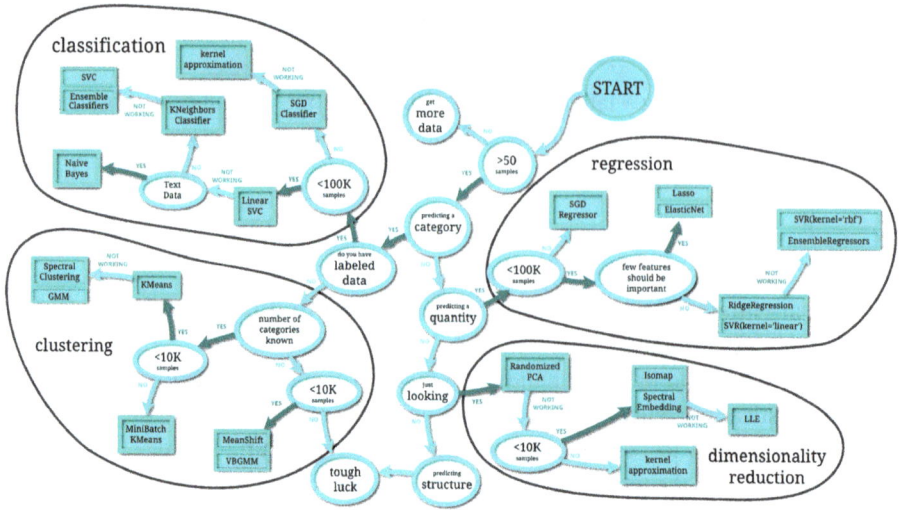

Figure 33.2: Choosing the right estimator: *Category -> Quantity -> Structure (Scikit-learn).*

scikit classifier comparison

This comparison has three datasets, with 100 points per dataset and 2 dimensions per point:
- Moons: Two interleaving half-circles
- Circles: A larger circle containing the smaller one
- Linear: A linearly separable dataset

The plots show training points in solid colors and testing points as semitransparent. The lower right shows the classification accuracy on the test set.

```
# scikit classifier comparison
# authors: Varoquaux, Müller, Grobler
# license: BSD 3 clause
# code: scikit-learn.org/.../plot classifier comparison
# activity: single example     # index: 33-1

# ax is the matplotlib object, dismissed for readability

from sklearn.model_selection import train_test_split
from sklearn.preprocessing import StandardScaler
from sklearn.datasets import make_moons, make_circles, make_classification
from sklearn.neural_network import MLPClassifier
from sklearn.neighbors import KNeighborsClassifier
```

```python
from sklearn.svm import SVC
from sklearn.gaussian_process import GaussianProcessClassifier
from sklearn.gaussian_process.kernels import RBF
from sklearn.tree import DecisionTreeClassifier
from sklearn.ensemble import RandomForestClassifier, AdaBoostClassifier
from sklearn.naive_bayes import GaussianNB
from sklearn.discriminant_analysis import QuadraticDiscriminantAnalysis

names = ["Nearest Neighbors", "Linear SVM", "RBF SVM",
         "Gaussian Process", "Decision Tree", "Random Forest",
         "Neural Net", "AdaBoost", "Naive Bayes", "QDA"]

classifiers = [
    KNeighborsClassifier(3),
    SVC(kernel="linear", C=0.025),
    SVC(gamma=2, C=1),
    GaussianProcessClassifier(1.0 * RBF(1.0)),
    DecisionTreeClassifier(max_depth=5),
    RandomForestClassifier(max_depth=5, n_estimators=10, max_features=1),
    MLPClassifier(alpha=1, max_iter=1000),
    AdaBoostClassifier(),
    GaussianNB(),
    QuadraticDiscriminantAnalysis()]

X, y = make_classification(n_features=2, n_redundant=0, n_informative=2,
                           random_state=1, n_clusters_per_class=1)
rng = np.random.RandomState(2)
X += 2 * rng.uniform(size=X.shape)
linearly_separable = (X, y)

datasets = [make_moons(noise=0.3, random_state=0),
            make_circles(noise=0.2, factor=0.5, random_state=1),
            linearly_separable]

# iterate over datasets
for ds_cnt, ds in enumerate(datasets):
    # preprocess dataset, split into training and test part
    X, y = ds
    X = StandardScaler().fit_transform(X)
    X_train, X_test, y_train, y_test =
        train_test_split(X, y, test_size=.4, random_state=42)

    x_min, x_max = X[:, 0].min() - .5, X[:, 0].max() + .5
    y_min, y_max = X[:, 1].min() - .5, X[:, 1].max() + .5
    xx, yy = np.meshgrid(np.arange(x_min, x_max, h),
                         np.arange(y_min, y_max, h))

    # Plot the training points
    ax.scatter(X_train[:, 0], X_train[:, 1], c=y_train, cmap=cm_bright,
               edgecolors=colors)
```

```
    # Plot the testing points
    ax.scatter(X_test[:, 0], X_test[:, 1], c=y_test,
            cmap=cm_bright, alpha=0.6, edgecolors=colors)

    # iterate over classifiers
    for name, clf in zip(names, classifiers):
        clf.fit(X_train, y_train)
        score = clf.score(X_test, y_test)

        Z = Z.reshape(xx.shape)
        ax.contourf(xx, yy, Z, cmap=cm, alpha=.8)
        # Plot the training points
        ax.scatter(X_train[:, 0], X_train[:, 1], c=y_train,
                cmap=cm_bright, edgecolors=colors)
        # Plot the testing points
        ax.scatter(X_test[:, 0], X_test[:, 1], c=y_test, cmap=cm_bright,
                edgecolors=colors, alpha=0.6)
        ax.text(xx.max() - .3, yy.min()
                        + .3, ('%.2f' % score).lstrip('0'),
                        size=15, horizontalalignment='right')

plt.tight_layout()
plt.show()
```

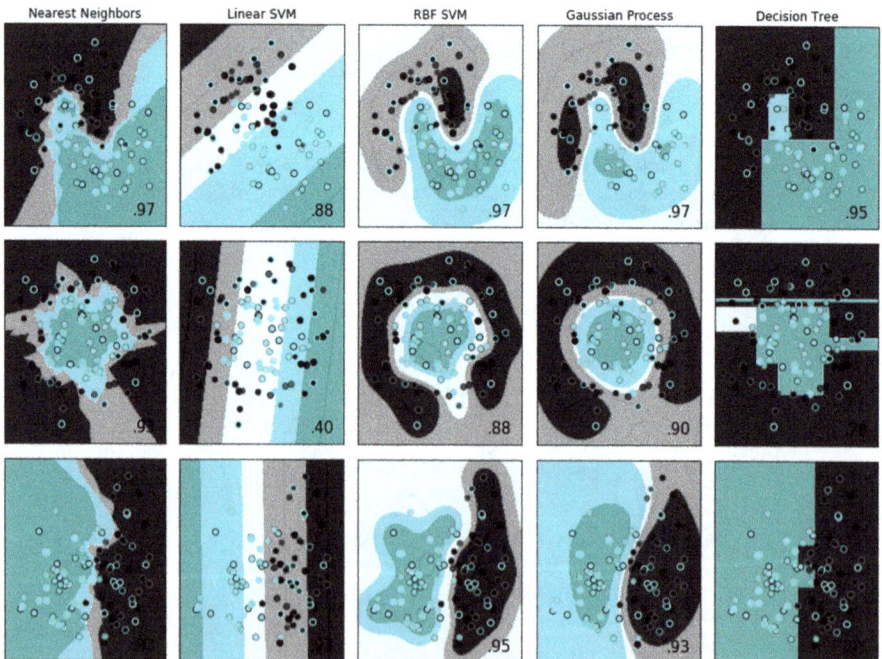

Figure 33.3a: Classifier comparison (scikit [4]).

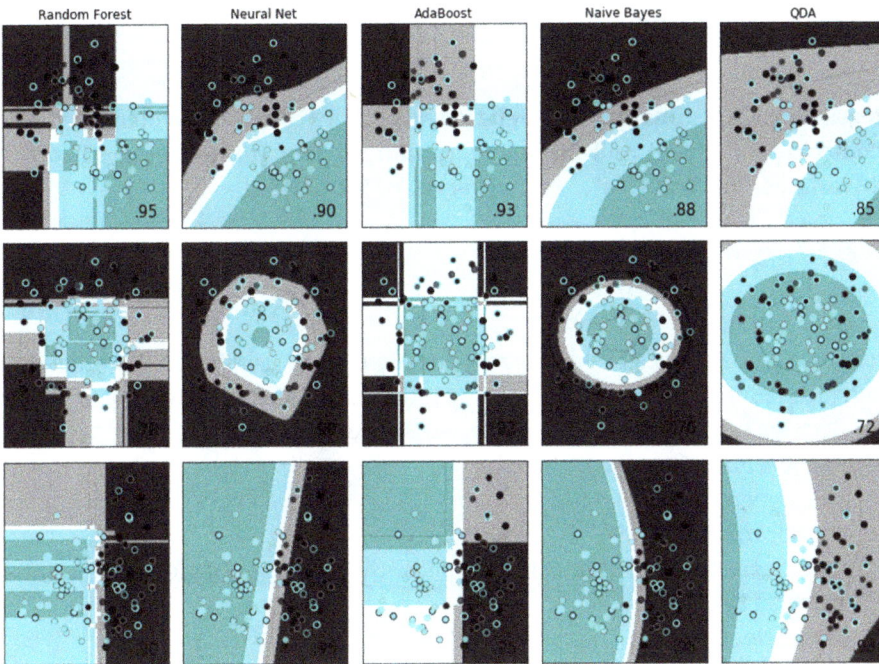

Figure 33.3b: (continued)

References

1. Timeline of artificial intelligence (Wikipedia) https://en.wikipedia.org/wiki/Timeline_of_artificial_intelligence.
2. Liu, H.; Lang, B. Machine Learning and Deep Learning Methods for Intrusion Detection Systems: A Survey. **2019**, 9, 4396.
3. Karpathy, A. Software 2.0 https://medium.com/@karpathy/software-2-0-a64152b37c35.
4. Classifier comparison – scikit-learn 0.22.1 documentation https://scikit-learn.org/stable/auto_examples/classification/plot_classifier_comparison.html.

34 Second generation: regression

Regression is estimating the relationships between a *dependent variable* ("outcome variable") and one or more *independent variables* ("predictors," "covariates," or "features").[1] It is a *supervised* ML pattern.

Important algorithms

Performing data science of the second generation means using algorithms that deal with probabilities. The most important algorithms are discussed here:
- Decision tree – classification and regression tree (CART)
- Random forest (bagging)
- Linear regression
- Logistic regression
- RidgeRegression SVR (kernel linear) or RBF-SVR (ensemble regressor)
- Bayesian linear regression
- Elastic Net Lasso
- Linear discriminant analysis (LDA)

Decision tree: classification and regression tree (CART)

```
# properties: simple and explainable; non-parametric model
# useCases: classify continuous dependent and categorical variables
```

The representation of the decision tree model is a binary tree. The algorithm is designed by answering either yes or no to questions with certain parameters. See Chapter 35 for more details.

Random forest (bagging)

```
# properties: fast, accurate
```

Random forests are basically the combination of tree predictors, where each tree depends on the values of a random vector that are sampled independently and with the same distribution for all the trees in the forest. This technique is easy to use as well as flexible because it can be both used for classification and regression tasks.

https://doi.org/10.1515/9783110629453-034

Difference between random forests and decision trees

- **A decision tree** is built on an entire dataset using all the features (variables) of interest. The accuracy keeps improving with more and more splits until overfitting.
- **Random forest** is a collection of decision trees (n_estimators, max_features). It randomly selects observations and specific features to build *multiple decision trees* and then *averages the results*[2].

Random Forest Simplified

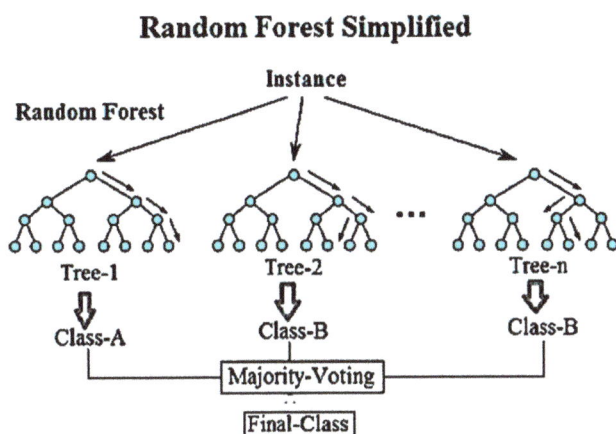

Figure 34.1: Random forest[3] *(Jagannath)*.

Examples in chemistry

- **Different strategies for the use of random forest in NMR spectra**[4]
- **A random forest model for predicting the crystallisability of organic molecules**[5]
- **A random forest model for the analysis of chemical descriptors for the elucidation of HIV-1 protease protein–ligand interactions**[5]

Linear regression

```
# properties: fast, linear
```

A linear regression estimates[6] the coefficients – or specific weights – (b, slope, a, intercept) of a linear equation with one or more variables. The variable to predict (y) is known as the dependent variable and the variable *used* (x) is the *independent variable*. The *most simple* form has a single regressor x, which has a relationship with a response y. *The relationship is a straight line of the form*, y=ax+b.

Different techniques can be used:
–   linear algebra solution for ordinary least squares (OLS)
–   gradient descent optimization

Python linear regression example

```
# Example form scikit⁷
# train
regressor = LinearRegression()
regressor.fit(X_train, y_train)

# predict
y_pred = regressor.predict(X_test)
```

Examples in chemistry

–   **Analytical chemistry.** Determine the best equation for calibration data to generate a calibration curve. Then a concentration of an analyte in a sample can be determined by comparing a measured value to that curve.
–   **Colorimetry.** X-axis variable is the concentration of a known solution. The y-axis variable is the measured absorbance.

Logistic regression

```
# properties: binary classification problems (problems with two class
values)
```

The prediction for the output is transformed using a nonlinear function called the *logistic function*. In a standard regression analysis, the dependent variable is binary. The logit classifier in the logistic regression is a mathematical form that estimates the variable between 0 and 1.

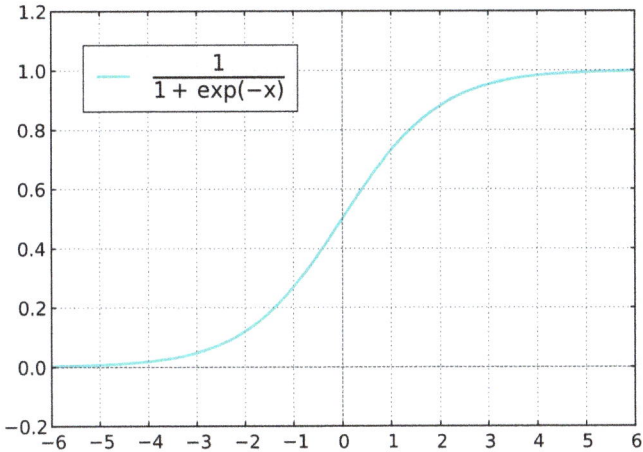

Figure 34.2: Logistic function.[8]

Python logistic regression example (Scikit)

```
from sklearn.datasets import load_iris
X, y = load_iris(return_X_y=True)

from sklearn.linear_model import LogisticRegression
clf = LogisticRegression(random_state=0).fit(X, y)

clf.predict(X[:2, :])
> array([0, 0])
clf.predict_proba(X[:2, :])
> array([[9.8...e-01, 1.8...e-02, 1.4...e-08],
>        [9.7...e-01, 2.8...e-02, ...e-08]])
clf.score(X, y)
> 0.97...
```

RidgeRegression SVR (kernel linear) or RBF-SVR (ensemble regressor)

```
# properties: predicting a quantity
# properties: [>50k ; <100k] samples
```

Bayesian linear regression

```
#properties: few samples, linear
```

The Bayesian approach is a *general way* of estimating statistical models. It can be applied to different models. "Naive" assumes that each input variable is independent.

Bayesian linear regression[9] assumes the responses are sampled from a *probability distribution* such as the normal (Gaussian) distribution. This implies that y is *not a single value*. Bayesian linear regression is an approach to linear regression in which statistical analysis is carried out with Bayesian inference. The algorithm can also be used for classification.

Python example of a Bayesian linear regression on lab data

```python
# Bayesian Linear Regression with pymc3
# author: Gressling, T
# license: MIT License        # code: github.com/gressling/examples
# activity: single example    # index: 34-1

import pymc3 as pm
from sklearn.model_selection import train_test_split

data = pd.read_csv('data.csv')
X = data.drop([Sample_ID', 'pH'], axis = 1)    # the features
Y = data['pH']

X_train, X_test, y_train, y_test = train_test_split(X, Y, test_size=0.2)
Formula = 'pH ~ ' + ' + '.join(
                ['%s' % variable for variable in X_train.columns[0:]])

with pm.Model() as normal_model:
    f = pm.glm.families.Normal()
    pm.GLM.from_formula(Formula, data=pd.concat([X_train, y_train],
                        axis=1), family=f)
    normal_trace = pm.sample(draws=2000, chains = 2, tune = 500)

plt.figure(figsize=(7, 7))
traceplot(normal_trace[100:])
plt.tight_layout()
```

Elastic Net Lasso

```
# properties: predict a quantity
# properties: <100k samples
# properties: features are important
```

The Lasso is a linear model that estimates sparse coefficients. It is useful in some contexts due to its tendency to prefer solutions with fewer non-zero coefficients, effectively reducing the number of features upon which the given solution is dependent.[10]

Python elastic net Lasso example

```
# Example form scikit
from sklearn import linear_model

reg = linear_model.Lasso(alpha=0.1)
reg.fit([[0, 0], [1, 1]], [0, 1])

reg.predict([[1, 1]])
> array([0.8])
```

Linear discriminant analysis

```
# properties: linear classification; two or more classes
# type: dimensionality reduction
```

LDA is a supervised classification technique that takes labels into consideration. It finds a linear combination of features that characterizes or separates classes of objects or events. A standard technique for LDA is to do a *projection* of the data from a high-dimensional space onto a perceivable subspace such that the data can be separated by visual inspection *(dimensionality reduction)*.

The resulting combination may be used as a linear classifier or more commonly, for dimensionality reduction before later classification. LDA is also closely related to principal component analysis and factor analysis in that they both look for linear combinations of variables that best explain the data.

LDA vs QDA

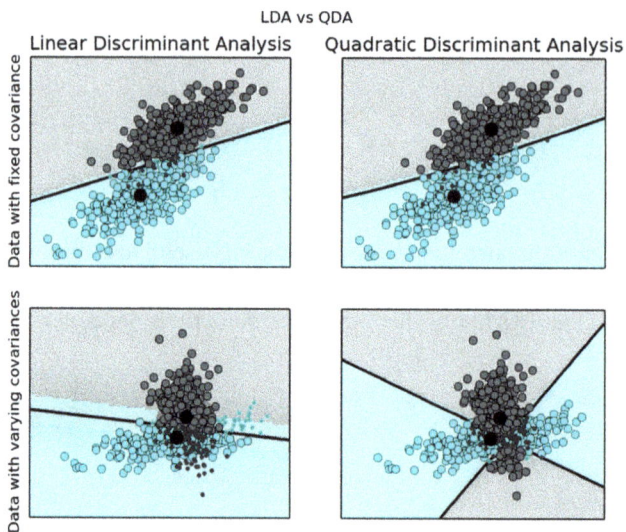

Figure 34.3: Linear discriminant analysis *(scikit [11])*.

Python linear discriminant analysis example (scikit[7])

```python
from sklearn.discriminant_analysis import LinearDiscriminantAnalysis

X = np.array([[-1, -1], [-2, -1], [-3, -2], [1, 1], [2, 1], [3, 2]])
y = np.array([1, 1, 1, 2, 2, 2])

clf = LinearDiscriminantAnalysis()
clf.fit(X, y)

print(clf.predict([[-0.8, -1]]))
> [1]
```

Example in chemistry

- **Search for complex chemical pathways using harmonic linear discriminant analysis[12]**

More algorithms

Despite the major algorithms this list contains, the implementations that may also be successful in regression questions under special circumstances:

Constrained linear regression

The least-squares method can confuse overshoots, false fields, etc. Restrictions are needed to reduce the variance of the line that is put in the data set: matching the linear regression model to models that can be L1 (LASSO) or L2 (Ridge Regression) or both (elastic regression).

SGD regressor

```
# predicting a quantity
# >100k samples
```

Ordinal regression

```
#properties: data in rank-ordered categories
```

Poisson regression

```
#properties: Predict event-counts
```

Fast forest quantile regression

```
#properties: predict a distribution
```

Neural network regression

This is discussed in the context of neural networks in Chapter 38.

References

1. Regression analysis (Wikipedia) https://en.wikipedia.org/wiki/Regression_analysis.
2. Difference between Random Forests and Decision Tree (StackExchange) https://stats. stackexchange.com/questions/285834/difference-between-random-forests-and-decision-tree.
3. Random forest (Wikimedia) https://commons.wikimedia.org/wiki/File:Random_forest_ diagram_complete.png.
4. Lovatti, B. P. O.; Nascimento, M. H. C.; Rainha, K. P.; Oliveira, E. C. S.; Neto, Á. C.; Castro, E. V. R.; Filgueiras, P. R. Different Strategies for the Use of Random Forest in NMR Spectra. J. Chemom. **2020**, 50, 729.https://onlinelibrary.wiley.com/doi/full-xml/10.1002/cem.3231.
5. Bhardwaj, R. M.; Johnston, A.; Johnston, B. F.; Florence, A. J. A Random Forest Model for Predicting the Crystallisability of Organic Molecules. CrystEngComm **2015**, 17 (23), 4272–4275. https://pubs.rsc.org/en/content/articlelanding/2015/ce/c4ce02403f.
6. Rawski, R. I.; Sanecki, P. T.; Kijowska, K. M.; Skitat, P. M.; Saletnik, D. E. Regression Analysis in Analytical Chemistry. Determination and Validation of Linear and Quadratic Regression Dependencies. S.Afr.j.chem. **2016**, 69. http://www.scielo.org.za/pdf/sajc/v69/27.pdf.
7. Chauhan, N. Beginner's guide to Linear Regression in Python with Scikit-Learn https://www. kdnuggets.com/2019/03/beginners-guide-linear-regression-python-scikit-learn.html.
8. Logistic function (Wikimedia) https://de.m.wikipedia.org/wiki/Datei:Mplwp_logistic_function. svg.
9. GLM: Linear regression – PyMC documentation https://docs.pymc.io/notebooks/GLM-linear. html.
10. scikit-learn: machine learning in Python, Documentation https://scikit-learn.org/stable/.
11. Linear and quadratic discriminant analysis – scikit-learn documentation https://scikit-learn. org/0.15/modules/lda_qda.html.
12. Rizzi, V.; Mendels, D.; Sicilia, E.; Parrinello, M. Blind Search for Complex Chemical Pathways Using Harmonic Linear Discriminant Analysis. J. Chem. Theory Comput. **2019**, 15 (8), 4507–4515 https://arxiv.org/pdf/1904.06276.pdf.

35 Decision trees

A decision tree[1] is a decision support tool that uses a tree-like model of decisions and their possible consequences, including chance event outcomes, resource costs, and utility. It is one way to display an algorithm that only contains conditional control statements.

Decision trees are the fundamental building block of gradient boosting machines and random forests, probably the two most popular machine learning models for structured data. Visualizing decision trees is a tremendous aid when learning how these models work and when interpreting models.[2,3]

Visualization example

```
# Decision tree visualization
# authors: Parr, Terence; Grover, Prince³
# License: MIT License          # code: github.com/parrt/dtreeviz
# activity: active (2019)       # index: 35-1


from animl.trees import *
from animl.viz.trees import *


diabetes = load_diabetes()


regr = tree.DecisionTreeRegressor(max_depth=2)
regr.fit(diabetes.data, diabetes.target)
# random sample from training
X = diabetes.data[np.random.randint(0, len(diabetes.data)),:]


viz = dtreeviz(regr,
               diabetes.data,
               diabetes.target,
               target_name='value',
               orientation ='LR',  # left-right orientation
               feature_names=diabetes.feature_names,
               X=X)  # need to give single observation for prediction


viz.view()
```

https://doi.org/10.1515/9783110629453-035

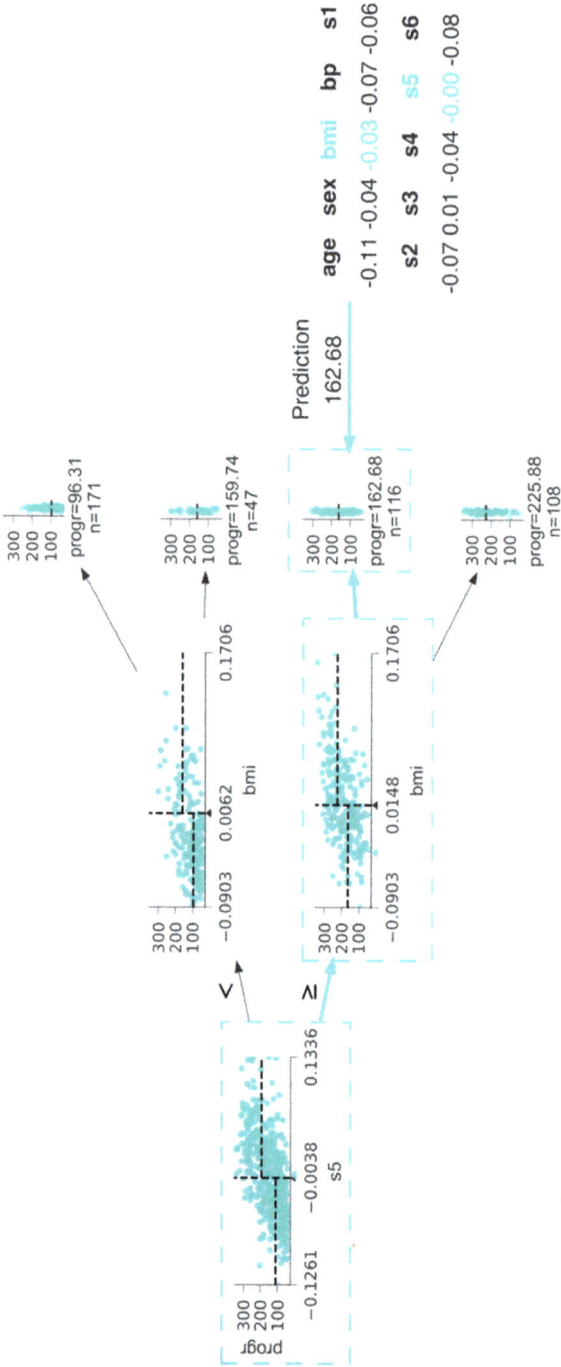

Figure 35.1: Decision tree example from AnIML *(Parr, Grover)*.

References

1. Decision tree https://en.wikipedia.org/wiki/Decision_tree.
2. Parr, T.; Grover, P. animl (GitHub) *https://github.com/darrenzeng2012/animl*.
3. How to visualize decision tree https://explained.ai/decision-tree-viz/index.html.

36 Second generation: classification

Classification is the problem[1] of identifying to which set of categories (subpopulations) a new observation belongs on the basis of a training set of data containing observations (or instances) whose category membership is known. It is an example of pattern recognition and is considered an instance of supervised learning, that is, learning where a training set of correctly identified observations is available.

An algorithm that implements classification is known as a classifier and sometimes also refers to the implemented mathematical function.

The corresponding unsupervised procedure is known as clustering and involves grouping data into categories based on some measure of inherent similarity or distance.

The observations are analyzed into a set of quantifiable properties: explanatory variables or features. These properties may variously be
- *categorical (e.g., "A", "B", "AB" or "O", for blood type),*
- *ordinal (e.g., "large", "medium" or "small"),*
- *integer-valued (e.g., the number of occurrences of a particular word in an email)*
- *real-valued (e.g., measurement of blood pressure)*

An example in chemistry is generating a two-class classification model for predicting toxicity[2].

Important algorithms

k-Nearest neighbors (*k*-NN)

k-Nearest neighbors (k-NN) is lazy learning as the function is only approximated locally and all the computations are deferred until classification.

The algorithm selects the k-nearest training samples for a test sample and then predicts the test sample with the major class amongst k-nearest training samples. The number of samples can be a user-defined constant (k-NN learning) or vary based on the local density of points (radius-based neighbor learning). Standard Euclidean distance is the most common distance.

```
#  k-nearest neighbors algorithm (k-NN)
# author: Gressling, T
# license: MIT License      # code: github.com/gressling/examples
# activity: single example   # index: 36-1
```

https://doi.org/10.1515/9783110629453-036

```
# (black dot, center) is test sample
# Classification to square or triangle?
```

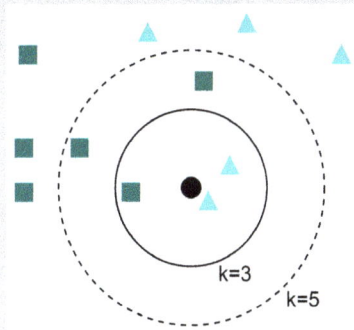

Figure 36.1: k-Nearest neighbors[3] *(Ajanki).*

```
case k = 3 (solid circle) -> assigned to triangle
                            # (2 triangles and 1 square in circle)
case k = 5 (dashed circle) -> assigned to square
                            # (3 squares and 2 triangles in circle)
```

Python logistic regression example (Scikit)

sklearn.neighbors provides functionality for unsupervised and supervised neighbors-based learning methods.[4]

```
from sklearn.neighbors import NearestNeighbors

X = np.array([[-1, -1], [-2, -1], [-3, -2], [1, 1], [2, 1], [3, 2]])

nbrs = NearestNeighbors(n_neighbors=2, algorithm='ball_tree').fit(X)
distances, indices = nbrs.kneighbors(X)

indices
> array([[0, 1],
>        [1, 0],
>        [2, 1],
>        [3, 4],
>        [4, 3],
>        [5, 4]])
```

```
distances
> array([[0.        , 1.        ],
>        [0.        , 1.        ],
>        [0.        , 1.41421356],
>        [0.        , 1.        ],
>        [0.        , 1.        ],
>        [0.        , 1.41421356]])


nbrs.kneighbors_graph(X).toarray()
> array([[1., 1., 0., 0., 0., 0.],
>        [1., 1., 0., 0., 0., 0.],
>        [0., 1., 1., 0., 0., 0.],
>        [0., 0., 0., 1., 1., 0.],
>        [0., 0., 0., 1., 1., 0.],
>        [0., 0., 0., 0., 1., 1.]])
```

Examples in chemistry

k-NN has been applied in chemistry since the early 1970s[5] to **classify:**
– molecular structures using their nuclear magnetic resonance (NMR) spectra
– elements of the periodic table with respect to their representative oxide
– hydrocarbons within three different classes using their mass spectra
– samples within a group of extraction kinetics using mid-infrared spectra
– polymer materials into four polymer classes

Use cases for **prediction:**
– structure–activity/property relationships (QSAR) studies to predict the volume of distribution at steady-state and clearance of antimicrobial agents in humans
– toxicity activity and anticonvulsant activity in different compounds
– predict melting points of organic molecules and drugs:

Melting point prediction employing k-NN algorithms and genetic parameter optimization[6]

A nearest neighbor approach was applied to 4119 structurally diverse organic molecules and 277 drug-like molecules to create a simple and reliable method for the prediction of melting points. The results gave valuable insights into the applicability of the "molecular similarity principle" that provides the basis for the kNN method in the field of property prediction.

Defining a novel k-NN approach to assess the applicability domain of a QSAR model for reliable predictions[7]

The algorithm was implemented using autoscaled Euclidean distances. The k-optimization procedure was carried out initially to decide upon an optimal k value and the training set contained 378 samples with 20% of the samples in the test set. The study proposes a descriptor-based applicability domain method for QSAR models, which exploits the k-NN principle to derive a heuristic decision rule within the descriptor space.

Naïve Bayes

```
# properties: easy to design
```

These classifiers are a family of *"probabilistic classifiers"* based on Bayes' theorem. There are *strong independence assumptions between the features*. It is called "Naive" because it makes the assumption that the occurrence of a certain feature is independent of the occurrence of other features. Naïve Bayes classifier does *not need any optimization* algorithms such as the gradient descent or maximum likelihood.

The classifiers are based on the simplest Bayesian network models. The algorithm aims to calculate the conditional probability of an object with a feature vector, which belongs to a particular class (probability distribution).

Python of Naïve Bayes

Four types of classes are available in scikit:
- **Gaussian**: features are in the dataset, which is normally distributed
- **Multinomial**: Used for document classification features; multinomial distributed
- **Complement**: imbalanced features
- **Bernoulli**: more than two or multiple features that are binary variables
- **Categorical** distributed data; each feature (by the index) has its own distribution

```
# Gaussian and complement Naïve Bayes (scikit)
# author: Gressling, T
# license: MIT License       # code: github.com/gressling/examples
# activity: single example   # index: 36-2

from sklearn.datasets import load_iris # iris example dataset

from sklearn.naive_bayes import GaussianNB
X, y = load_iris(return_X_y=True)
```

```
X_train, X_test, y_train, y_test =
          train_test_split(X, y, test_size=0.5, random_state=0)

gnb = GaussianNB()
y_pred = gnb.fit(X_train, y_train).predict(X_test)
print("Number of mislabeled points out of a total %d points : %d"
...       % (X_test.shape[0], (y_test != y_pred).sum()))

> Number of mislabeled points out of a total 75 points : 4
```

```
# Complement Naïve Bayes (scikit)
from sklearn.naive_bayes import ComplementNB

rng = np.random.RandomState(1)
X = rng.randint(5, size=(6, 100))
y = np.array([1, 2, 3, 4, 5, 6])

clf = ComplementNB()
clf.fit(X, y)
ComplementNB()

print(clf.predict(X[2:3]))
> [3]
```

Examples in chemistry

Naïve Bayes Classification Using 2D Pharmacophore Feature Triplet Vectors[8]

Molecules are described using a vector where each element in the vector contains the number of times a particular triplet of atom-based features separated by a set of topological distances occurs. Using the feature, triplet vectors, it is possible to generate naïve Bayes classifiers that predict whether molecules are likely to be active against a given target (or family of targets).

Classification of Metal Binders by Naïve Bayes Classifier on the Base of Molecular Fragment Descriptors and Ensemble Modeling[9]

Calculating two-class classification for organic molecules that are able to bind various metal cations in water: The modeling was performed on 30 data sets, each corresponding to a particular metal using the Naïve Bayes method and ISIDA fragment descriptors. The ligands were classified into weak and strong binders.

Novel naïve Bayes classification models for predicting the chemical Ames mutagenicity[10]

The prediction of drug candidates for mutagenicity was modeled. In addition, four molecular descriptors (Apol, No. of H donors, Num-Rings and Wiener) related to muta-genicity and representative substructures of mutagens (aromatic nitro, hydroxylamine, nitroso, aromatic amine, ...) produced by fingerprints were identified.

SVM/SVC: Support Vector Machine and classifier

Support Vector Machine is a supervised learning technique that represents the datasets as points. The main goal of SVM is to construct a hyperplane that divides the datasets into different categories and the hyperplane should be at the maximum margin from the various categories.

This algorithm helps in removing the over-fitting nature of the samples and provides better accuracy. A hyperplane is selected to best separate the points in the input variable space by their class, either class 0 or class 1. In two-dimensions, you can visualize this as a line. The algorithm finds the coefficients that result in the best separation of the classes by the hyperplane.

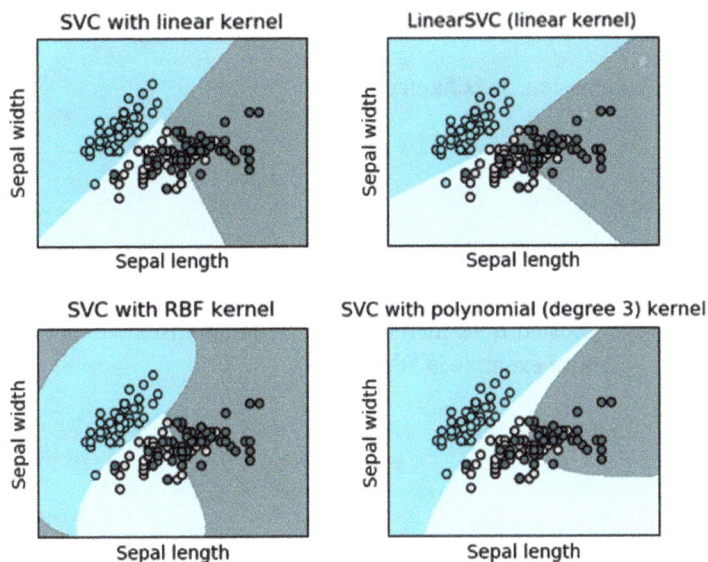

Figure 36.2: Support Vector Machine[4] *(Iris dataset)*.

Python SVC (Scikit)

```
# Support Vector Machine SVM (scikit)
# author: Gressling, T
# license: MIT License        # code: github.com/gressling/examples
# activity: single example    # index: 36-3

from sklearn import svm
X = [[0, 0], [1, 1]]
y = [0, 1]
clf = svm.SVC()
clf.fit(X, y)

clf.predict([[2., 2.]])
> array([1])
```

Examples in chemistry

–   **Applications of Support Vector Machines in Chemistry**[11]

Random Forest

–   Already discussed in regression, see Chapter 34.

AdaBoost (boosting)

Ensemble technique to create a strong classifier from a number of weak classifiers: Boosting is an algorithm that puts together prediction powers of two or more estimators to increase robustness. Boosted trees incrementally build an ensemble by training each new instance. A typical example is AdaBoost.

Other algorithms

Linear SVM/SVC

```
# properties: <100k samples
(-> alternative) Naiive Bayes    <100k samples, Text Data
(-> alternative) K-Neighbors     <100k samples
                    (-> alternative) SVC Ensemble Classifier
```

SGD Classifier

```
# properties: >100k samples
         (-> alternative) kernel approximation    >100k samples
```

References

1. Statistical classification (Wikipedia) https://en.wikipedia.org/wiki/Statistical_classification.
2. Kohtarou, Y. Generating two-class classification model for predicting chemical toxicity (Patent) https://patents.google.com/patent/US7725413.
3. k-nearest neighbors algorithm (Wikimedia) https://en.wikipedia.org/wiki/K-nearest_neighbors_algorithm#/media/File:KnnClassification.svg.
4. scikit-learn: machine learning in Python, Documentation https://scikit-learn.org/stable/.
5. Medina, J. L. V. Reliability of Classification and Prediction in K-Nearest Neighburs https://www.tdx.cat/bitstream/handle/10803/127108/Tesis%20Joe_Luis_Villa_Medina.pdf.
6. Nigsch, F.; Bender, A.; van Buuren, B.; Tissen, J.; Nigsch, E.; Mitchell, J. B. O. Melting point prediction employing k-nearest neighbor algorithms and genetic parameter optimization https://pubs.acs.org/doi/abs/10.1021/ci060149f.
7. Sahigara, F.; Ballabio, D.; Todeschini, R.; Consonni, V. Defining a novel k-nearest neighbours approach to assess the applicability domain of a QSAR model for reliable predictions https://www.ncbi.nlm.nih.gov/pmc/articles/PMC3679843/.
8. Watson, P. Naïve Bayes classification using 2D pharmacophore feature triplet vectors http://pubs.acs.org/doi/abs/10.1021/ci7003253.
9. Solov'ev, V.; Tsivadze, A.; Marcou, G.; Varnek, A. Classification of Metal Binders by Naïve Bayes Classifier on the Base of Molecular Fragment Descriptors and Ensemble Modeling https://onlinelibrary.wiley.com/doi/pdf/10.1002/minf.201900002.
10. Zhang, H.; Kang, Y. -L.; Zhu, Y. -Y.; Zhao, K. -X.; Liang, J. -Y.; Ding, L.; Zhang, T. -G.; Zhang, J. Novel naïve Bayes classification models for predicting the chemical Ames mutagenicity https://linkinghub.elsevier.com/retrieve/pii/S0887-2333(17)30037-1.
11. Ivanciuc, O. Applications of Support Vector Machines in Chemistry http://www.ivanciuc.org/Files/Reprint/Ivanciuc_Applications_of_Support_Vector_Machines_in_Chemistry.pdf.

37 Second generation: clustering and dimensionality reduction

Cluster analysis[1] is the task of grouping a set of objects in such a way that objects in the same group are more similar to each other than to those in other groups (clusters). Cluster analysis discovers structures in unlabeled data.

It can be achieved by various algorithms that differ significantly in their understanding of what constitutes a cluster and how to efficiently find them. Clustering can be formulated as a multiobjective optimization problem.

Cluster analysis as such is not an automatic task, but an iterative process of knowledge discovery or interactive multiobjective optimization that involves trial and failure.

Important algorithms

K-means clustering

```
# useCase: number of categories known, <10k samples
```

K-means clustering is a method that is commonly used to automatically partition a dataset into k groups. The algorithm proceeds by selecting the k initial cluster centers and then iteratively filtering them as each instance is assigned to its closest cluster center, while each cluster center is updated to the mean of its constituent.

And finally, the algorithm converges when there is no further change in the assignment of instances to clusters. This method is popular for cluster analysis in data mining.

Example using K-means

```
# Clustering using K-Means
# author: Roy[2], Jacques
# License: (permission by author)
# code: github.com/jacquesroy/...notebook W002
# activity: active (2020)          # index: 37-1

# Read data, extract 6 columns, binary values
# data = pd.read_csv(...)
X = data.iloc[:,2:17].values
y = data.iloc[:,1].values # RESULT column
```

https://doi.org/10.1515/9783110629453-037

```
# Encoding categorical data before split
from sklearn.preprocessing import LabelEncoder, StandardScaler
labelencoder_X_0 = LabelEncoder()
X[:,0] = labelencoder_X_0.fit_transform(X[:,0])
labelencoder_X_1 = LabelEncoder()
X[:,1] = labelencoder_X_1.fit_transform(X[:,1])
...

# Feature scaling; all the values in a standardized range
sc = StandardScaler()
X_scaled = sc.fit_transform(X)

k=10
model = KMeans(n_clusters=k)
kmeans = model.fit(X_scaled)

# Finding the optimal K Using the elbow method
from scipy.spatial.distance import cdist
distortions = []
K = range(2,15)
for k in K :
    kmeanModel = KMeans(n_clusters=k).fit(X_scaled)
    distortions.append(sum(
            np.min(cdist(X_scaled, kmeanModel.cluster_centers_,
            'euclidean'
            ),
            axis=1)) / X_scaled.shape[0])
...
plt.show()
```

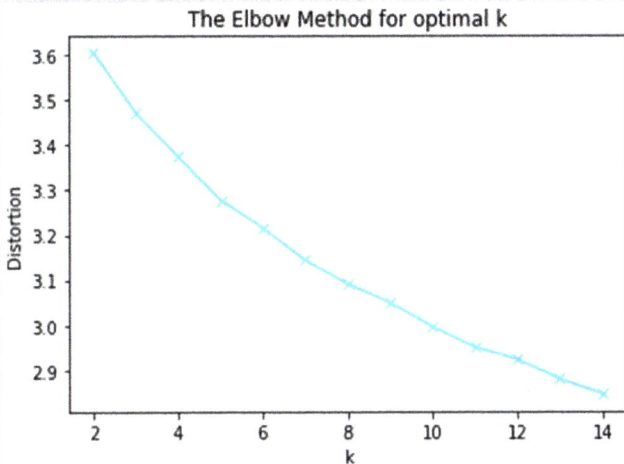

Figure 37.1: K-means elbow method.

```
# show K Means Cluster
vals=[0] * k
for i in kmeans.labels_ :
    vals[i] = vals[i] + 1

# Distribution between clusters
print(vals)
# [194, 37, 195, 244, 202, 132, 218, 262, 96, 219]
```

Examples in chemistry

The algorithm is used,[3] for example, to select a subset from a larger set of molecules, using Morgan fingerprints or in virtual screening (QSAR).

MiniBatch K-means

```
# useCase: number of categories known, >10k samples
```

Subsets are used in each iteration of the original K-means algorithm.

MeanShift VBGMM

```
# useCase: <10k samples
```

Dimensionality reduction

Cluster analysis is a method of unsupervised learning where the goal is to discover groups in the data; the groups are not known in advance (although you may know the number of groups). That is what "unsupervised" means here. There are a whole bunch of variations, but the two main types are hierarchical and k-means.

  PCA is a method of data reduction. It aims to reduce a large number of variables to a (much) smaller number, while losing as little information as possible.

Principal component analysis (PCA) and singular value decomposition (SVD)

The following examples are just a short introduction; more of principal component analysis (PCA) and singular value decomposition (SVD) is discussed later in the book, in the chemometric part as well as in Chapter 74.

```
# properties: simple and explainable
# useCases: classify continuous dependent and categorical variables
```

PCA forms the basis for multivariate data analysis. PCA is not a single algorithm – different ways of calculation are possible. PCA is a statistical method that converts a set of observations of *possibly correlated* variables into a set of values of *linearly uncorrelated* variables. This method is helpful in evaluating the minimum number of factors for the maximum variance in the data.

Small PCA example in Python

When a component analysis is applied to a dataset, it identifies the combination of attributes (principal components or directions in the feature space) that account for the most variance in the data. Here, we plot the samples on the first two principal components of the famous Iris dataset.

```
# PCA 2D projection (scikit)
# author: (scikit)
# license:  BSD new
# code: scikit-learn.org/.../decomposition/plot_pca
# activity: active (2020)      # index: 37-2

from sklearn.decomposition import PCA

# load
iris = datasets.load_iris()
X = iris.data
y = iris.target
target_names = iris.target_names

# model, fit
pca = PCA(n_components=2)
X_r = pca.fit(X).transform(X)

# plot
plt.figure()
for color, i, target_name in zip(colors, [0, 1, 2], target_names):
```

```
    plt.scatter(X_r[y == i, 0], X_r[y == i, 1],
                color=color, alpha=.8, lw=lw, label=target_name)
plt.legend(loc='best', shadow=False, scatterpoints=1)
plt.show()
```

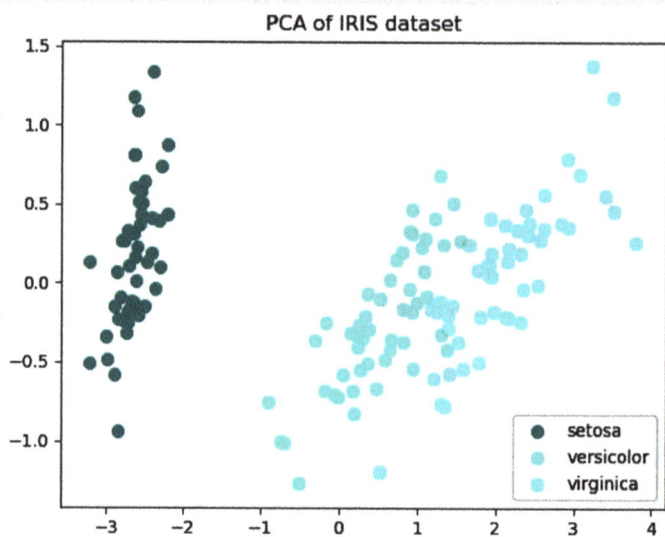

Figure 37.2: PCA of Iris dataset *(Scikit)*.

PCA examples in chemistry

PCA is one of the most important and powerful methods in chemometrics, the basic building block.[4,5] From the correlation of samples with fluorescence intensities and application of PCA on NMR spectra to use in clustering of chromatic data, all fields are covered. To investigate the intrinsic molecular properties of a series of compounds, the chemical properties (observables) for the compounds need to be collected. It is important that the properties which are chosen are relevant to the application and represent the compound. PCA analysis is one of the basic classes of algorithms. Read more about chemometrics in Chapters 77 and 78.

SVD - Singular value decomposition

It is widely used in statistics, where it is related to PCA and to correspondence analysis, in signal processing and pattern recognition. Several SVD implementations are discussed in this book.

ICA – Independent component analysis

A statistical technique for revealing hidden factors that underlie data sets, ICA is related to PCA, but it is a much more powerful technique capable of finding the underlying factors of sources when the classic methods fail.

Anomaly detection

One-class SVM

```
# useCase: >100 features, aggressive boundary
```

PCA-based anomaly detection

```
# properties: fast
```

References

1. Wikipedia contributors. Cluster analysis https://en.wikipedia.org/wiki/Cluster_analysis.
2. Roy, J. byte-size-data-science Tutorial (GitHub and YouTube) https://github.com/jacquesroy/byte-size-data-science.
3. Walters, P. K-means Clustering http://practicalcheminformatics.blogspot.com/2019/01/k-means-clustering.html.
4. Kucheryavskiy, S. PCA: Principal Component Analysis in mdatools: Multivariate Data Analysis for Chemometrics https://rdrr.io/cran/mdatools/man/pca.html.
5. B.y., C. PCA: The Basic Building Block of Chemometrics http://www.intechopen.com/books/analytical-chemistry/pca-the-basic-building-block-of-chemometrics.

38 Third generation: deep learning models (ANN)

Artificial neural networks[1] (ANN) or *connectionist systems* are computing systems vaguely inspired by the biological neural networks that constitute animal brains. Such systems *"learn"* to perform tasks by considering examples, generally without being programmed with task-specific rules.

Shallow models and deep learning

Figure 33.1 contains two main classes of machine learning. In the previous chapters, the shallow models (second-generation AI) were discussed. This chapter contains the deep learning part.

Neural network taxonomies

An ANN[2] is based on a collection of connected units or nodes called artificial neurons, which loosely models the neurons in a biological brain. Each connection, like the synapses in a biological brain, can transmit a signal to other neurons. An artificial neuron receives a signal, processes it, and can then signal neurons connected to it. Figure 38.1 shows a chart of neural networks. Table 38.1 contains the most important functions. Each ANN topology has its own characteristics and use-cases:

Table 38.1: ANN topologies and usage scenarios *(Gressling)*.

Algorithms	Suitable data types	Supervised or unsupervised	Functions
Autoencoder	Raw data; feature vectors	Unsupervised	Feature extraction; feature reduction; denoising
RBM	Feature vectors	Unsupervised	Feature extraction; feature reduction; denoising
DBN	Feature vectors	Supervised	Feature extraction; classification
DNN	Feature vectors	Supervised	Feature extraction; classification
CNN	Raw data; feature vectors; matrices	Supervised	Feature extraction; classification
RNN, Transformers[3], LSTM	Raw data; feature vectors; sequence data	Supervised	Feature extraction; classification
GAN	Raw data; feature vectors	Unsupervised	Data augmentation; adversarial training

https://doi.org/10.1515/9783110629453-038

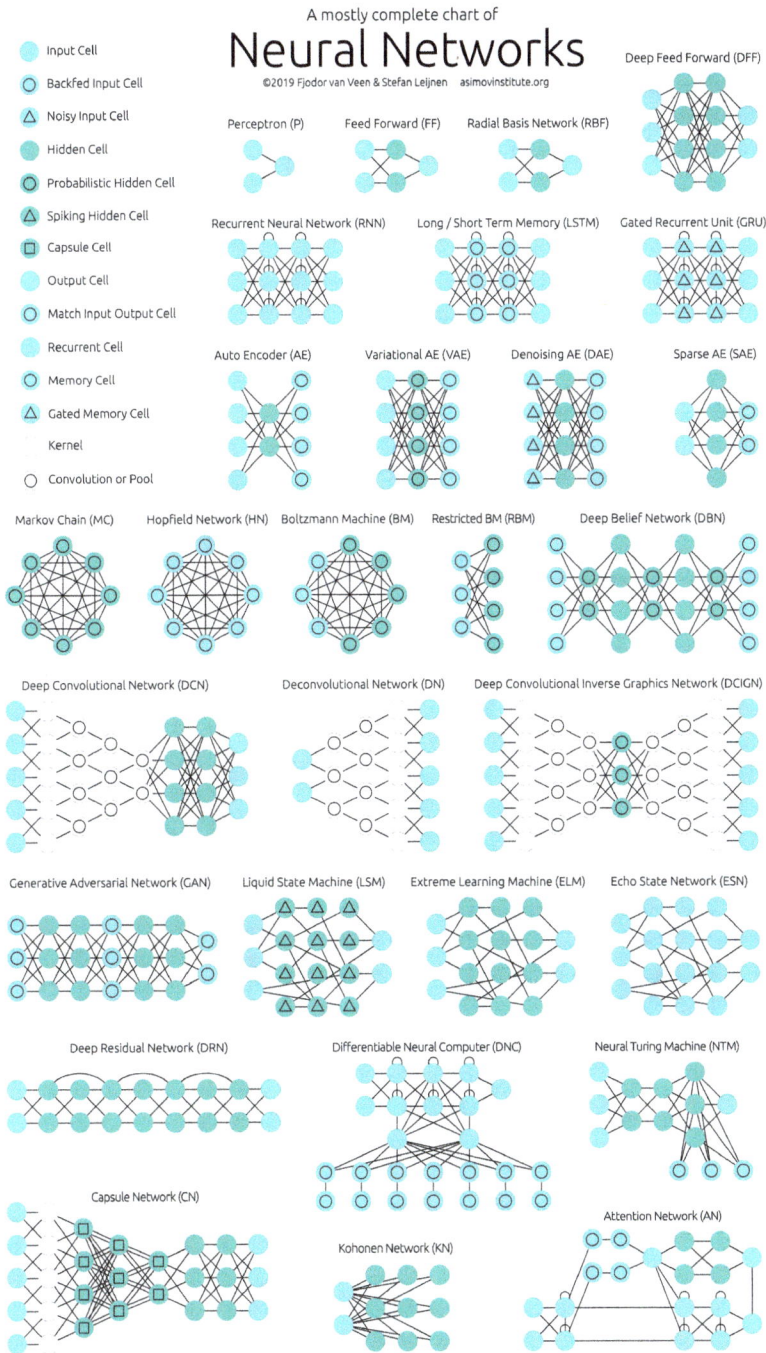

Figure 38.1: Chart of neural networks *(The Asimov Institute⁴).*

Design questions
(i)   How many hidden neurons can be used?
(ii) How many training pairs should be used?
(iii) Which training algorithm can be used?
(iv) What neural network architecture should be used?

CNN

Convolution neural networks are deep ANNs which are used to classify images, cluster
them by similarity and perform object recognition within scenes. They are algorithms
that can identify molecules, symbols, faces, and many other aspects of visual data.

```
# Sequence classification with 1D convolutions (Keras)
# author: Chollet, François (Et al.)
# license: MIT License          # code: keras.io/getting-started/examples
# activity: active (2020)        # index: 38-1

from keras.models import Sequential
from keras.layers import Dense, Dropout
from keras.layers import Embedding
from keras.layers import Conv1D, GlobalAveragePooling1D, MaxPooling1D

seq_length = 64
model = Sequential()
model.add(Conv1D(64, 3, activation='relu', input_shape=(seq_length, 100)))
model.add(Conv1D(64, 3, activation='relu'))
model.add(MaxPooling1D(3))
model.add(Conv1D(128, 3, activation='relu'))
model.add(Conv1D(128, 3, activation='relu'))
model.add(GlobalAveragePooling1D())
model.add(Dropout(0.5))
model.add(Dense(1, activation='sigmoid'))

model.compile(loss='binary_crossentropy',
              optimizer='rmsprop',
              metrics=['accuracy'])

model.fit(x_train, y_train, batch_size=16, epochs=10)
score = model.evaluate(x_test, y_test, batch_size=16)
```

RNN

Recurrent neural networks are specially used for processing sequential data such as sound, time series, or written natural languages. This technique differs from the feed-forward networks because they include a feedback loop.

```
# Character-level RNN to classify words (PyTorch)
# author: Robertson, Sean
# license:  BSD License              # code: pytorch.org/tutorials
# activity: active (2020)            # index: 38-2

import torch.nn as nn

class RNN(nn.Module):
    def __init__(self, input_size, hidden_size, output_size):
        super(RNN, self).__init__()
        self.hidden_size = hidden_size
        self.i2h = nn.Linear(input_size + hidden_size, hidden_size)
        self.i2o = nn.Linear(input_size + hidden_size, output_size)
        self.softmax = nn.LogSoftmax(dim=1)

    def forward(self, input, hidden):
        combined = torch.cat((input, hidden), 1)
        hidden = self.i2h(combined)
        output = self.i2o(combined)
        output = self.softmax(output)
        return output, hidden

    def initHidden(self):
        return torch.zeros(1, self.hidden_size)

n_hidden = 128
rnn = RNN(n_letters, n_hidden, n_categories)
```

LSTM

The selection of the number of hidden layers and the number of memory cells in LSTM probably depends on the application domain and context where you want to apply this LSTM.

```
# Sequence classification with LSTM (Keras)
# author: Chollet, François (Et al.)
# license: MIT License               # code: keras.io/getting-started/examples
# activity: active (2020)            # index: 38-3
```

```
from keras.models import Sequential
from keras.layers import Dense, Dropout
from keras.layers import Embedding
from keras.layers import LSTM

max_features = 1024

model = Sequential()
model.add(Embedding(max_features, output_dim=256))
model.add(LSTM(128))
model.add(Dropout(0.5))
model.add(Dense(1, activation='sigmoid'))

model.compile(loss='binary_crossentropy',
              optimizer='rmsprop',
              metrics=['accuracy'])

model.fit(x_train, y_train, batch_size=16, epochs=10)
score = model.evaluate(x_test, y_test, batch_size=16)
```

GAN, DCGAN, and deep topologies

A collection[5] of Keras implementations of Generative Adversarial Networks (GANs) are suggested in various research papers.

Transformers

Transformers[3] provide state-of-the-art general-purpose architectures (BERT, GPT-3, RoBERTa, XLM, DistilBert, XLNet, CTRL…) for Natural Language Understanding and Natural Language Generation with over 32+ pretrained models in 100+ languages and deep interoperability between TensorFlow 2.0 and PyTorch.

Metrics to evaluate

Use tools like comet.ml or UBER magnifold.

Accuracy, precision, and recall[6]: Accuracy is the proportion of true results among the total number of cases examined. Precision is a valid choice of evaluation metric when we want to be very sure of our prediction. Recall: what proportion of actual positives is correctly classified?
-   F1 score: precision/recall tradeoff
-   Log loss/binary cross-entropy

- Categorical cross-entropy
- AUC. AUC is the area under the ROC curve:

ROC

A receiver-operating characteristic curve, or ROC curve, is a plot that illustrates the diagnostic ability of a binary classifier system. The ROC curve is created by outlining the true positive rate against the false-positive rate at various threshold settings.

MoleculeNet benchmark

This library[7] is specially designed for testing *machine learning methods* and *featurizations of* molecular properties. It curates a number of dataset collections and creates a suite of software that implements many known featurizations and algorithms. It is built upon multiple public databases. The full collection currently includes over 700,000 compounds tested on a range of different properties. MoleculeNet[8] tests the performances of various machine learning models with different featurizations on the datasets (detailed descriptions here), with all results reported in AUC-ROC, AUC-PRC, RMSE, and MAE scores.

Extended-Connectivity Fingerprints

Molecule is decomposed into segments of variable sizes, all originated from heavy atoms (C, N, O).
All segment are then assigned with unique identifiers, which are hashed together into a fixed length binary fingerprint.

Deterministic

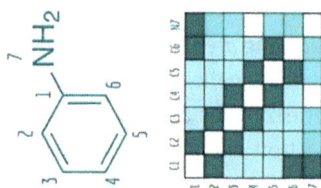

Weave

With the same feature vectors for atoms as Graph convolutions featurizer, Weave featurizer elaborates the neighbour list as a matrix of pair feature vectors, each representing the connectivity and distance between a pair of atoms.

Variable

Graph Convolutions

Molecule is represented by a neighbout list and a set of initial feature vectors, each corresponding to a single atom.. Feature vector summarizes the atom's local chemical environment, including atom-types, hybridization types and valence structures.

Variable

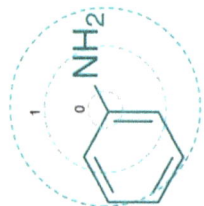

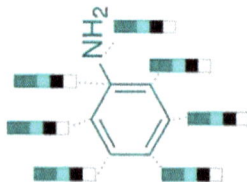

Grid Featurizer

Grid Featurizer, initially built for PDBbind, relies on detailed structures of protein–ligand pair to summarize inter-molecular forces. It incorporates fingerprints of both proteins and ligands, as well as an enumeration of salt bridges, hydrogen bonding, etc.

Deterministic   3D Coordinates

Figure 38.2: Featurizations contained in molecule.ai (Pande-Group, Stanford).

References

1. Artificial neural network (Wikipedia) https://en.wikipedia.org/wiki/Artificial_neural_network.
2. Types of artificial neural networks (Wikipedia) https://en.wikipedia.org/wiki/Types_of_artificial_neural_networks.
3. Transformers (GitHub) https://github.com/huggingface/transformers.
4. Van Veen, F.; Leijnen, S. The Neural Network Zoo – The Asimov Institute http://www.asimovinstitute.org/neural-network-zoo/.
5. Linder-Norén, E. Keras-GAN (GitHub) https://github.com/eriklindernoren/Keras-GAN.
6. The 5 Classification Evaluation Metrics Every Data Scientist Must Know – (KDnuggets) https://www.kdnuggets.com/2019/10/5-classification-evaluation-metrics-every-data-scientist-must-know.html.
7. Wu, Z.; Ramsundar, B.; Feinberg, E. N.; Gomes, J.; Geniesse, C.; Pappu, A. S.; Leswing, K.; Pande, V. MoleculeNet: A Benchmark for Molecular Machine Learning http://arxiv.org/abs/1703.00564.
8. moleculenet.ai (DeepChem package on GitHub) http://moleculenet.ai/.

39 Third generation: SNN – spiking neural networks

Spiking neural networks (SNN) are the next generation of neural networks. They do not react to each stimulus, but rather accumulate inputs until they reach a threshold potential and generate a "*spike.*" Because of their very nature, SNNs cannot be trained like current generation artificial neural networks using gradient descent.[1]

The future of SNNs, therefore, remains unclear. On the one hand, they are the natural successors of our current neural networks, but on the other, they are quite far from being practical tools for most tasks.[2] The neuron model is based on the *simple model of the spiking neuron.*[3]

Python implementation

```
# SNN in Tensorflow (shortened version)
# author: author: Corvoysier, David³
# license:  GNU GPL v2.0
# code: kaizou.org/simulating-spiking-neurons-with-tensorflow
# activity: active (2020)  # index: 39-1


# Evaluate membrane potential
dv_op = tf.where(has_fired_op,
                 tf.zeros(self.v.shape),
                 tf.subtract(
                     tf.add_n([tf.multiply(
                         tf.square(v_reset_op), 0.04),
                     tf.multiply(v_reset_op, 5.0),
                     tf.constant(140.0, shape=[self.n]),
                         i_op]),
                 self.u))

# Evaluate membrane recovery
du_op = tf.where(has_fired_op,
                 tf.zeros([self.n]),
                 tf.multiply(self.A,
                 tf.subtract(
                     tf.multiply(self.B, v_reset_op), u_reset_op)))

# Increment membrane potential
v_op = tf.assign(self.v,
                 tf.minimum(
                     tf.constant(self.SPIKING_THRESHOLD,
                         shape=[self.n]),
```

https://doi.org/10.1515/9783110629453-039

```
                    tf.add(v_reset_op,
                        tf.multiply(dv_op, self.dt))))
# Decrease in membrane recovery
u_op = tf.assign(self.u,
             tf.add(u_reset_op, tf.multiply(du_op, self.dt)))
```

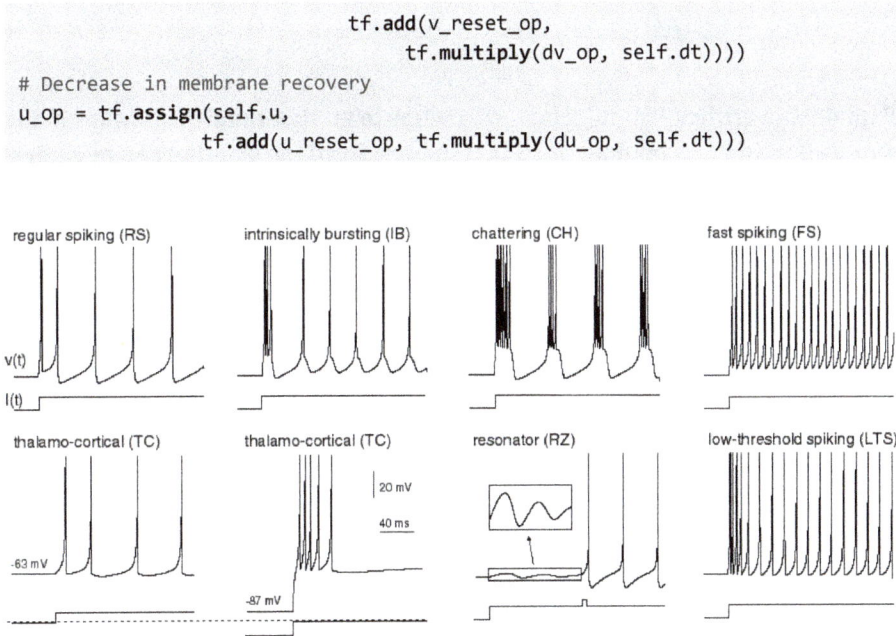

Figure 39.1: Spiking neurons[4] *(Izhikevich)*.

Implementations

– SNN experiments on GitHub[5]
– Spiking neural network simulator, basic SNN propagating spikes between layers of neurons[6]

References

1. Identify Repeating Patterns using Spiking Neural Networks in Tensorflow http://www.kaizou.org/2018/07/stdp-tensorflow/.
2. Soni, D. Spiking Neural Networks, the Next Generation of Machine Learning https://towardsdatascience.com/spiking-neural-networks-the-next-generation-of-machine-learning-84e167f4eb2b.
3. Simulating spiking neurons with Tensorflow http://www.kaizou.org/2018/07/simulating-spiking-neurons-with-tensorflow/.
4. Izhikevich, E. M. Simple Model of Spiking Neurons https://www.izhikevich.org/publications/spikes.htm.
5. de Azambuja, R. SNN-Experiments (GitHub) https://github.com/ricardodeazambuja/SNN-Experiments.
6. Strefford, M. Spiking-Neural-Network (GitHub) https://github.com/markstrefford/Spiking-Neural-Network.

40 xAI: eXplainable AI

Explainable artificial intelligence (xAI) refers to methods and techniques in the application of AI technology such that the results of the solution can be understood by human experts.

It contrasts with the concept of the "black box" in machine learning where even their designers cannot explain why the AI arrived at a specific decision. XAI is an implementation of the social right to explanation[1] and enables "third-wave AI systems."[2]

On a technical level, in ML models, there is a frequent tension between model performance and interpretability. Also, the optimal location in the solution space is highly application dependent.

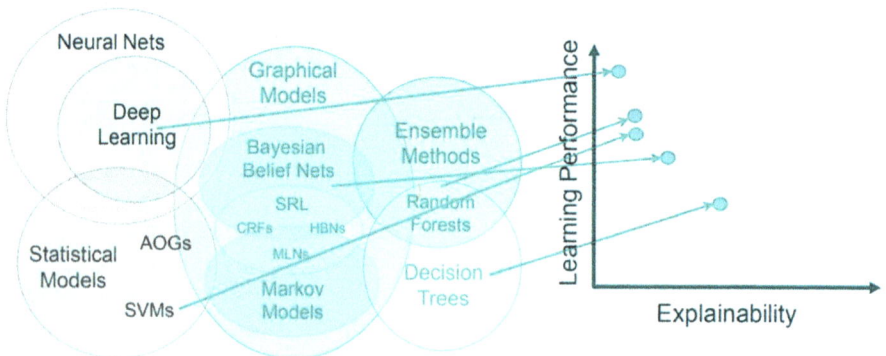

Figure 40.1: Explainability versus model[3] *(DARPA)*.

There are also external reasons[3] for an explainability demand:
- an incorrect model gets pushed
- incoming data is corrupted
- incoming data changes and no longer resembles datasets used during training

How does it work?

1. Given is a model's *complex* decision function f (unknown). It's a **black box.**
2. Now, **one new data point** is the instance being explained.
3. Takes many sample instances from the **original data** and gets predictions (using f, the black box).
4. **Weighs them** by the **proximity to the instance being explained.**
5. The **weights** now show the **relevance of the features** that are important for the explanation of the black box.

https://doi.org/10.1515/9783110629453-040

LIME library

Local interpretable model-agnostic explanations (LIME[4]) is an actual method to gain greater transparency on what is happening inside an algorithm. When the number of dimensions is high, maintaining local fidelity for such models becomes increasingly hard.[5] LIME solves the task of finding a model that approximates the original model locally.

```python
# Explainable AI with LIME
# author: Ribeiro, Marco Tulio Correia (Et al.)
# license: BSD 2 clause          # code: github.com/marcotcr/Lime/MNIST
# activity: active (2020)        # index: 40-1

import matplotlib.pyplot as plt
from sklearn.datasets import fetch_mldata
from sklearn.pipeline import Pipeline
from sklearn.ensemble import RandomForestClassifier
from sklearn.preprocessing import Normalizer


# 1. Data
mnist = fetch_mldata('MNIST original')


# 2. Prepare
# pipeline for processing the images
# flatten the image back to 1d vectors
class PipeStep(object):
    def __init__(self, step_func):
        self._step_func=step_func
    def fit(self,*args):
        return self
    def transform(self,X):
        return self._step_func(X)

makegray_step = PipeStep(Lambda img_list: [rgb2gray(img) for img in img_list])
flatten_step = PipeStep(Lambda img_list: [img.ravel() for img in img_list])


# 3. Train
# RandomForest Classifier
from sklearn.model_selection import train_test_split
X_train, X_test, y_train, y_test = train_test_split(X_vec, y_vec,
                                        train_size=0.55)

simple_rf_pipeline = Pipeline([
    ('Make Gray', makegray_step),
    ('Flatten Image', flatten_step),
    ('RF', RandomForestClassifier())
                             ])
```

```python
simple_rf_pipeline.fit(X_train, y_train)

# 4. Explain
from lime import lime_image
from lime.wrappers.scikit_image import SegmentationAlgorithm
explainer = lime_image.LimeImageExplainer(verbose = False)
segmenter = SegmentationAlgorithm('quickshift',
                    kernel_size=1, max_dist=200, ratio=0.2)

explanation = explainer.explain_instance(X_test[0],
        classifier_fn = simple_rf_pipeline.predict_proba,
        top_labels=10, hide_color=0,
        num_samples=10000, segmentation_fn=segmenter)

# 5. Show
fig, m_axs = plt.subplots(2,5, figsize = (12,6))
for i, c_ax in enumerate(m_axs.flatten()):
    temp, mask = explanation.get_image_and_mask(i, positive_only=True,
            num_features=1000, hide_rest=False, min_weight = 0.01 )
    c_ax.imshow(label2rgb(mask,X_test[0], bg_label = 0), interpolation =
    'nearest')
    c_ax.set_title('Positive for {}\nActual {}'.format(i, y_test[0]))
    c_ax.axis('off')
```

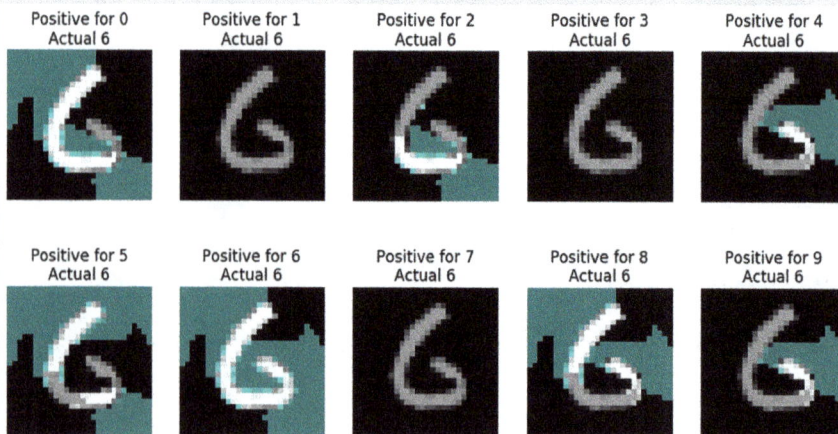

Figure 40.2: Positive regions for digits 0–9 (colored) compared to digit 6 *(Ribeiro)*.

There will be many use cases in chemistry because model evaluation and the detection of underrepresented/imbalanced data is a core AI problem. An example is described in the paper "Reliable and explainable machine learning methods for accelerated material discovery." [6]

References

1. Explainable artificial intelligence (Wikipedia) https://en.wikipedia.org/wiki/Explainable_ artificial_intelligence.
2. Explainable Artificial Intelligence https://www.darpa.mil/program/explainable-artificial-intelligence.
3. Explainable Artificial Intelligence (XAI) DARPA (CC0) https://www.darpa.mil/attachments/ XAIProgramUpdate.pdf.
4. Ribeiro, M. T. C. lime (GitHub) https://github.com/marcotcr/lime.
5. Ribeiro, M. T.; Singh, S.; Guestrin, C. "Why Should I Trust You?": Explaining the Predictions of Any Classifier. In *Proceedings of the 22nd ACM SIGKDD International Conference on Knowledge Discovery and Data Mining, San Francisco, CA, USA, August 13–17, 2016*; 2016; pp. 1135–1144.
6. Kailkhura, B.; Gallagher, B.; Kim, S.; Hiszpanski, A.; Han, Y.-J. Reliable and Explainable Machine Learning Methods for Accelerated Material Discovery. **2019**.

41 Crystallographic data

A crystalline[1] solid is a solid material whose constituents (such as atoms, molecules, or ions) are arranged in a highly ordered microscopic structure, forming a crystal lattice that extends in all directions. Crystals are usually identifiable by their geometrical shape – flat faces with specific, characteristic orientations.

ICSD: inorganic structure database

The ICSD[2] is a chemical database founded in 1978. It is now maintained by FIZ Karlsruhe in Europe and the U.S. National Institute of Standards and Technology. It contains information on all inorganic crystal structures published since 1913, including pure elements, minerals, metals, and intermetallic compounds (with atomic coordinates). ICSD is updated twice a year. The data is not freely available.

Query an ICSD structure dataset with aiida

This example shows how to obtain structure data with the aiida[3,4] (Automated Interactive Infrastructure and Database for Computational Science) package:

```
# Query ICSD crystallographic data (aiida)
# author: aiida team (ECOLE POLYTECHNIQUE FEDERALE DE LAUSANNE)
# license: MIT License
# code: aiida.readthedocs.io/projects/aiida-core
# activity: active (2020)          # index: 41-1


from aiida.tools.dbimporters.plugins.icsd import IcsdDbImporter

importer = IcsdDbImporter(server="http://ICSDSERVER.com/", host=
"127.0.0.1")

cif_nr_list = ["50542", "617290", "35538"]
query_results = importer.query(id=cif_nr_list)
for result in query_results:
    print(result.source['db_id'])
    aiida_structure = result.get_aiida_structure()
    # do something with the structure
```

The following export formats[5] are available:
- xsf
- xyz

https://doi.org/10.1515/9783110629453-041

- cif (export to CIF format, without symmetry reduction, i.e. always storing the structure as P1 symmetry)
- tcod (extension to the CIF format: supports symmetry reduction and typically adds in the CIF file, a number of additional information)

CSD: the Cambridge Structural Database

The Cambridge Structural Database (CSD) was established in 1965. It is a repository for small-molecule organic and metal-organic crystal structures that contain over one million structures from x-ray and neutron diffraction analyses. CCDC compiles the data-base and provides access.[6] Currently, it has about 9 GB of data.

The CSD Python API is a commercial package, and a Python API[7] is available.

References

1. Crystal (Wikipedia) https://en.wikipedia.org/wiki/Crystal.
2. Inorganic Crystal Structure Database (Wikipedia) https://en.wikipedia.org/wiki/Inorganic_Crystal_Structure_Database.
3. aiida-core (GitHub) https://github.com/aiidateam/aiida-core.
4. Pizzi, G.; Cepellotti, A.; Sabatini, R.; Marzari, N.; Kozinsky, B. AiiDA: Automated Interactive Infrastructure and Database for Computational Science. *Comput. Mater. Sci.* **2016**, 111, 218–230. https://doi.org/10.1016/j.commatsci.2015.09.013.
5. Export data nodes to various formats – AiiDA 0.11.2 documentation https://aiida-core.readthedocs.io/en/v0.11.2/datatypes/functionality.html.
6. The Cambridge Crystallographic Data Centre (CCDC) https://www.ccdc.cam.ac.uk/.
7. Dave Bardwell; Andrew Maloney; Ilenia Giangreco; Seth Wiggin; Pickard, Frank. CSD Python API Forum - The Cambridge Crystallographic Data Centre (CCDC) https://www.ccdc.cam.ac.uk/forum/csd_python_api/.

42 Crystallographic calculations

A space group is the symmetry group of a configuration in space, usually, in three dimensions where there are 219 distinct types or 230 if chiral copies are considered distinct. Space groups are also studied in dimensions other than 3, where they are sometimes called Bieberbach groups and are discrete cocompact groups of isometries of an oriented Euclidean space. In crystallography, space groups are also called the crystallographic or Fedorov groups and represent a description of the symmetry of the crystal.

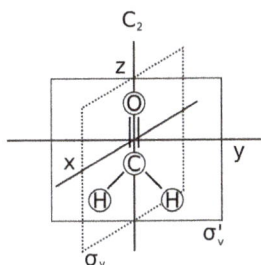

Figure 42.1: Symmetry elements of formaldehyde[1]. C_2 is a twofold rotation axis. σ_v and σ_v' are two nonequivalent reflection planes.

Space group analyzer (matgenb)

Pymatgen provides many analyses functions for structures. Some common ones are given below.[2] The library is part of the Materials Virtual Lab[3] at the University of California San Diego's Department of NanoEngineering; see Chapter 54.

```
# Symmetry analyzer (pymatgen)
# author: materialsvirtuallab
# license: BSD 3-Clause   # code: matgenb...org/.../Basic-functionality
# activity: active (2020)   # index: 42-1


# Determining the symmetry
from pymatgen.symmetry.analyzer import SpacegroupAnalyzer
finder = SpacegroupAnalyzer(<YOUR_STRUCTURE>)
print("The spacegroup is {}".format(finder.get_space_group_symbol()))
> The spacegroup is Pmm2
```

We also have an extremely powerful structure matching tool. It returns a mapping, which maps s1 and s2 onto each other. Strict element fitting is also available. Let us

https://doi.org/10.1515/9783110629453-042

create two structures that are the same topologically but with different elements and one lattice is larger.[2]

```
# Crystallographic structure matcher
# author: materialsvirtuallab
# license: BSD 3-Clause   # code: matgenb.materialsvirtuallab.org
# activity: active (2020)  # index: 42-2

from pymatgen.analysis.structure_matcher import StructureMatcher

s1 = mg.Structure(lattice, ["Cs", "Cl"], [[0, 0, 0], [0.5, 0.5, 0.5]])
s2 = mg.Structure(mg.Lattice.cubic(5),
                  ["Rb", "F"], [[0, 0, 0], [0.5, 0.5, 0.5]])
m = StructureMatcher()
print(m.fit_anonymous(s1, s2))
> true
```

Crystallography of materials

This illustrates various symmetry and crystallography concepts and computations,[4] and probability and statistics and their application to science and engineering.[5,6]

```
# Generate a structure from spacegroup (symmetry)
# author: Ong, Shyue Ping
# license: BSD 2   # code: github/shyuep/miworkshop/Materials API
# activity: single example  # index: 42-3
# Using pymatgen and the Materials API

from pymatgen import MPRester, Structure, Lattice, Element

perovskite = Structure.from_spacegroup(
        "Pm-3m", Lattice.cubic(3),
        ["Sr", "Ti", "O"],
        [[0,0,0], [0.5, 0.5, 0.5], [0.5, 0.5, 0]]
)
print(perovskite)

> Full Formula (Sr1 Ti1 O3)
> Reduced Formula: SrTiO3
> abc   :   3.000000   3.000000   3.000000
> angles:  90.000000  90.000000  90.000000
> Sites (5)
```

```
>   #  SP   abc
> ---  ----  ---   ---   ---
>   0  Sr000
>   1  Ti0.5  0.5  0.5
>   2  0 0.5  0.5  0
>   3  0 00.5  0.5
>   4  0 0.5  00.5
```

```
new_structure = perovskite.copy()
new_structure.apply_strain([0, 0, 0.1])
new_structure["Sr"] = "Ba"
print(new_structure)
```

```
> Full Formula (Ba1 Ti1 O3)
> Reduced Formula: BaTiO3
> abc   :   3.000000   3.000000   3.300000
> angles:  90.000000  90.000000  90.000000
> Sites (5)
>   #  SP   abc
> ---  ----  ---   ---   ---
>   0  Ba000
>   1  Ti0.5  0.5  0.5
>   2  0 0.5  0.5  0
>   3  0 00.5  0.5
>   4  0 0.5  00.5
```

```
# We can determine the spacegroup of this new structure.
# We get a tetragonal space group as expected.
```

```
print(new_structure.get_space_group_info())
> ('P4/mmm', 123)
```

Crystals

Crystals[7] is a library to manipulate abstract crystals with programming in python. Crystals helps with reading crystallographic files (like .cif and .pdb), provides access to atomic positions and scattering utilities, and allows for symmetry determination. Installation is standard with `pip install crystals` or `conda install -c conda-forge crystals`.

```
# data structure for crystallography (crystals)
# author: de Cotret, Laurent P. René⁸,⁹
# license: BSD 3    # code: crystals.readthedocs.io
# activity: active (2020)   # index: 42-4

from crystals import Crystal
vo2 = Crystal.from_database('vo2-m1')
print(vo2)
> Crystal object with following unit cell:
> Atom O  @ (0.90, 0.79, 0.80)
> Atom O  @ (0.90, 0.71, 0.30)
> Atom O  @ (0.61, 0.31, 0.71)
> Atom O  @ (0.39, 0.69, 0.29)
> Atom O  @ (0.61, 0.19, 0.21)
> Atom O  @ (0.10, 0.29, 0.70)
> Atom O  @ (0.10, 0.21, 0.20)
> Atom O  @ (0.39, 0.81, 0.79)
> Atom V  @ (0.76, 0.03, 0.97)
> Atom V  @ (0.76, 0.48, 0.47)
> ... omitting 2 atoms ...
> Lattice parameters:
> a=5.743Å..., b=4.517Å..., c=5.375Å...
> Î±=90.000Å°, Î²=122.600Å°, Î³=90.000Å°
> Chemical composition:
> O: 66.667%
> V: 33.333%
> Source:
> (...omitted...)\crystals\cifs\vo2-m1.cif
```

References

1. Andel. Formaldehyde symmetry elements (Wikimedia CC0) https://commons.wikimedia.org/ wiki/File:Formaldehyde_symmetry_elements.svg.
2. matgenb – Basic functionality http://matgenb.materialsvirtuallab.org/2013/01/01/ Basic-functionality.html.
3. Materials Virtual Lab http://materialsvirtuallab.org/.
4. Ong, S. P. NANO106 – Crystallography of Materials https://github.com/materialsvirtuallab/ nano106.
5. Ong, S. P. CENG/NANO 114; Probability and Statistics and their Application to Science and Engineering https://github.com/materialsvirtuallab/ceng114.
6. Ong, S. P. miworkshop https://github.com/shyuep/miworkshop.
7. crystals https://pypi.org/project/crystals/.
8. René de Cotret, L. P.; Otto, M. R.; Stern, M. J.; Siwick, B. J. An Open-Source Software Ecosystem for the Interactive Exploration of Ultrafast Electron Scattering Data. *Adv Struct Chem Imaging* **2018**, *4*, 11.
9. crystals : data structure for crystallography (1.0.0 documentation) https://crystals.readthedocs. io/en/master/.

43 Chemical kinetics and thermochemistry

Thermochemistry

Thermochemistry[1] is the study of heat energy that is associated with chemical reactions and/or physical transformations. A reaction may release or absorb energy, and a phase change may do the same, such as in melting and boiling. Thermochemistry focuses on these energy changes, particularly on the system's energy exchange with its surroundings.

    Thermochemistry is useful in predicting reactant and product quantities throughout the course of a given reaction. It coalesces the concepts of thermodynamics with the concept of energy in the form of chemical bonds.

Auxi

The Auxi toolkit helps metallurgical process engineers with their day-to-day tasks. Many of the calculations in thermochemistry require calculations such as molar masses, conversion of one compound to another using stoichiometry, enthalpy calculations, heat transfer calculations, mass balances, energy balances, etc. Auxi provides many of these calculations from within Python.

```
# Thermochemical calculations in metallurgy
# author: Zietsman², Johan (Et al.)
# license: GNU Lesser General Public License v3.0
# code: github.com/Ex-Mente/auxi.0
# activity: active (2019) (0.3.6)                    # index: 43-1


from auxi.tools.chemistry import thermochemistry as thermo


m_ZrO2 = 2.34
Cp_ZrO2 = thermo.Cp("ZrO2[S1]", 893.5, m_ZrO2)
print("The Cp of 2.34 kg of ZrO2[S1] at 893.5 °C is", Cp_ZrO2, "kWh/K.")
> The Cp of 2.34 kg of ZrO2[S1] at 70 °C is 0.0004084615851157184 kWh/K.


# Adding Material to a Package - A Compound Mass
from auxi.modelling.process.materials.thermo import Material


ilmenite = Material("Ilmenite", "./materials/ilmenite.txt")
reductant = Material("Reductant", "./materials/reductant.txt")
mix = Material("Mix", "./materials/mix.txt")


ilma_package = ilmenite.create_package("IlmeniteA", 300.0, 1.0, 25.0)
```

https://doi.org/10.1515/9783110629453-043

```
ilma_package += ("TiO2[S1]", 150.0)
print(ilma_package)

> ================================================================
> MaterialPackage
> ================================================================
> Material              Ilmenite
> Mass                  4.50000000e+02 kg
> Amount                5.40632064e+00 kmol
> Pressure              1.00000000e+00 atm
> Temperature           3.84927151e+02 °C
> Enthalpy              -1.14449836e+03 kWh
> ----------------------------------------------------------------
> Compound Details
> Formula          Mass            Mass Fraction   Mole Fraction
> ----------------------------------------------------------------
> Al2O3[S1]        3.48725349e+00  7.74945219e-03  6.32625154e-03
> CaO[S]           6.61375661e-02  1.46972369e-04  2.18151669e-04
> Cr2O3[S]         2.40500241e-02  5.34444979e-05  2.92683040e-05
> ...
```

Chemical kinetics

Chemical kinetics,[3] also known as reaction kinetics, is the branch of physical chemistry that is concerned with understanding the rates of chemical reactions. It is to be contrasted with thermodynamics, which deals with the direction in which a process occurs, but in itself tells nothing about its rate.

Chemical kinetics includes investigations of how *experimental conditions influence the speed of a chemical reaction* and yield information about the reaction's mechanism and transition states.

ChemPy[4,5] is a Python package useful in chemistry (mainly physical/inorganic/analytical chemistry). Currently, it includes:
-   Chemical kinetics (ODE solver front-end)
-   Rate expressions (and convenience fitting routines)
-   Solver for equilibria (including multiphase systems)

Physical chemistry:
-   Debye–Hückel expressions
-   Arrhenius and Eyring equation
-   Einstein–Smoluchowski equation

```python
# Chemical kinetics (differential equations, ChemPy)
# author: Dahlgren, Björn
# license: BSD 2-Clause "Simplified" License
# code: github.com/bjodah/chempy
# activity: active (2020)                    # index: 43-2


from chempy import ReactionSystem  # The rate constants below are arbitrary
rsys = ReactionSystem.from_string("""2 Fe+2 + H2O2 -> 2 Fe+3 + 2 OH-; 42
    2 Fe+3 + H2O2 -> 2 Fe+2 + O2 + 2 H+; 17
    H+ + OH- -> H2O; 1e10
    H2O -> H+ + OH-; 1e-4""")  # "[H2O]" = 1.0 (actually 55.4 at RT)

from chempy.kinetics.ode import get_odesys
odesys, extra = get_odesys(rsys)

from collections import defaultdict
import numpy as np
tout = sorted(np.concatenate((np.linspace(0, 23), np.logspace(-8, 1))))
c0 = defaultdict(float, {'Fe+2': 0.05, 'H2O2': 0.1,
                         'H2O': 1.0, 'H+': 1e-2, 'OH-': 1e-12})
result = odesys.integrate(tout, c0, atol=1e-12, rtol=1e-14)

import matplotlib.pyplot as plt
fig, axes = plt.subplots(1, 2, figsize=(12, 5))
for ax in axes:
    ...
_ = ax.set_xlabel('Time')
_ = ax.set_ylabel('Concentration')
_ = axes[1].set_ylim([1e-13, 1e-1])
```

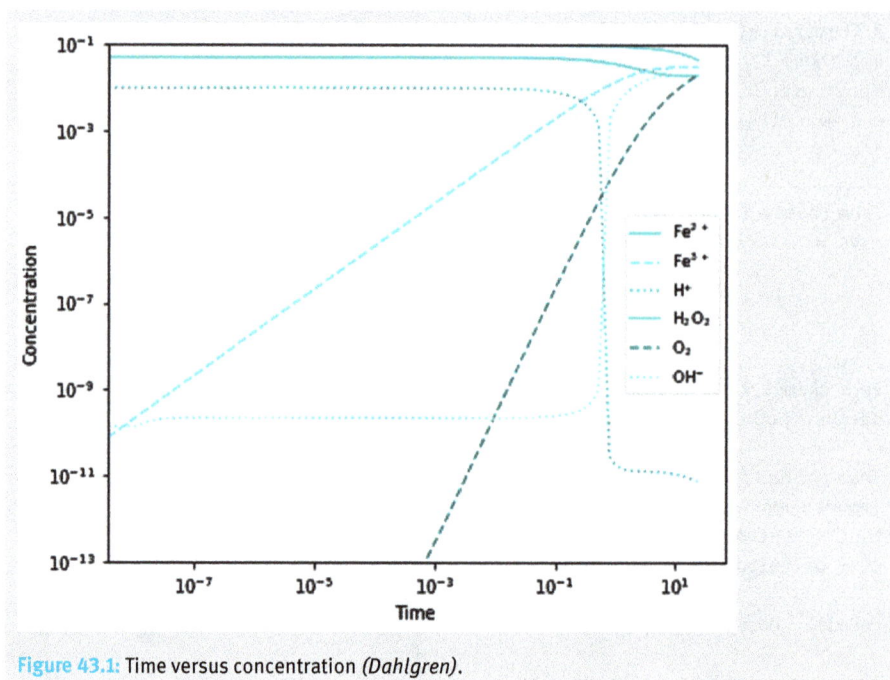

Figure 43.1: Time versus concentration *(Dahlgren)*.

References

1. Thermochemistry (Wikipedia) https://en.wikipedia.org/wiki/Thermochemistry.
2. Zietsman, J.; Sandrock, C.; Kok, C.; Jantzen, T. D.; Grewar, M. auxi – Open Source Python Package for Metallurgical Calculations https://www.researchgate.net/project/auxi-Open-Source-Python-Package-for-Metallurgical-Calculations.
3. Chemical kinetics (Wikipedia) https://en.wikipedia.org/wiki/Chemical_kinetics.
4. Dahlgren, B. chempy (GitHub) https://github.com/bjodah/chempy.
5. Dahlgren, B. ChemPy: A package useful for chemistry written in Python https://www.theoj.org/joss-papers/joss.00565/10.21105.joss.00565.pdf.

44 Reaction paths and mixtures

In engineering, physics, and chemistry, the study of transport phenomena[1] concerns the exchange of mass, energy, charge, momentum, and angular momentum between the observed and studied systems. While it draws from fields as diverse as continuum mechanics and thermodynamics, it places a heavy emphasis on the commonalities between the topics covered. Mass, momentum, and heat transport, all share a very similar mathematical framework.

Examples with Cantera

Cantera[2] is an open-source suite for problems involving chemical kinetics, thermodynamics, and/or transport processes. It can be used for applications including combustion, detonations, electrochemical energy conversion and storage, fuel cells, batteries, aqueous electrolyte solutions, plasmas, and thin film deposition.

This index lists examples included with the Cantera[3] Python module:

Thermodynamics

- critical state properties for the fluids for which Cantera has built-in liquid/vapor equations of state
- isentropic, adiabatic flow
- Rankine vapor power cycle
- "equilibrium" and "frozen" sound speeds for a gas

```
# Air and methane mixed in stoichiometric proportions
# author: Cantera Developers
# license: Copyright (c) 2016-2018, Cantera Developers (Open Source)
# also parts licensed with: BSD 3-Clause "New" or "Revised" License
# code: github.com/Cantera/cantera-jupyter
# activity: active (2020)        # index: 44-1

import cantera as ct
gas = ct.Solution('gri30.xml')

# Stream A (air)
A = ct.Quantity(gas, constant='HP')
A.TPX = 300.0, ct.one_atm, 'O2:0.21, N2:0.78, AR:0.01'
# Stream B (methane)
B = ct.Quantity(gas, constant='HP')
B.TPX = 300.0, ct.one_atm, 'CH4:1'
```

https://doi.org/10.1515/9783110629453-044

```
# Set the molar flow rates corresponding to stoichiometric reaction,
# CH4 + 2 O2 -> CO2 + 2 H2O
A.moles = 1
nO2 = A.X[A.species_index('O2')]
B.moles = nO2 * 0.5

# Compute the mixed state
M = A + B
print(M.report())
# Show that this state corresponds to stoichiometric combustion
M.equilibrate('TP')
print(M.report())
```

Kinetics

- Uses species and reaction objects to extract a reaction sub mechanism
- Dynamically manipulating chemical mechanisms

```
# Viewing a reaction path diagram
# author: Cantera Developers
# license: Copyright (c) 2016-2018, Cantera Developers (Open Source)
# also parts licensed with: BSD 3-Clause "New" or "Revised" License
# code: github.com/Cantera/cantera-jupyter
# activity: active (2020)        # index: 44-2
import cantera as ct

gas = ct.Solution('gri30.xml')
gas.TPX = 1300.0, ct.one_atm, 'CH4:0.4, O2:1, N2:3.76'
r = ct.IdealGasReactor(gas)
net = ct.ReactorNet([r])
T = r.T
while T < 2400:
    net.step(1.0)
    T = r.T

element = 'N'
diagram = ct.ReactionPathDiagram(gas, element)
diagram.title = 'Reaction path diagram following {0}'.format(element)
diagram.label_threshold = 0.1
diagram.show_details = True
diagram.write_dot(dot_file)
print(diagram.get_data())
```

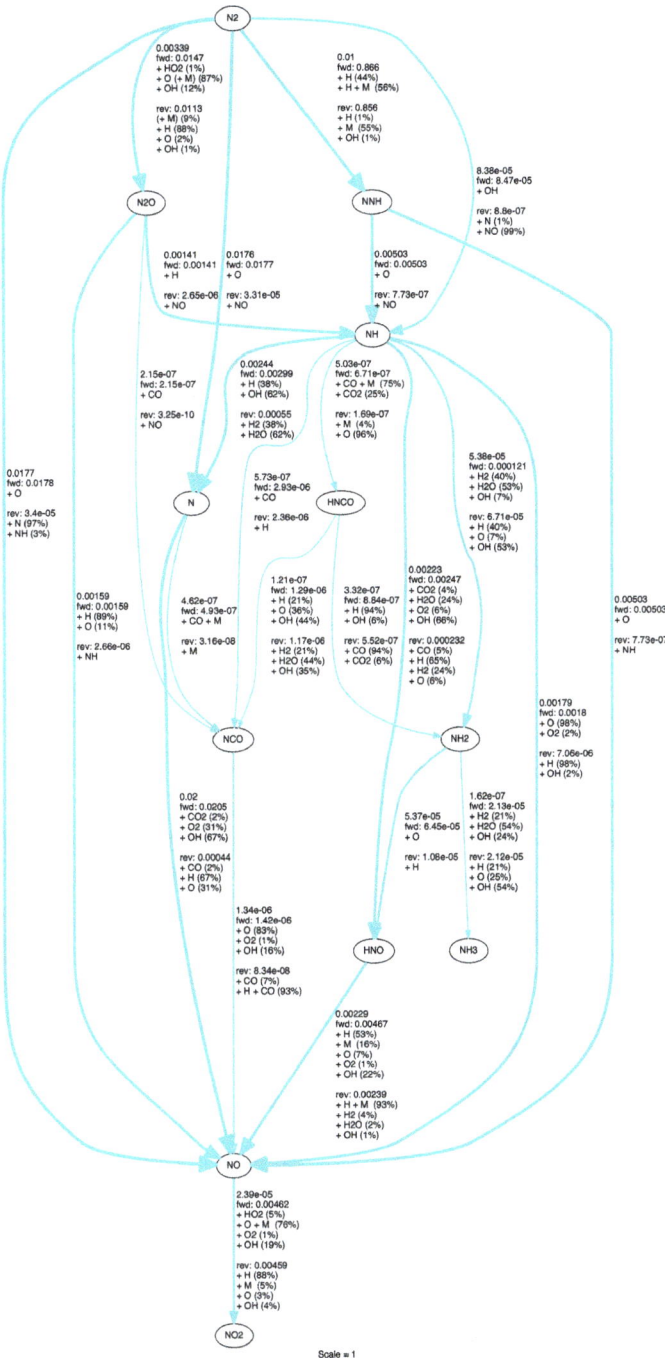

Figure 44.1: NOx generation routes on a reaction path with Cantera4 *(Salehi)*.

Transport

- Dusty gas transport model
- Viscosity

Reactor networks

- A combustor: Two separate streams, one, pure methane and the other, air, both at 300 K and 1 atm, flow into an adiabatic combustor
- Constant pressure ignition problem where the governing equations
- Mixing two streams
- Continuously stirred tank reactor (CSTR) with steady inputs
- Plug-flow reactor with a simulation of a Lagrangian fluid particle and a simulation of a chain of reactors
- Reactors separated by a piston
- Adiabatic kinetic simulations

One-dimensional flames and multiphase mixtures

- Freely propagating, premixed hydrogen flat flame
- Opposed-flow ethane/air diffusion flame
- Extinction point of a counterflow diffusion flame
- Equilibrium example with charged species in the gas phase and multiple condensed phases.

Surface chemistry

- Catalytic combustion of methane on platinum
- Simulating growth of a diamond film
- Model of a solid oxide fuel cell

References

1. Transport phenomena (Wikipedia) https://en.wikipedia.org/wiki/Transport_phenomena.
2. MACCCR-Meeting-2019 (GitHub) https://github.com/Cantera/MACCCR-Meeting-2019.
3. Cantera 2.3.0 documentation https://cantera.github.io/docs/sphinx/html/index.html.
4. Salehi, M. NOx generation routes on a reaction path example https://groups.google.com/forum/#!topic/cantera-users/rLFTgjpHO_M.

45 The periodic table of elements

Alternative periodic tables are tabulations of chemical elements that differ in their organization from the traditional depiction of the periodic system. Over a thousand have been devised.[1] Major periodic trends include electronegativity, ionization energy, electron affinity, atomic radii, ionic radius, metallic character, and chemical reactivity.[2]

Mendeleev

Mendeleev[3] by Lukasz Mentel is a Python API to access a query on atomic properties for elements in the periodic table. It contains a database and can access various properties of elements, ions, and isotopes. The purpose is to get parameters for calculations in an easy way. This list contains an excerpt of the properties that are included[3]:

```
# Periodic table of elements
# author: Mentel, Lukasz
# license: MIT License (MIT)      # code: github.com/Lmmentel/mendeleev
# activity: active (2020)         # index: 45-1

from mendeleev import element
si = element('Si')
si
> Element(
>      abundance_crust=282000.0,
>      abundance_sea=2.2,
>      annotation='',
>      atomic_number=14,
>      atomic_volume=12.1,
>      atomic_weight=28.085,
>      atomic_weight_uncertainty=None,
...
>      covalent_radius_pyykko_double=107.0,
>      covalent_radius_pyykko_triple=102.0,
>      covalent_radius_slater=110.00000000000001,
>      cpk_color='#daa520',
>      density=2.33,
>      econf='[Ne] 3s2 3p2',
>      electron_affinity=1.3895211,
>      en_allen=11.33
...
```

https://doi.org/10.1515/9783110629453-045

Periodic/Bokeh

Also, the Bokeh[4] visualization library provides some data. Within the package bokeh.
sampledata.periodic_table, the periodic table of elements as a tabular display of the
chemical elements is available. The data can be arranged by atomic number, electron
configuration, and recurring chemical properties.

```
# Periodic table in Bokeh
# authors: Anaconda, Inc., and Bokeh Contributors
# license: BSD 3-Clause      # code: github.com/bokeh
# activity: active (2020)    # index: 45-2

from bokeh.sampledata import periodic_table

periodic_table.elements["atomic mass"] =
                periodic_table.elements["atomic mass"].astype(str)
elements = periodic_table.elements[periodic_table.elements["group"] != "-"]
reversed(sorted(set(elements["period"])))]
colormap = {
    "alkali metal"        : "#a6cee3",
    "alkaline earth metal" : "#1f78b4",
    "halogen"             : "#fdbf6f",
    "metal"               : "#b2df8a",
    "metalloid"           : "#33a02c",
    "noble gas"           : "#bbbb88",
    "nonmetal"            : "#baa2a6",
    "transition metal"    : "#e08e79"
}

p = figure(title="Periodic Table", tools=TOOLS,
    x_range=group_range, y_range=period_range)
...
show(p)
```

Periodic Table (omitting LA and AC Series)

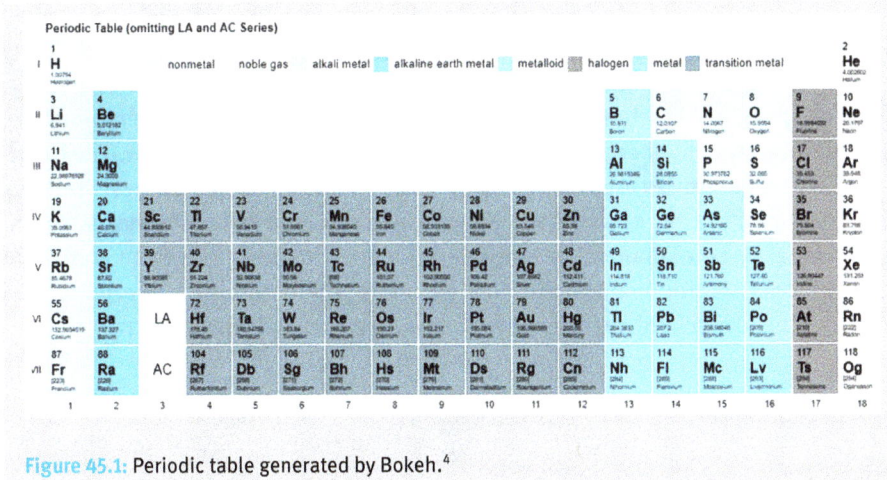

Figure 45.1: Periodic table generated by Bokeh.[4]

References

1. Alternative periodic tables (Wikipedia) https://en.wikipedia.org/wiki/Alternative_periodic_tables.
2. Periodic trends (Wikipedia) https://en.wikipedia.org/wiki/Periodic_trends.
3. Mentel, L. M. mendeleev – A Python resource for properties of chemical elements, ions and isotopes https://mendeleev.readthedocs.io/en/stable/.
4. periodic – Bokeh documentation https://docs.bokeh.org/en/latest/docs/gallery/periodic.html.

46 Applied thermodynamics

Thermodynamics[1] is a branch of physics that deals with heat and temperature, and their relation to energy, work, radiation, and properties of matter. The behavior of these quantities is governed by the four laws of thermodynamics. The laws provide a quantitative description using measurable macroscopic physical quantities but may be explained in terms of microscopic constituents by statistical mechanics. Thermodynamics applies to a wide variety of topics in science and engineering, especially physical chemistry and chemical engineering.

PyTherm

Iuri Segtovich[2] has created some educational Jupyter notebooks for *applied thermodynamics*. These examples are not based on a library; they use standard Python mathematics.

```
# Isotherms of the van der Waals Equation of State
# author: Segtovich³, Iuri
# license: GPL v3
# code: github.com/iurisegtovich/PyTherm-...-thermodynamics
# activity: single example # index: 46-1

R = 8.3144598 # gas constant
# pure component critical point of hexane:
Tc = 507.5 # Tc (K)
Pc = 30.1*(10**5) # Pc (Pa)
#  variables a and b for the EoS parameters
a = 27/64*((Tc**2)*(R**2)/Pc)
b = (R*Tc)/(8*Pc)

def Pressure_van_Der_waals(T, Vm):
    P = R*T/(Vm-b) - a/(Vm**2)
    return P

fig = plt.figure(1)
for T in np.array([.1,.25,.5,.75,1,2])*Tc:
  Vmi = np.logspace(np.log10(b*1.01),2+np.log10(R*Tc/Pc),100)
  plt.semilogx(Vmi, Pressure_van_Der_waals(T, Vmi),label=str(T)+' K')
    plt.ylabel(r'$P (\mathrm{Pa})$')
    plt.xlabel(r'$\bar{V} (\mathrm{m^{3}mol^{-1}})$')
    plt.title('vdW\'s EoS isotherms \n (for n-hexane)')
    plt.ylim(-Pc,Pc*3)
    plt.xlim(b,(R*Tc/Pc)*100)

plt.show()
```

https://doi.org/10.1515/9783110629453-046

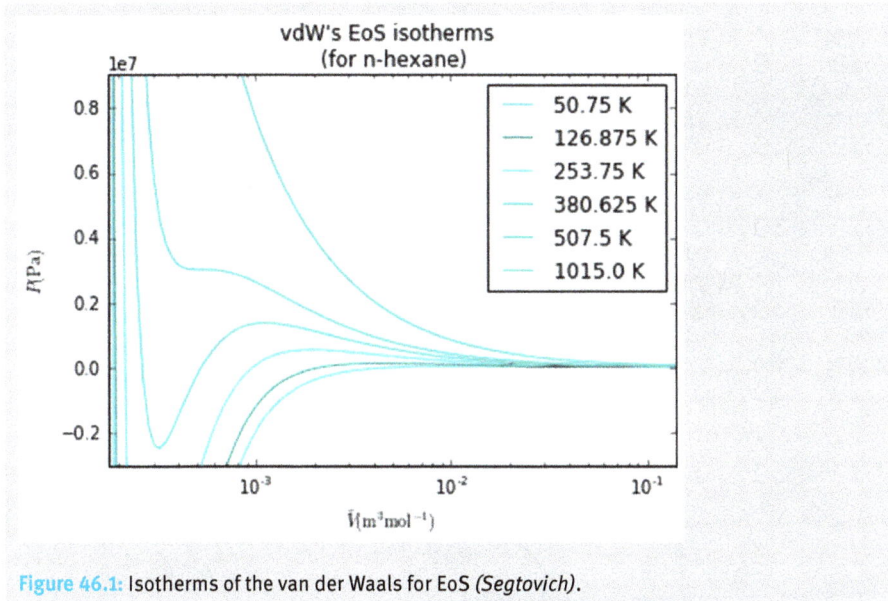

Figure 46.1: Isotherms of the van der Waals for EoS *(Segtovich)*.

- Calculations of molar volume and saturation pressure of a pure substance
- Analytical solutions of the van der Waals equation of state
- Natural gas hydrate models and algorithms: methane hydrates
- Excess thermodynamics properties, relations, and models
- Azeotropy in vapor–liquid phase equilibrium
- Liquid–liquid phase equilibrium flash algorithm
- Calculation of adsorption for pure fluids by equation of state

References

1. Thermodynamics https://en.wikipedia.org/wiki/Thermodynamics.
2. Segtovich, I. AwesomeThermodynamics https://github.com/iurisegtovich/
   AwesomeThermodynamics.
3. Segtovich, I. PyTherm-applied-thermodynamics (GitHub) https://github.com/iurisegtovich/
   PyTherm-applied-thermodynamics.

Material science

47 Material informatics

Materials informatics is a field of study that applies the principles of informatics to materials science and engineering to understand the use, selection, development, and discovery of materials. This is an emerging field, with a goal of achieving high-speed and robust acquisition, management, analysis, and dissemination of diverse materials data with the intention of greatly reducing the time and risk required to develop, produce, and deploy a new material, which generally takes longer than 20 years.[1]

Material informatics[2] includes combinatorial chemistry, process modeling, materials property databases, and materials data management. By gathering appropriate metadata, the value of each individual data point can be expanded.

For example, for exploring quaternary compounds (~10^{13} materials) it is obvious that a new paradigm for discovery is needed. This can be by accelerating the design of inorganic materials and the route to new materials, using machine learning.

Analysis of structure–property–processing relationships

Materials Knowledge Systems (MKS) is a data science approach for solving multiscale materials science problems. It uses techniques from physics, machine learning, regression analysis, signal processing, and spatial statistics to create processing–structure–property relationships.

MKS in Python (PyMKS[3]) is a set of tools and examples written in Python that provides high-level access to the MKS framework.

```
# Quantify Microstructures using 2-Point Statistics (PyMKS)
# author: MATIN materials research group
# license: MIT License
# code: materialsinnovation.github.io/pymks
# activity: active (2020)          # index: 47-1


from pymks.datasets import make_microstructure

X_1 = make_microstructure(n_samples=1, grain_size=(25, 25))
X_2 = make_microstructure(n_samples=1, grain_size=(95, 15))
X = np.concatenate((X_1, X_2))

from pymks.tools import draw_microstructures
draw_microstructures(X)
```

https://doi.org/10.1515/9783110629453-047

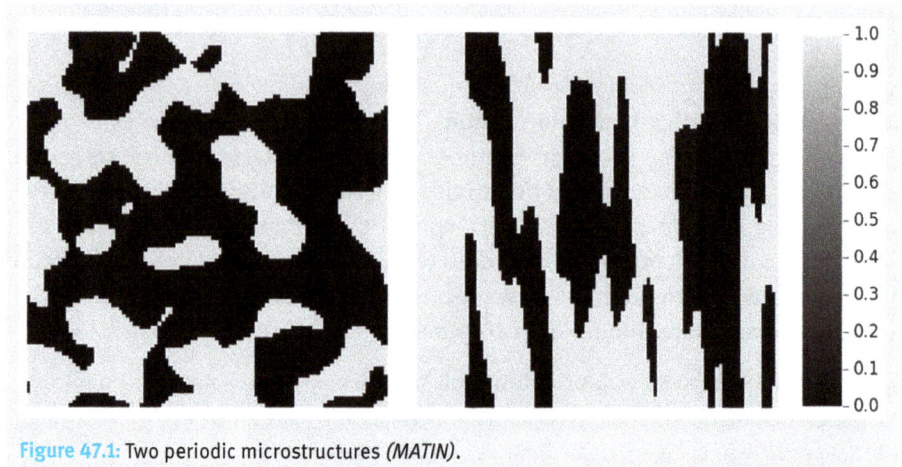

Figure 47.1: Two periodic microstructures *(MATIN)*.

References

1. Materials informatics (Wikipedia) https://en.wikipedia.org/wiki/Materials_informatics.
2. Isayev, O.; Tropsha, A.; Curtarolo, S. *Materials Informatics: Methods, Tools, and Applications*; John Wiley & Sons, 2019.
3. Wheeler, D.; Brough, D.; Fast, T.; Kalidindi, S.; Reid, A. PyMKS: Materials Knowledge System in Python https://materialsinnovation.github.io/pymks/.

48 Molecular dynamics workflows

Molecular dynamics (MD) is a computer simulation method for studying the *physical movements* of atoms and molecules. The atoms and molecules are allowed to interact for a fixed period of time, giving a view of the dynamic evolution of the system.[1]

MD use either quantum mechanics (QM), or molecular mechanics (MM), or a mixture of both to calculate forces, which are then used to solve Newton's laws of motion to examine the time-dependent behavior of systems. The result of a MD simulation is a trajectory that describes how the position and velocity of particles vary with time.[2]

MD difference to MM

MM and MD are related, as both are based on the same classical force fields[3].
- MD is modeling of molecular motions
- MM implements "static" energy minimization methods (i.e., potential energy surfaces)

MD difference to QM

- QM has scale limitation: volume (size) and shape (boundary conditions) of corresponding simulation boxes (3-d structures, 2-d grids, 1-d chains, or 0-d points)
- QM is problem-dependent: phase transitions (e.g., magnetic) or quantum mechanically (magnetism is a quantum phenomenon)

Hybrid QM – MM

The hybrid QM MM approach is a simulation method that combines the strengths of the QM (accuracy) and MM (speed). An advantage of the combination is their efficiency.[4]

Python toolchain examples

FESetup simulation setup and workflow

FESetup[5] automates the setup of relative alchemical free energy simulations such as thermodynamic integration and free energy perturbation. FESetup can also be used

https://doi.org/10.1515/9783110629453-048

for general simulation setup ("equilibration") through an abstract MD engine (AMBER, GROMACS, NAMD, etc.).[6]

```
# Workflow of a free energy simulations (FeSETUP)
# author: Löffler, Hannes H
# license: GNU General Public License v2.0    # code: github.com/halx/FESetup
# activity: on hold? (2016)                   # index: 48-1


# force field sub type, water model, divalent ions, MD engine
amber = prepare.ForceField('amber', 'ff14SB', 'tip3p', 'cm', [], 'amber')


# make protein
protein_file = os.path.join(top, 'thrombin/protein/2ZC9/protein.pdb')
protein_wd = os.path.join(top, '_protein', '2ZC9')
name = '2ZC9'
protein = amber.Protein(name, protein_file)


# create model
model = ModelConfig(name)
protein.charge = float(model['charge.total'])
protein.amber_top = model['top.filename']
protein.amber_crd = model['crd.filename']
protein.orig_file = model['crd.original']
if 'box.dimensions' in model:
    protein.box_dims = model['box.dimensions']
```

OVITO toolchain and visualization

A workflow in MD MM also contains visualization of particle-based simulations. The following OVITO[7] example loads an atomic structure from a simulation file, selects all hydrogen atoms, deletes them, and writes the results back to an output file, enabling visualization.

```
# Ovito pipeline / ASE Example
# author: Stukowski⁸, Alexander (ovito.org)
# author: Janssen, Jan (MPI für Eisenforschung)
# license:  GNU General Public License v3.0
# code: github.com/chilammps/Ovito
# code: github.com/jan-janssen/ovito3-example
# activity: on hold? (2015)        # index: 48-2


from ovito.io import import_file, export_file
from ovito.modifiers import SelectTypeModifier, DeleteSelectedModifier
```

```
# 1 - data wrangling example
pipeline = import_file('input_file.xyz')
pipeline.modifiers.append(
     SelectTypeModifier(property='Particle Type', types={'H'}) )
pipeline.modifiers.append( DeleteSelectedModifier() )
export_file(pipeline, 'output_file.xyz', 'xyz')


# 2 - visualization example
from ase.build import bulk
import nglview
from ovito.io.ase import ase_to_ovito
from ovito.pipeline import Pipeline, StaticSource
from ovito.modifiers import CreateBondsModifier,
     CommonNeighborAnalysisModifier


# ASE
cu_atoms = bulk('Cu', 'fcc', a=3.6)
atoms = cu_atoms.repeat([4,4,4])


# NGLview
view = nglview.show_ase(atoms)
view.add_spacefill(radius_type='vdw', color_scheme='element', scale=0.5)
view.remove_ball_and_stick()
view
> NGLWidget()


# 3 - Ovito pipeline example
pipeline = Pipeline(source = StaticSource(data = ase_to_ovito(atoms)))
pipeline.modifiers.append(CommonNeighborAnalysisModifier(
          mode=CommonNeighborAnalysisModifier.Mode.AdaptiveCutoff))
output = pipeline.compute()
output.attributes
> {'CommonNeighborAnalysis.counts.OTHER': 0, 'CommonNeighborAnalysis.counts.
FCC': 64,
> 'CommonNeighborAnalysis.counts.HCP': 0, 'CommonNeighborAnalysis.counts.
BCC': 0,
> 'CommonNeighborAnalysis.counts.ICO': 0}
```

MD simulations with QM/MM and adaptive neural networks

The computational cost of QM/MM calculations during MD simulations can be reduced using semiempirical QM/MM methods (with lower accuracy). A neural network method like QM/MM-NN can predict the potential energy difference between semiempirical and ab initio QM/MM approaches.[9]

In the QM/MM method, the whole system is divided into a QM subsystem containing the active site and an MM subsystem containing the rest.

References

1. Molecular dynamics (Wikipedia) https://en.wikipedia.org/wiki/Molecular_dynamics.
2. Computational chemistry (Wikipedia) https://en.wikipedia.org/wiki/Computational_chemistry.
3. Srivastava, A. What is the difference between molecular mechanics and molecular dynamics? https://www.researchgate.net/post/What_is_the_difference_between_molecular_mechanics_and_molecular_dynamics.
4. QM/MM (Wikipedia) https://en.wikipedia.org/wiki/QM/MM.
5. Loeffler, H. H.; Michel, J.; Woods, C. FESetup: Automating Setup for Alchemical Free Energy Simulations. *J. Chem. Inf. Model.* **2015**, *55* (12), 2485–2490. https://doi.org/10.1021/acs.jcim.5b00368.
6. Loeffler, H. FESetup (GitHub) https://github.com/halx/FESetup.
7. OVITO's Python interface – Scientific visualization and analysis software for atomistic simulation data https://www.ovito.org/docs/current/python/introduction/running.php.
8. Stukowski, A. Visualization and Analysis of Atomistic Simulation Data with OVITO–the Open Visualization Tool. *Modell. Simul. Mater. Sci. Eng.* **2009**, *18* (1), 015012. https://doi.org/10.1088/0965-0393/18/1/015012.
9. Shen, L.; Yang, W. Molecular Dynamics Simulations with Quantum Mechanics/Molecular Mechanics and Adaptive Neural Networks. *J. Chem. Theory Comput.* **2018**, *14* (3), 1442–1455. https://doi.org/10.1021/acs.jctc.7b01195.

49 Molecular mechanics

Molecular mechanics[1] uses classical mechanics to model molecular systems. The Born–Oppenheimer approximation is assumed valid, and the potential energy of all systems is calculated as a function of the nuclear coordinates, using force fields.

Molecular mechanics can be used to study molecule systems ranging in size and complexity from small to large biological systems or material assemblies with many thousands to millions of atoms.

The following assumptions are used in calculations:
- Each atom is simulated as one particle.
- Each particle is assigned a
  - radius (van der Waals),
  - polarizability,
  - constant net charge (derived from quantum calculations or experiment).
- Bonded interactions are treated as springs with an equilibrium distance to the bond length.

The prototypical application of molecular mechanics is *energy minimization*. That is, the force field is used as an optimization criterion, and the (local) minimum searched by an appropriate algorithm. Molecular mechanics formulation is a vast simplification and only an approximation. Verify through quantum calculations on small systems that the molecular mechanics' parameters describe that process in a satisfactory manner.

Force field

In the context of molecular modeling, a force field[2] (a special case of energy functions or interatomic potentials; not to be confused with the force field in classical physics) refers to the functional form and parameter sets used to calculate the potential energy of a system of atoms or coarse-grained particles in molecular mechanics and molecular dynamics simulations. The parameters of the energy functions may be derived from experiments in physics or chemistry, calculations in quantum mechanics, or both.

Open force field consortium

Open force field[3,4] is an academic-industry consortium to develop open biomolecular force fields. The Python toolkit is for the development and application of modern molecular

https://doi.org/10.1515/9783110629453-049

mechanics force fields, based on direct chemical perception and statistical parameteriza-
tion methods. The toolkit currently covers two main areas:
- Tools for using the SMIRKS native open force field (SMIRNOFF) specification
- Tools for direct chemical environment perception and manipulation

Example using openforcefield with Python

```
$ conda config --add channels omnia --add channels conda-forge
$ conda install openforcefield
```

```
# Working with SMIRNOFF parameter sets (OpenForceField)
# author: OpenForceField Team
# license: MIT License
# code: github.com/openforcefield
# activity: active (2020)          # index: 49-1

# Load a molecule into the openforcefield Molecule object
from openforcefield.topology import Molecule
from openforcefield.utils import get_data_file_path
sdf_file_path = get_data_file_path('molecules/ethanol.sdf')
molecule = Molecule.from_file(sdf_file_path)

# Create an openforcefield Topology object from the molecule
from openforcefield.topology import Topology
topology = Topology.from_molecules(molecule)

# Load the smirnoff99Frosst SMIRNOFF force field definition
from openforcefield.typing.engines.smirnoff import ForceField
forcefield = ForceField('forcefields/smirnoff99Frosst.offxml')

# Create an OpenMM system representing the molecule with SMIRNOFF-applied
parameters
openmm_system = forcefield.create_openmm_system(topology)

# Load a SMIRNOFF small molecule forcefield for alkanes, ethers, and alcohols
forcefield = ForceField('forcefields/Frosst_AlkEthOH_parmAtFrosst.offxml')
```

Run the simulation (with OpenMM) as follows:

```
from simtk import openmm, unit

# Propagate the System with Langevin dynamics.
time_step = 2*unit.femtoseconds
temperature = 300*unit.kelvin  # simulation temperature
```

```
friction = 1/unit.picosecond  # collision rate
integrator = openmm.LangevinIntegrator(temperature, friction, time_step)
# Length of the simulation
num_steps = 1000  # number of integration steps to run
simulation = openmm.app.Simulation(omm_topology, system, integrator)

[...]

# Run the simulation
simulation.step(num_steps)
```

If successful, the directory contains a trajectory file that can be visualized and a
data.csv file including potential energy, density, and temperature of each frame.

OpenMM

OpenMM is a high-performance toolkit[5] for molecular simulation. It can be used as a
library, with bindings for Python. The code is open source and actively maintained on
Github, and licensed under MIT and LGPL. It supports AMBER and CHARMM force fields.

```
# Automate the use of AMBER force fields with GAFF (OpenMM)
# author: OpenMM Team
# License: MIT License    # code: https://github.com/openmm/openmm
# activity: active (2020) # index: 49-2

# Define the keyword arguments to feed to ForceField
from simtk import unit
from simtk.openmm import app
forcefield_kwargs = { 'constraints' : app.HBonds, 'rigidWater' : True,
'removeCMMotion' : False, 'hydrogenMass' : 4*unit.amu }

# Initialize a SystemGenerator using General AMBER force field (GAFF)
# provides parameters for small organic molecules to facilitate simulations
# of drugs and small molecule ligands
from openmmforcefields.generators import SystemGenerator
system_generator = SystemGenerator(forcefields=['amber/ff14SB.xml',
                    'amber/tip3p_standard.xml'],
                    small_molecule_forcefield='gaff-2.11',
                    forcefield_kwargs=forcefield_kwargs,
                    cache='db.json')

system = system_generator.create_system(openforcefield_topology)
# Alternatively, create an OpenMM System from an OpenMM Topology object and
a list of openforcefield Molecule objects
system = system_generator.create_system(openmm_topology,
                                  molecules=molecules)
```

References

1. Molecular mechanics (Wikipedia) https://en.wikipedia.org/wiki/Molecular_mechanics.
2. Force field (chemistry) (Wikipedia) https://en.wikipedia.org/wiki/Force_field_(chemistry).
3. Open Force Field Toolkit https://open-forcefield-toolkit.readthedocs.io/en/latest/.
4. newpixcom. Open Force Fields https://openforcefield.org/.
5. OpenMM http://openmm.org/.

50 VASP

The Vienna ab initio simulation package (VASP[1]) is a computer program for atomic scale materials modeling, e.g. electronic structure calculations and quantum-mechanical molecular dynamics, from first principles. It uses[2] either Vanderbilt pseudopotentials or the projector augmented wave method and a plane wave basis set.

The basic methodology is density functional theory (DFT) but the code also allows the use of post-DFT corrections such as hybrid functionals mixing DFT and Hartree-Fock exchange, many-body perturbation theory (the GW method) and dynamical electronic correlations, within the random phase approximation.

VASP installation

Local builds of DFT codes can be useful. It is easy to install via Docker[3,4] for VASP, also with GPU support. A license can be obtained by request, example:

```
$ docker build -t vasp        # code: github.com/ajjackson/vasp-docker
```

Python interfaces to VASP

Many different solutions are available. Some are part of a bigger library like `ase.calculators.vasp` (discussed in Chapter 55) and some are stand-alone. Here are some examples of handling VASP.

Usage in matgenb/pymatgen

The vaspio_set module[5] provides a means to obtain a complete set of VASP input files for performing calculations. Several useful presets based on the parameters used in the Materials Project are provided.[6]

```
# VASP in pymatgen
# author: pymatgen team
# license: MIT License
# code: matgenb.materialsvirtuallab.org/Basic-functionality
# activity: active (2020)      # index: 50-1
```

https://doi.org/10.1515/9783110629453-050

```
# Create example structure
lattice = mg.Lattice.cubic(4.2)
structure = mg.Structure(lattice, ["Cs", "Cl"],
                        [[0, 0, 0], [0.5, 0.5, 0.5]])

# Convenient IO to various formats
structure.to(filename="POSCAR")
structure.to(filename="CsCl.cif")

from pymatgen.io.vasp import Poscar
poscar = Poscar.from_file("POSCAR")
structure = poscar.structure

# obtain a complete set of VASP input files
from pymatgen.io.vasp.sets import MPRelaxSet
v = MPRelaxSet(structure)
v.write_input("MyInputFiles")
```

PytLab/VASPy

VASPy[7] is a pure Python library designed to make it easy and quick to manipulate VASP files. You can use VASPy to manipulate VASP files in command lines or write your own python scripts to process VASP files and visualize VASP data. This is an example for plotting ELF[8] (electron localization function):

```
# Plotting the electron localization function (PytLab/VASPy)
# author: Shao Zhengjiang
# license: MIT License            # code: github.com/PytLab/VASPy
# activity: on hold? (2016) (0.7.0)  # index: 50-2

from vaspy.electro import ElfCar
a = ElfCar()
a.plot_contour()
a.plot_mcontour()   # Plot coutour using mlab(with Mayavi installed)
a.plot_contour3d()
a.plot_field()      # Plot scalar field
```

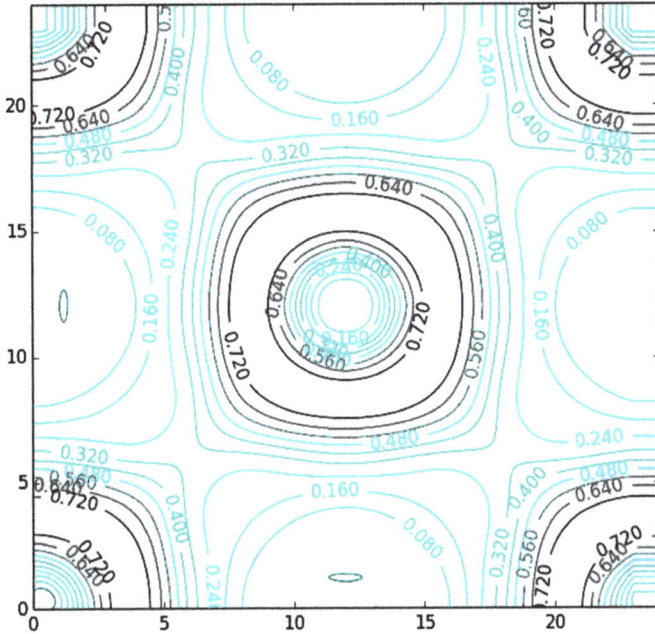

Figure 50.1: ELF contour plot *(Zhengjiang).*

vasp (Python)

`vasp.py` provides threads, pool, and multiprocessing for interacting with VASP[9,10]:

```
# Single, Threads, Pool and Multiprocessing with vasp.py
# author: John Kitchin
# License: GNU Free Documentation License    # code: github.com/jkitchin/vasp
# activity: on hold? (2016) (0.7.0)          # index: 50-3

from vasp import Vasp
from ase import Atom, Atoms
import Queue
import threading

class Worker(threading.Thread):
    def __init__(self, queue):
        self.queue = queue
        threading.Thread.__init__(self)
        self.daemon = True
```

```python
    def run(self):
        while true:
            item = self.queue.get()
            f, energies, index = item
            a = 3.6
            atoms = Atoms([Atom('Cu',(0, 0, 0))],
                        cell=0.5 * a * f * np.array([[1.0, 1.0, 0.0],
                                                    [0.0, 1.0, 1.0],
                                                    [1.0, 0.0, 1.0]]))
            calc = Vasp('mp/queue-Cu-{}'.format(index),
                        xc='pbe', encut=300, kpts=[6, 6, 6],
                        isym=2, atoms=atoms)
            energies[index] = calc.potential_energy
            q.task_done()

# Setup our queue
q = Queue.Queue()
num_threads = 2
for i in range(num_threads):
    Worker(q).start()

# Return evenly spaced numbers over a specified interval:
factors = np.linspace(0.9, 1.1, 10)

energies = [None for f in factors]
for i, f in enumerate(factors):
    q.put([f, energies, i])

q.join()
# print(energies)
```

References

1. VASP – Vienna Ab initio Simulation Package https://www.vasp.at/.
2. Vienna Ab initio Simulation Package (Wikipedia) https://en.wikipedia.org/wiki/Vienna_Ab_initio_Simulation_Package.
3. Hampel, A. VASP/ESPRESSO Dockerfiles https://github.com/materialstheory/Dockerfiles.
4. Jackson, A. J. vasp-docker (GitHub) https://github.com/ajjackson/vasp-docker.
5. pymatgen documentation https://pymatgen.org/.
6. matgenb – Basic functionality http://matgenb.materialsvirtuallab.org/2013/01/01/Basic-functionality.html.
7. Zhengjiang, S. VASPy (GitHub) https://github.com/PytLab/VASPy.
8. ELFCAR – Vaspwiki https://www.vasp.at/wiki/index.php/ELFCAR.
9. Kitchin, J. vasp (gitHub) https://github.com/jkitchin/vasp.
10. DFT blog http://kitchingroup.cheme.cmu.edu/dft-book/.

51 Gaussian (ASE)

Gaussian[TM,1] is a computational chemistry software package released in 1970. It has been continuously updated, since then. The name originates from the use of Gaussian orbitals to speed up molecular electronic structure calculations – as opposed to using Slater-type orbitals. This choice was made to improve performance on the limited computing capacities of the then-current computer hardware for Hartree–Fock calculations.

The current version of the program is Gaussian[TM] 16; it was later licensed out of Carnegie Mellon University, and since 1987 has been developed and licensed by Gaussian[TM], Inc.[2]

ASE[3,4] is described in Chapter 55.

Interfacing with Gaussian[TM]

This example requires a Hessian calculation followed by optimization to a saddle point ("transition state optimization"; the Hessian matrix is the matrix of second derivatives of the energy with respect to geometry, so a force calculation):

```
# GAUSSIAN™ Wrapper (ASE), transition state optimization
# author: ASE-developers
# license: LGPLv2.1+          # code: fysik.dtu.dk/ase/.../calculators/gaussian
# activity: active (2020)     # index: 51-1


from ase.calculators.gaussian
    import Gaussian, GaussianOptimizer, GaussianIRC

atoms = ...
# Optimize to a saddle point
calc_opt = Gaussian(label='opt', ...)
opt = GaussianOptimizer(atoms, calc_opt)
opt.run(fmax='tight', steps=100, opt='calcfc,ts')
tspos = atoms.positions.copy()


# Do a vibrational frequency calculation and store the Hessian in a
# checkpoint file, for use in subsequent IRC calculations
atoms.calc = Gaussian(label='sp', chk='sp.chk', freq='')
atoms.get_potential_energy()


# Perform IRC in the "forwards" direction
calc_irc_for = Gaussian(label='irc_for', chk='irc_for.chk',
                        oldchk='sp.chk', ...)
```

https://doi.org/10.1515/9783110629453-051

```
irc_for = GaussianIRC(atoms, calc_irc_for)
irc_for.run(direction='forward', steps=20, irc='rcfc')  # reuses Hessian

# Perform IRC in the "reverse" direction
# First, restore TS positions
atoms.positions[:] = tspos
calc_irc_rev = Gaussian(label='irc_rev', chk='irc_rev.chk',
                oldchk='sp.chk', ...)
irc_rev = GaussianIRC(atoms, calc_irc_rev)
irc_rev.run(direction='reverse', steps=20, irc='rcfc')
```

References

1. Gaussian (software) (Wikipedia) https://en.wikipedia.org/wiki/Gaussian_(software).
2. Interfacing to Gaussian 16 http://gaussian.com/interfacing/.
3. About – ASE documentation https://wiki.fysik.dtu.dk/ase/about.html.
4. ase (GitLab) https://gitlab.com/ase/ase.

52 GROMACS

GROMACS – Groningen Machine for Chemical Simulations – is a versatile package to perform molecular dynamics, that is, simulate the Newtonian equations of motion for systems with hundreds to millions of particles.[1,2]

Running GROMACS

Because the system is not available as a binary setup, there are two possibilities for installation: Use a docker image or setup with compiling the sources from the ground up.

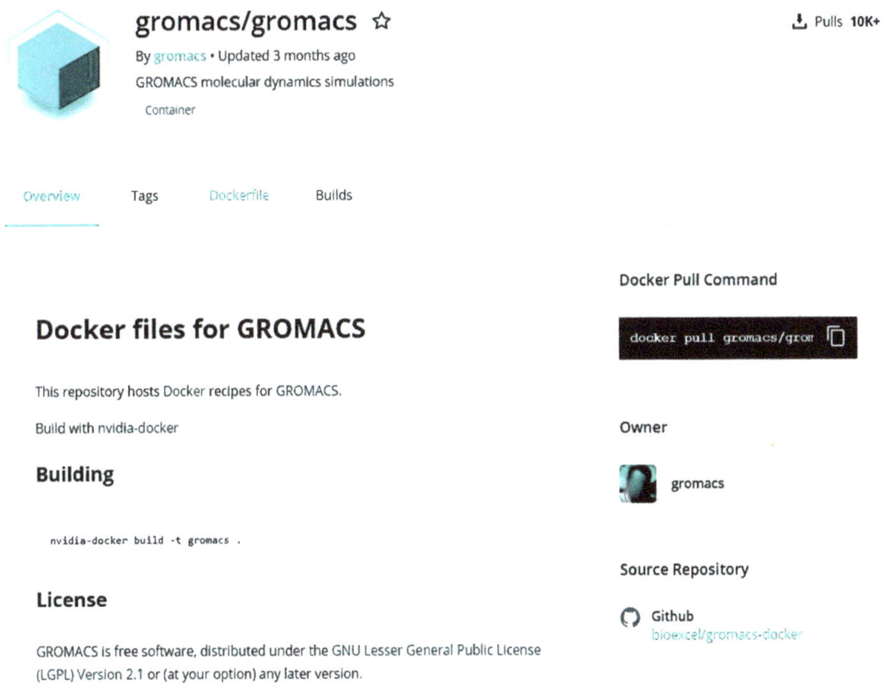

gromacs/gromacs ☆

By gromacs • Updated 3 months ago
GROMACS molecular dynamics simulations
Container

↓ Pulls 10K+

Overview   Tags   Dockerfile   Builds

Docker files for GROMACS

This repository hosts Docker recipes for GROMACS.

Build with nvidia-docker

Building

```
nvidia-docker build -t gromacs .
```

License

GROMACS is free software, distributed under the GNU Lesser General Public License (LGPL) Version 2.1 or (at your option) any later version.

Docker Pull Command

```
docker pull gromacs/grom
```

Owner

gromacs

Source Repository

○ Github
bioexcel/gromacs-docker

Figure 52.1: Docker/KITEMATIC GROMACS additional information *(Screenshot by Gressling)*.

https://doi.org/10.1515/9783110629453-052

Using the command line:

```
$ docker pull gromacs/gromacs
Using default tag: latest
latest: Pulling from gromacs/gromacs
976a760c94fc: Pull complete
c58992f3c37b: Pull complete
0ca0e5e7f12e: Pull complete
f2a274cc00ca: Pull complete
708a53113e13: Pull complete
371ddc2ca87b: Pull complete
f81888eb6932: Pull complete
19dbd9dd59d6: Downloading  33.76MB/507.4MB
e55ccb8a3273: Download complete
2c56d49a1521: Download complete
6ee19bb70913: Downloading  8.074MB/69.66MB
```

Workflow

The process[3] has four steps:
- Setup (loading data, solvation, i.e., addition of water and ions)
- Energy minimization of the protein
- Equilibration of the solvent around the protein
- Production simulation, which produces the trajectory

Runtime example

The Oak Ridge National Laboratory's summit contains 9216 IBM processors, each with 22 computing cores and 27,000 Tesla V100 accelerators. According to Nvidia, the GPUs run the GROMACS software, which simulates molecular dynamic processes.[4]

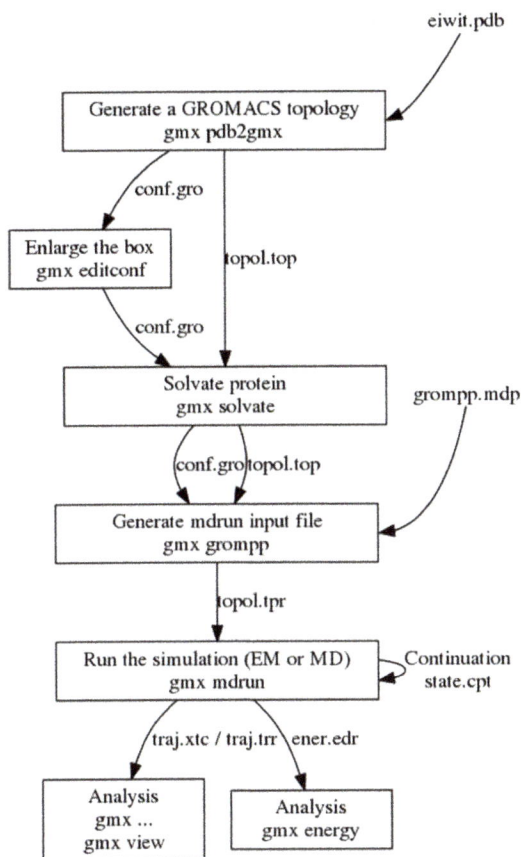

Figure 52.2: Gromacs workflow[2] *(Gromacs)*.

Python with GromacsWrapper

GromacsWrapper[5] is a Python package that wraps system calls to Gromacs tools into classes and so into Python scripts. This uses Python's better error handling and superior data structures instead of shell calls for Gromacs.

Python example

```
# Gromacs Wrapper
# author: Beckstein, Oliver (Et al.)
# license: GPL 3          # code: github.com/BecksteinLab/GromacsWrapper
# activity: active (2019)  # index: 52-1

# Given a PDB file 1iee.pdb, set up and run a simple simulation
# (assuming all other input files at hand such as the MDP files)
import gromacs
gromacs.pdb2gmx(f="1ake.pdb", o="protein.gro", p="topol.top",
                ff="oplsaa", water="tip4p")
gromacs.editconf(f="protein.gro", o="boxed.gro",
                 bt="dodecahedron", d=1.5, princ=True,
                 input="Protein")
gromacs.solvate(cp="boxed.gro", cs="tip4p", p="topol.top",
                o="solvated.gro")
gromacs.grompp(f="emin.mdp", c="solvated.gro", p="topol.top",
               o="emin.tpr")
gromacs.mdrun(v=True, deffnm="emin")
gromacs.grompp(f="md.mdp", c="emin.gro", p="topol.top", o="md.tpr")
gromacs.mdrun(v=True, deffnm="md")
```

References

1. Gromacs http://www.gromacs.org/About_Gromacs.
2. User guide – GROMACS 2018 documentation (LGPL 2.1 License) http://manual.gromacs.org/documentation/2018/user-guide/index.html.
3. Michael Papili. GROMACS TUTORIAL: Your first Simulation Made Easy! https://www.youtube.com/watch?v=rYZ1p5lXNyc.
4. Heise. Die weltweit schnellsten Supercomputer rechnen an Lösungen gegen Coronavirus https://news.developer.nvidia.com/oak-ridge-national-laboratory-coronavirus-research/.
5. Beckstein, O. GromacsWrapper – a Python framework for Gromacs https://pythonhosted.org/GromacsWrapper/.

53 AMBER, NAMD, and LAMMPS

AMBER

Assisted Model Building with Energy Refinement (AMBER[1]) is a family of force fields for molecular dynamics of biomolecules originally developed by Peter Kollman's group at the University of California, San Francisco. AMBER[2] is also the name of the molecular dynamics software package that simulates these force fields.

NAMD

NAnoscale Molecular Dynamics (NAMD)[3,4] is a parallel molecular dynamics code designed for high-performance simulation of large biomolecular systems. Based on Charm++ parallel objects, NAMD scales to hundreds of cores for typical simulations and beyond 500,000 cores for the largest simulations. NAMD uses the popular molecular graphics program VMD for simulation setup and trajectory analysis but is also file compatible with AMBER, CHARMM, and X-PLOR. NAMD is distributed free of charge with source code.

To run any MD simulation, NAMD requires at least four files:
- Protein data bank (pdb) stores atomic coordinates and/or velocities.
- Protein structure file (psf) stores structural information of the protein (bonding interactions).
- Force field parameter file defines bond strengths, equilibrium lengths, and so on. CHARMM, X-PLOR, AMBER, and GROMOS are four types of force fields.
- a configuration file

Interface via PyNAMD

PyNAMD is a package to provide an interface to **NAMD** input and output. The focus is on energy-based output:[5]

```
# Output energies from NAMD (pyNAMD)
# author: Radak, Brian
# license: MIT License # code: github.com/radakb/pynamd/parse-log
# activity: active (2019)      # index: 53-1

import pynamd
l = pynamd.NamdLog("00001.log", "00002.log")
```

https://doi.org/10.1515/9783110629453-053

```
#log.energy can easily be converted into a pandas dataframe
import pandas as pd
df =  pd.DataFrame(l.energy)

#convert time to nanoseconds
outputEnergies_ns = l.info['outputEnergies']*l.info['timestep']/1000/1000
df['t [ns]']=df.index*outputEnergies_ns

mean = lf["TOTAL"].mean()
df["TOTAL_rolling"] = df["TOTAL"].rolling(window=20,center=True).mean()
```

MDAnalysis

A Python package[6,7] for the analysis of molecular dynamics simulations is MDAnalysis, supporting among many MD software types also **AMBER**. A typical usage pattern is to iterate through a trajectory and analyze coordinates for every frame. In the following example, the end-to-end distance of a protein and the radius of gyration of the backbone atoms are calculated:

```
# Analyze coordinates in a trajectory (MDAnalysis)
# author: MDAnalysis team
# license: GPL-2              # code: github.com/MDAnalysis/.../examples/
# activity: active (2020)     # index: 53-2

import MDAnalysis
from MDAnalysis.tests.datafiles import PSF, DCD   # test trajectory
import numpy.linalg

u = MDAnalysis.Universe(PSF,DCD)  # always start with a Universe
# can access via segid (4AKE) and atom name
# we take the first atom named N and the last atom named C
nterm = u.select_atoms('segid 4AKE and name N')[0]
cterm = u.select_atoms('segid 4AKE and name C')[-1]

bb = u.select_atoms('protein and backbone')  # a selection (AtomGroup)

for ts in u.trajectory:     # iterate through all frames
    r = cterm.position - nterm.position # end-to-end vector from atom positions
    d = numpy.linalg.norm(r)  # end-to-end distance
    rgyr = bb.radius_of_gyration()  # method of AtomGroup
    print("frame = {0}: d = {1} A, Rgyr = {2} A".format(
          ts.frame, d, rgyr))
```

MDTraj

MDTraj[8] is a python library that allows users to manipulate molecular dynamics (MD) trajectories and performs a variety of analyses, including root mean square deviation (RMSD), solvent-accessible surface area, and hydrogen bonding. Read, write, and analyze MD trajectories with only a few lines of Python code.

The library supports molecular dynamics trajectory file formats, including RCSB pdb, GROMACS (xtc and trr), CHARMM/NAMD dcd, AMBER binpos, AMBER NetCDF, AMBER mdcrd, TINKER arc, and MDTraj HDF5.

```
# Baker-Hubbard Hydrogen Bond Identification (MDTraj)
# author: MDTraj team
# license: LGPL v2.1       # code: github.com/mdtraj
# activity: active (2020)  # index: 53-3


import mdtraj as md

t = md.load_pdb('http://www.rcsb.org/pdb/files/2EQQ.pdb')
print(t)
> <mdtraj.Trajectory with 20 frames, 423 atoms, 28 residues, without unit-
cells>
hbonds = md.baker_hubbard(t, periodic=False)
label = lambda hbond : '%s -- %s' % (t.topology.atom(hbond[0]),
t.topology.atom(hbond[2]))
for hbond in hbonds:
    print(label(hbond))
> GLU1-N -- GLU1-OE2
> GLU1-N -- GLU1-OE1
> GLY6-N -- SER4-O
...
a_distances = md.compute_distances(t, hbonds[:, [0,2]], periodic=False)

# plot
color = itertools.cycle(['r', 'b', 'gold'])
for i in [2, 3, 4]:
    plt.hist(da_distances[:, i], color=next(color), label=label(hbonds[i]),
    alpha=0.5)
plt.legend()
plt.ylabel('Freq');
plt.xlabel('Donor-acceptor distance [nm]')
```

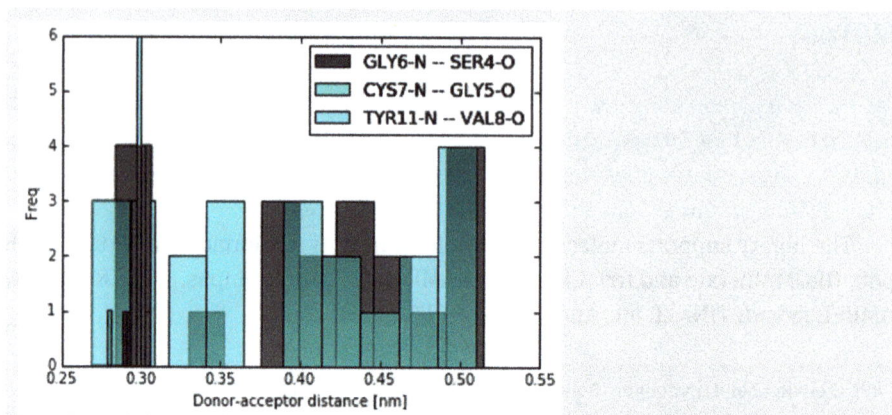

Figure 53.1: Frequency versus donor–acceptor distance *(MDTraj)*.

LAMMPS

LAMMPS[9] is a molecular dynamics program (Large-Scale Atomic/Molecular Massively Parallel Simulator) with a focus on materials modeling. LAMMPS uses neighbor lists (Verlet lists) to keep track of nearby particles. They are optimized for systems with particles that repel at *short distances* so that the local density of particles never grows too large.[10] So LAMMPS is a particle simulator at the atomic, meso, or continuum scale.

pysimm

Python Simulation Interface for Molecular Modeling (pysimm[11]) is an open-source object-oriented Python package for molecular simulations. It handles data organization for particles, force field parameters, and simulation settings so you can focus on developing your simulation workflow.[12]

This example demonstrates some features of pysimm, including structure generation, force field parameter assignment, and **LAMMPS[9]** simulation. Requires LAMMPS to be installed (e.g., with `$ docker pull lammps/lammps`).

```
# Energy minimization of a polymer system using LAMMPS
# author: Demidov, Alexander; Fortunato, Michael E.
# license: MIT License          # code: github.com/polysimtools/pysimm
# activity: hold? (2018)        # index: 53-4


# Creating a polyethylene-co-styrene chain using the random walk application
from pysimm import system, lmps, forcefield
from pysimm.apps.random_walk import copolymer
```

```
from pysimm.models.monomers.dreiding.pe import monomer as pe_monomer
from pysimm.models.monomers.dreiding.ps import monomer as ps_monomer

# Creating polyethylene and polystyrene monomers
pe = pe_monomer()
ps = ps_monomer()
# Creating a reference to a forcefield object
f = forcefield.Dreiding()
# Deriving Gasteiger partial charges for both of our monomers.
pe.apply_charges(f, charges='gasteiger')
ps.apply_charges(f, charges='gasteiger')
# Creating a copolymer chain
# The force field objects retrieved from the forcefield.
pe.pair_style = 'lj'
ps.pair_style = 'lj'
polymer = copolymer([pe, ps], 10, pattern=[1, 1], forcefield=f)

# Write xyz, yaml, lammps data, and chemdoodle json file formats.
polymer.write_xyz('polymer.xyz')
polymer.write_yaml('polymer.yaml')
polymer.write_lammps('polymer.lmps')
polymer.write_chemdoodle_json('polymer.json')

# run LAMMPS simulation
sim = lmps.Simulation(polymer, log= 'steps.log')
sim.add_min(min_style = 'fire', name = 'min_fire', etol = 1.0e-5, ftol = 1.0e-5)
sim.add_md(ensemble='nvt', timestep=0.5)
sim.run()
```

pyiron

A project for computational materials science is pyiron.[13,14] It is an environment for implementing, testing, and running simulations in materials development.
- Atomic structure objects – compatible with the Atomic Simulation Environment (ASE)
- Atomistic simulation codes – like LAMMPS and VASP
- Feedback loops – to construct dynamic simulation life cycles</DL>

```
# LAMMPS integration with pyiron
# author: Neugebauer, Jörg (MPI Eisenforschung)
# license: BSD 3-Clause "New" # code: github.com/pyiron/pyiron
# activity: active (2020)     # index: 53-5
```

```
from pyiron.project import Project
pr = Project(path='material_project')

basis = pr.create_ase_bulk('Al', cubic=True)
supercell_3x3x3 = basis.repeat([3, 3, 3])
supercell_3x3x3.plot3d()
job = pr.create_job(job_type=pr.job_type.Lammps, job_name='Al_T800K')
```

References

1. AMBER (Wikipedia) https://en.wikipedia.org/wiki/AMBER.
2. The Amber Molecular Dynamics Package http://ambermd.org/.
3. NAMD – Scalable Molecular Dynamics https://www.ks.uiuc.edu/Research/namd/.
4. Rudack, T.; Ribeiro, J. V.; Barragan, A.; Lihan, M.; Shahoei, R.; Zhang, Y. QwikMD – Easy Molecular Dynamics with NAMD and VMD.
5. Radak, B. pynamd (GitHub) https://github.com/radakb/pynamd.
6. MDAnalysis (PyPI) https://pypi.org/project/MDAnalysis/.
7. Michaud-Agrawal, N.; Denning, E. J.; Woolf, T. B.; Beckstein, O. MDAnalysis: A Toolkit for the Analysis of Molecular Dynamics Simulations. *J. Comput. Chem.* **2011**, *32* (10), 2319–2327. https://doi.org/10.1002/jcc.21787.
8. McGibbon, R. T.; Beauchamp, K. A.; Harrigan, M. P.; Klein, C.; Swails, J. M.; Hernández, C. X.; Schwantes, C. R.; Wang, L.-P.; Lane, T. J.; Pande, V. S. MDTraj: A Modern Open Library for the Analysis of Molecular Dynamics Trajectories. *Biophys. J.* **2015**, *109* (8), 1528–1532. https://doi.org/10.1016/j.bpj.2015.08.015.
9. Plimpton, S. LAMMPS Molecular Dynamics Simulator https://lammps.sandia.gov/.
10. LAMMPS (Wikipedia) https://en.wikipedia.org/wiki/LAMMPS.
11. Fortunato, M. E.; Colina, C. M. Pysimm: A Python Package for Simulation of Molecular Systems. *SoftwareX* **2017**, *6*, 7–12. https://doi.org/10.1016/j.softx.2016.12.002.
12. pysimm https://pysimm.org/.
13. About – pyiron 0.2.2 documentation https://pyiron.org/.
14. Janssen, J.; Surendralal, S.; Lysogorskiy, Y.; Todorova, M.; Hickel, T.; Drautz, R.; Neugebauer, J. Pyiron: An Integrated Development Environment for Computational Materials Science. *Comput. Mater. Sci.* **2019**, *163*, 24–36. https://doi.org/10.1016/j.commatsci.2018.07.043.

54 Featurize materials

The Materials Genome Initiative is a multiagency initiative designed to create a new era of policy, resources, and infrastructure that supports U.S. institutions in the effort to discover, manufacture, and deploy advanced materials twice as fast, at a fraction of the cost.

The Materials Virtual Lab[1] is a group of scientists at the University of California San Diego's Department of NanoEngineering dedicated to the application of first-principles calculations and informatics to accelerate materials design. Their research includes the development of novel thermodynamic analyses of quantum mechanical calculations and the investigation of structure–property relationships.

To use the Materials Project python login to the Materials Project and generate an API key.

maml: convert materials into features

maml[2] (MAterials Machine Learning) is a python package[3] for materials machine learning. It aims to convert materials (crystals and molecules) into features. In addition to the common compositional, site, and structural features, it provides fine-grain local environment features. Also, it is possible to use ML to learn relationships between features and targets. (The software supports sklearn and Keras models.)

Example of Spectral Neighbor Analysis Potential (SNAP)

This example demonstrates that machine-learned Spectral Neighbor Analysis Potential (SNAP) models can yield improvements even over well-established, high-performing embedded atom methods. (The SNAP potential comes with the installation of lammps: `conda install -c conda-forge/label/cf202003 lammps`):

```
# Spectral Neighbor Analysis Potential (SNAP) for Ni-Mo Binary Alloy
# author: Xiang-Guo Li
# license: BSD 3              # code: github.com/materialsvirtuallab/maml
# activity: active (2020)     # index: 54-1


from maml import ModelWithSklearn
from maml.describer import BispectrumCoefficients
from sklearn.linear_model import LinearRegression
from maml.apps.pes import SNAPotential
```

https://doi.org/10.1515/9783110629453-054

```
element_profile = {'Mo': {'r': 0.5, 'w': 1}, 'Ni': {'r': 0.5, 'w': 1}}
describer = BispectrumCoefficients(cutoff = 5.0, twojmax = 6,
   element_profile = element_profile, quadratic = True, pot_fit = True)
model = ModelWithSklearn(describer=describer, model=LinearRegression())
qsnap = SNAPotential(model=model)
qsnap.train(train_structures, train_energies, train_forces,
sample_weight=weights)
df_orig, df_predict = qsnap.evaluate(test_structures=train_structures,
                                     test_energies=train_energies,
                                     test_forces=train_forces,
                                     test_stresses=None)
```

The Pymatgen library

The Pymatgen[4] (Python Materials Genomics) library is an open-source Python library for material analysis. Main features are highly flexible classes for the representation of element, site, molecule, and structure objects.

Example: interface reaction

Obtain information about interface reactions between two solid substances in contact and plot reaction energy as a function of mixing ratio:

```
# Interface reaction of LiCoO₂ and Li₃PS₄
# author: pymatgen team
# license: MIT License
# code: matgenb.materialsvirtuallab.org/Interface-Reactions
# activity: active (2020)  # index: 54-2


from pymatgen.ext.matproj import MPRester
from pymatgen.analysis.interface_reactions import InterfacialReactivity
from pymatgen.analysis.phase_diagram
            import PhaseDiagram, GrandPotentialPhaseDiagram
from pymatgen import Composition, Element


mpr = MPRester(<<API Key>>)


reactant1 = 'LiCoO2'
reactant2 = 'Li3PS4'
comp1 = Composition(reactant1)
comp2 = Composition(reactant2)
elements = [e.symbol for e in comp1.elements + comp2.elements]
elements.append(open_el)
```

```
elements = list(set(elements)) # Remove duplicates
entries = mpr.get_entries_in_chemsys(elements)
pd = PhaseDiagram(entries)

# For an open system, include the grand potential phase diagram.
if grand:
    # Get the chemical potential of the pure substance.
    mu = pd.get_transition_chempots(Element(open_el))[0]
    # Set the chemical potential in the elemental reservoir.
    chempots = {open_el: relative_mu + mu}
    # Build the grand potential phase diagram
    gpd = GrandPotentialPhaseDiagram(entries, chempots)
    # Create InterfacialReactivity object.
    interface = InterfacialReactivity(
        comp1, comp2, gpd, norm=True, include_no_mixing_energy=True,
        pd_non_grand=pd, use_hull_energy=False)
else:
    interface = InterfacialReactivity(
        comp1, comp2, pd, norm=True, include_no_mixing_energy=False,
        pd_non_grand=None, use_hull_energy=False)
plt = interface.plot()
```

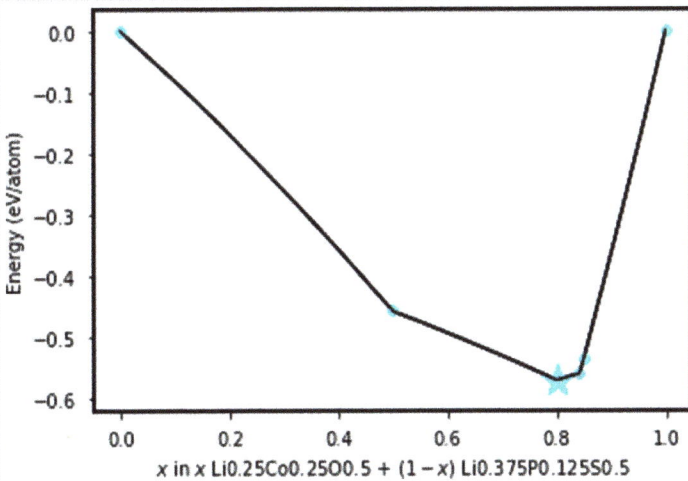

Figure 54.1: Energy for $Li_xCo_yO_z+Li_aPb_bS_c$.

Plotting a pourbaix diagram

The Materials Project REST interface includes functionality to construct Pourbaix diagrams[4-6] from computed entries:

```
# Pourbaix diagram for the Cu-O-H system (pymatgen)
# author: Montoya, Joseph
# license: MIT License
# code: matgenb.materialsvirtuallab.org/Pourbaix-Diagram
# activity: active (2020)  # index: 54-3


from pymatgen import MPRester
from pymatgen.analysis.pourbaix_diagram
        import PourbaixDiagram, PourbaixPlotter

mpr = MPRester(<<API Key>>)

# Get all pourbaix entries corresponding to the Cu-O-H chemical system.
entries = mpr.get_pourbaix_entries(["Cu"])
# Construct the PourbaixDiagram object

pbx = PourbaixDiagram(entries)

# The PourbaixAnalyzer includes a number of useful functions for determining
# stable species and stability of entries relative to a given
# pourbaix facet (i.e. as a function of pH and V).

entry = [e for e in entries if e.entry_id == 'mp-1692'][0]
print("CuO's potential energy per atom relative to the most",
      "stable decomposition product is {:0.2f} eV/atom".format(
          pbx.get_decomposition_energy(entry, pH=7, V=-0.2)))

plotter = PourbaixPlotter(pbx)
plotter.get_pourbaix_plot().show()
```

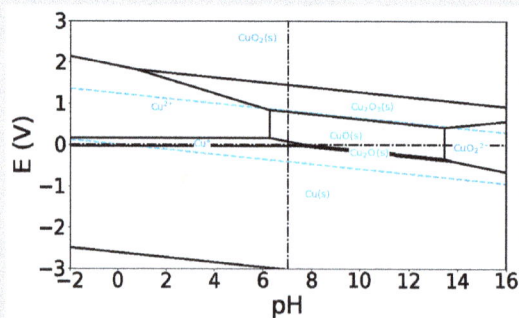

Figure 54.2: Pourbaix diagram for the Cu-O-H system *(Montoya).*

References

1. Materials Virtual Lab http://materialsvirtuallab.org/.
2. maml – MAterials Machine Learning (GitHub) https://github.com/materialsvirtuallab/maml.
3. maml documentation https://guide.materialsvirtuallab.org/maml/modules.html.
4. pymatgen documentation https://pymatgen.org/.
5. Singh, A. K.; Zhou, L.; Shinde, A.; Suram, S. K.; Montoya, J. H.; Winston, D.; Gregoire, J. M.; Persson, K. A. Electrochemical Stability of Metastable Materials. *Chem. Mater.* **2017**, *29*, 10159–10167.
6. Persson, K. A.; Waldwick, B.; Lazic, P.; Ceder, G. Prediction of Solid-Aqueous Equilibria: Scheme to Combine First-Principles Calculations of Solids with Experimental Aqueous States. *Phys. Rev. B Condens. Matter* **2012**, *85*, 235438.

55 ASE and NWChem

ASE

The Atomic Simulation Environment (ASE[1,2]) is a set of tools and Python modules for setting up, manipulating, running, visualizing, and analyzing atomistic simulations. The code is freely available under the GNU LGPL license.

ASE provides interfaces to different codes through Calculators that are used together with the central Atoms object and the many available algorithms in ASE.

NWChem

The Pacific Northwest National Laboratory (PNNL) develops the NWChem[3,4] package, which is an ab initio computational chemistry software package including quantum chemical and molecular dynamics functionality. It is designed to run on high-performance parallel supercomputers as well as conventional workstation clusters.

It has a built-in python interface but is not accessible from outside. An Educational Community License (ECL) is available. Also, this software can be used by docker environment:

```
$ docker pull nwchemorg/nwchem-qc
```

Interfacing NWChem

Optimize the geometry of a water molecule using the PBE (Perdew–Burke–Ernzerhof) density functional:

```
# PBE (Perdew-Burke-Ernzerhof) density from NWChem (ASE)
# author: ASE-developers
# license: LGPLv2.1+
# code: fysik.dtu.dk/ase/.../calculators/nwchem.html
# activity: active (2020)     # index: 55-1


# requires NWChem to be installed
from ase import Atoms
from ase.optimize import BFGS
```

https://doi.org/10.1515/9783110629453-055

```
from ase.calculators.nwchem import NWChem
from ase.io import write

# data
h2 = Atoms('H2', positions=[[0, 0, 0],[0, 0, 0.7]])
h2.calc = NWChem(xc='PBE')
opt = BFGS(h2, trajectory='h2.traj')

# simulation
opt.run(fmax=0.02)
> BFGS:    0  19:10:49    -31.435229    2.2691
> BFGS:    1  19:10:50    -31.490773    0.3740
> BFGS:    2  19:10:50    -31.492791    0.0630
> BFGS:    3  19:10:51    -31.492848    0.0023
write('H2.xyz', h2)
h2.get_potential_energy()  # ASE's units are eV and Ang
> -31.492847800329216
```

Nudged elastic band calculations

Calculating surface diffusion energy barriers using the nudged elastic band (NEB)
method:

```
# PBE (Perdew–Burke–Ernzerhof) density from NWChem (ASE)
# author: ASE-developers
# license: LGPLv2.1+          # code: fysik.dtu.dk/ase/.../diffusion
# activity: active (2020)     # index: 55-2

from ase.build import fcc100, add_adsorbate
from ase.constraints import FixAtoms
from ase.calculators.emt import EMT
from ase.optimize import QuasiNewton

# 1 - build data: 2x2-Al(001) surface with 3 layers + Au atom adsorbed
slab = fcc100('Al', size=(2, 2, 3))
add_adsorbate(slab, 'Au', 1.7, 'hollow')
slab.center(axis=2, vacuum=4.0)
mask = [atom.tag > 1 for atom in slab] # Fix second and third layers
slab.set_constraint(FixAtoms(mask=mask))
slab.set_calculator(EMT()) # Use EMT potential

# 2 - Run: initial state and final state:
qn = QuasiNewton(slab, trajectory='initial.traj')
```

```
qn.run(fmax=0.05)
slab[-1].x += slab.get_cell()[0, 0] / 2
qn = QuasiNewton(slab, trajectory='final.traj')
qn.run(fmax=0.05)

# 3 - NEB calculation
from ase.io import read
from ase.constraints import FixAtoms
from ase.calculators.emt import EMT
from ase.neb import NEB
from ase.optimize import BFGS

initial = read('initial.traj')
final = read('final.traj')
constraint = FixAtoms(mask=[atom.tag > 1 for atom in initial])
images = [initial]
for i in range(3):
    image = initial.copy()
    image.set_calculator(EMT())
    image.set_constraint(constraint)
    images.append(image)

images.append(final)
neb = NEB(images)
neb.interpolate()
qn = BFGS(neb, trajectory='neb.traj')
qn.run(fmax=0.05)
# Visualize the results with $ ase gui neb.traj@-5:

# 4 - Calculate barrier and plot
import matplotlib.pyplot as plt
from ase.neb import NEBTools
from ase.io import read

images = read('neb.traj@-5:')
nebtools = NEBTools(images)
# Get the barrier without any interpolation between highest images.
Ef, dE = nebtools.get_barrier(fit=False)
# Get the actual maximum force at this point in the simulation.
max_force = nebtools.get_fmax()

fig = nebtools.plot_band()
fig.savefig('diffusion-barrier.png')
```

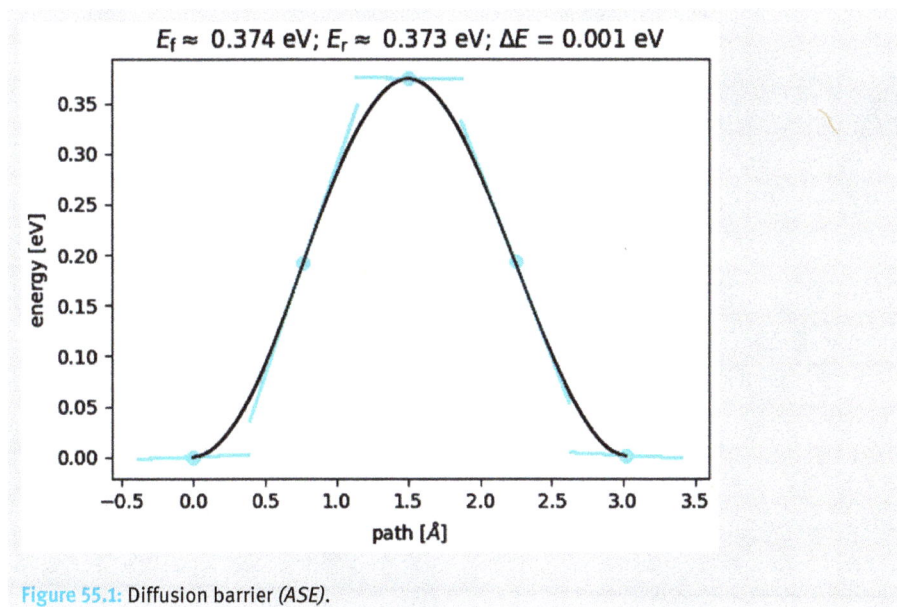

Figure 55.1: Diffusion barrier *(ASE)*.

References

1. About – ASE documentation https://wiki.fysik.dtu.dk/ase/about.html.
2. ase (GitLab) https://gitlab.com/ase/ase.
3. Valiev, M.; Bylaska, E. J.; Govind, N.; Kowalski, K.; Straatsma, T. P.; Van Dam, H. J. J.; Wang, D.; Nieplocha, J.; Apra, E.; Windus, T. L.; et al. NWChem: A Comprehensive and Scalable Open-Source Solution for Large Scale Molecular Simulations. *Comput. Phys. Commun.* **2010**, *181*, 1477. https://doi.org/10.1016/j.cpc.2010.04.018.
4. NWChem http://www.nwchem-sw.org/index.php/Main_Page.

Organic chemistry

56 Visualization

Visualization[1] is any technique using images, diagrams, or animations to communicate a message. Visualization through visual imagery is an *effective way to communicate* ideas. The use of visual representations to transfer knowledge between two persons aims to improve the transfer of knowledge. Visual analytics focuses on human interaction with visualization systems as part of a larger process of data analysis.

There are four main types of visualization for molecules, where two are within the Python context:
- Non-Jupyter: VR – virtual reality
- Non-Jupyter: Rich and interactive
- Jupyter: NGLView and py3Dmol
- Jupyter: Classic depiction examples (like in RDKit)

Virtual reality

GRASP, RasMol, PyMol, and UCSF Chimera are widely used software tools for molecular visualization with general functions for structural analysis. Autodesk Molecular Viewer implemented a VR environment with a Cardboard VR device. Till 2020, there is no professional or usable VR[2,3] environment available that can be used in visualizing molecules or structures that are scriptable with Python or moreover can be used in Jupyter notebooks.

Non-Jupyter: rich function, "fat client" viewers

Standalone visualization packages like VMD or PyMOL can be used to view molecular graphics. They are powerful and rich (via plugins) but proprietary and monolithic means they have to be installed depending on the operating system:
- **VMD**[4,5] is a molecular graphics program designed for the display and analysis of molecular Assemblies.[4] It is a software running locally and supports all OS.[5]
- **PyMOL**[6,7] is an open-source molecular visualization system created by Warren Lyford DeLano. It is currently commercialized by Schrödinger, Inc.[7] An educational license is available. It is only running in Jupyter with an XML-RPC server active.

https://doi.org/10.1515/9783110629453-056

Jupyter: NGLView (interactive)

The library is an *interactive* molecular graphics package for Jupyter notebooks.[8] It allows interactive viewing of molecular structures as well as trajectories from molecular dynamic simulations. The Viewer is embedded in a widget to provide WebGL accelerated molecular graphics.[9] A number of convenience functions are available to quickly display data from the file system, RCSB PDB, simpletraj, and from objects of analysis libraries mdtraj, pytraj, mdanalysis, ParmEd, rdkit, HTMD, and biopython.

```
# NGLView - WebGL Viewer for Molecular Visualization
# author: Rose, Alexander S.
# License: MIT License        # code: github.com/arose/nglview
# activity: on hold? (2015)   # index: 56-1


# !pip install nglview
import nglview
view = nglview.show_pdbid("3pqr")   # from RCSB PDB
# add component from url
view.add_component('rcsb://1tsu.pdb')

# not using default representation
view = nv.show_file('your.pdb', default=False)
view.center()
view.add_rope()
view
```

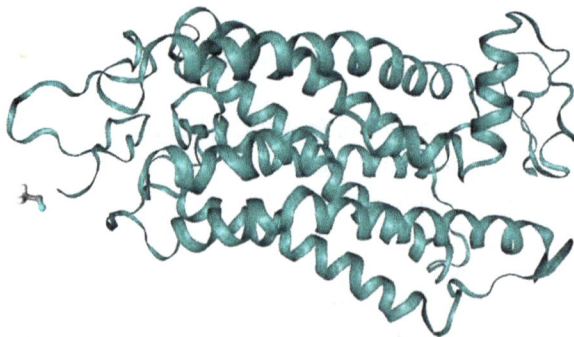

Figure 56.1: NGLView for a trajectory within Jupyter *(Rose)*.

```
# Representations
view.add_representation('cartoon', selection='protein')
view.add_surface(selection="protein", opacity=0.3)
view.add_cartoon(selection="protein", color='blue')
view.add_licorice('ALA, GLU')
# update parameters for ALL cartoons of component 0 (default)
view.update_cartoon(opacity=0.4, component=0)


# Add special components
# Density volumes (MRC/MAP/CCP4, DX/DXBIN, CUBE)
# Or adding derived class of 'nglview.Structure'
view.add_component('my.ccp4')
# adding new trajectory, Trajectory is a special case of component
# traj could be a 'pytraj.Trajectory', 'mdtraj.Trajectory', 'MDAnalysis.
Universe',
# 'parmed.Structure', 'htmd.Molecule' or derived class of 'nglview.Trajectory'
view.add_trajectory(traj)


view.clear_representations()
```

Jupyter: py3Dmol (interactive)

Py3DMol[10,11] is a python package for dependency-free molecular visualization in iPython notebooks. Objects from MDAnalysis, MDTraj, OpenBabel, and CClib can be visualized and manipulated directly in a notebook. The backend visualization library, 3DMol.js,[12] is included, so no additional libraries are necessary – visualizations will function in any modern browser using javascript and WebGL.

```
# py3Dmol - Viewer for Molecular Visualization
# author: Koes, David
# license: MIT License       # code: github.com/avirshup/py3dmol
# activity: on hold? (2015)  # index: 56-2


#!pip install py3Dmol
import py3Dmol
p=py3Dmol.view(query='cid:702')
p.setStyle({'stick': {'radius': .1}, 'sphere': {'scale': 0.25}})
p.show();
> [see Figure 56-2a]


m = Chem.MolFromSmiles('COc1ccc2c(c1)[nH]c(n2)[S@@](=O)Cc1ncc(c(c1C)OC)C')
AllChem.EmbedMultipleConfs(m,
                  useExpTorsionAnglePrefs=True,useBasicKnowledge=True)
```

```
p = py3Dmol.view(width=800,height=800)
p.addModel(Chem.MolToMolBlock(m),'sdf')
p.setStyle({'stick':{}})
p.setBackgroundColor('0xeeeeee')
p.zoomTo()
p.show()
> [Figure 56-2b]

p=py3Dmol.view(query='pdb:1ycr')
p.setStyle({'stick': {'radius': .2}, 'sphere': {'scale': 0.2}})
p.show();
> [Figure 56-2c]
```

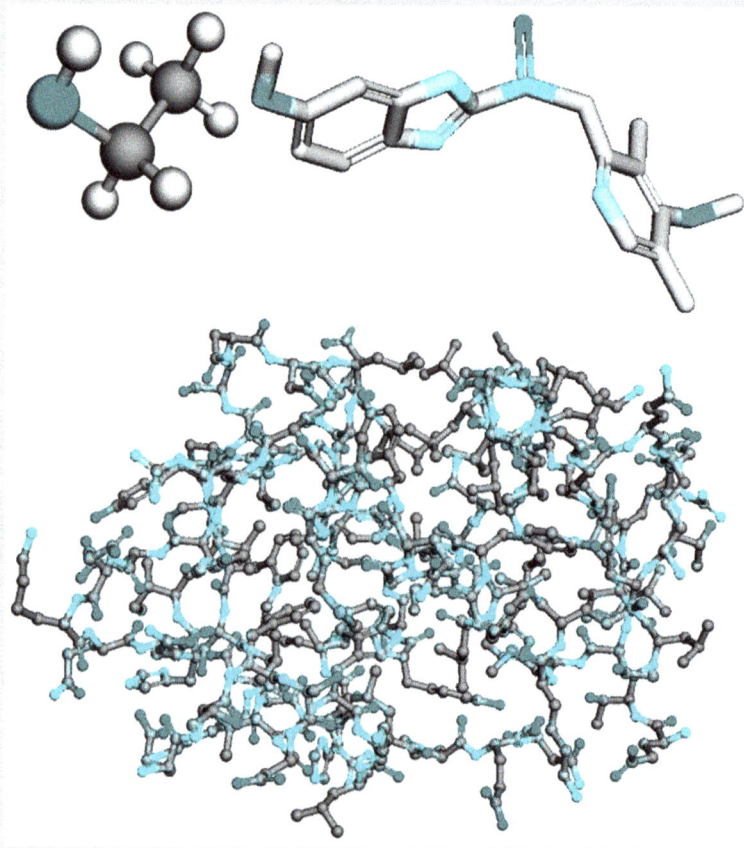

Figure 56.2: Three examples (a, b, and c) of py3Dmol *(Koes)*.

Jupyter: RDKit

RDKit by Greg Landrum is discussed in deep in the next chapters. A first example is listed here.

Depiction example: align molecules using a fixed substructure

When displaying a large number of (a few hundred) structures on a page, it is easier to identify the differences between the structures when the pictures are aligned[13] (other examples: Indigo and OpenEye). The structures are the database results from searching with a given SMARTS query, and the common SMARTS substructures are aligned in each depiction.

```
# RDKit Depiction examples
# author: Dahlke, Andrew
# license: CC-BY-SA
# code: ctr.fandom.com/.../Align_depiction_fixed_substructure
# activity: single example (2013)  # index: 56-3

from rdkit import Chem, Geometry
from rdkit.Chem import AllChem

# read in the first 16 molecules
suppl = Chem.SDMolSupplier('benzodiazepine.sdf')
first16 = [suppl[x] for x in range(16)]

# do the substructure matching
patt = Chem.MolFromSmarts(
        "[#7]~1~[#6]~[#6]~[#7]~[#6]~[#6]~2~[#6]~[#6]~[#6]~[#6]~[#6]12")
matchVs= [x.GetSubstructMatch(patt) for x in first16]

# compute the reference coordinates
AllChem.Compute2DCoords(first16[0])
coords = [first16[0].GetConformer().GetAtomPosition(x) for x in matchVs[0]]
coords2D = [Geometry.Point2D(pt.x,pt.y) for pt in coords]

# now generate coords for the other molecules using that reference
for molIdx in range(1,16):
    mol = first16[molIdx]
    coordDict={}
    for i,coord in enumerate(coords2D):
        coordDict[matchVs[molIdx][i]] = coord
    AllChem.Compute2DCoords(mol,coordMap=coordDict)
# generate 100x100 images for each molecule
```

```
from rdkit.Chem import Draw
imgs = [Draw.MolToImage(x,size=(100,100)) for x in first16]
# and use PIL to combine them
import Image
img = Image.new('RGBA',(400,400))
for i in range(16):
img.paste(imgs[i],(100*(i%4),100*(i//4),100*(i%4+1),100*(i//4+1)))
img
```

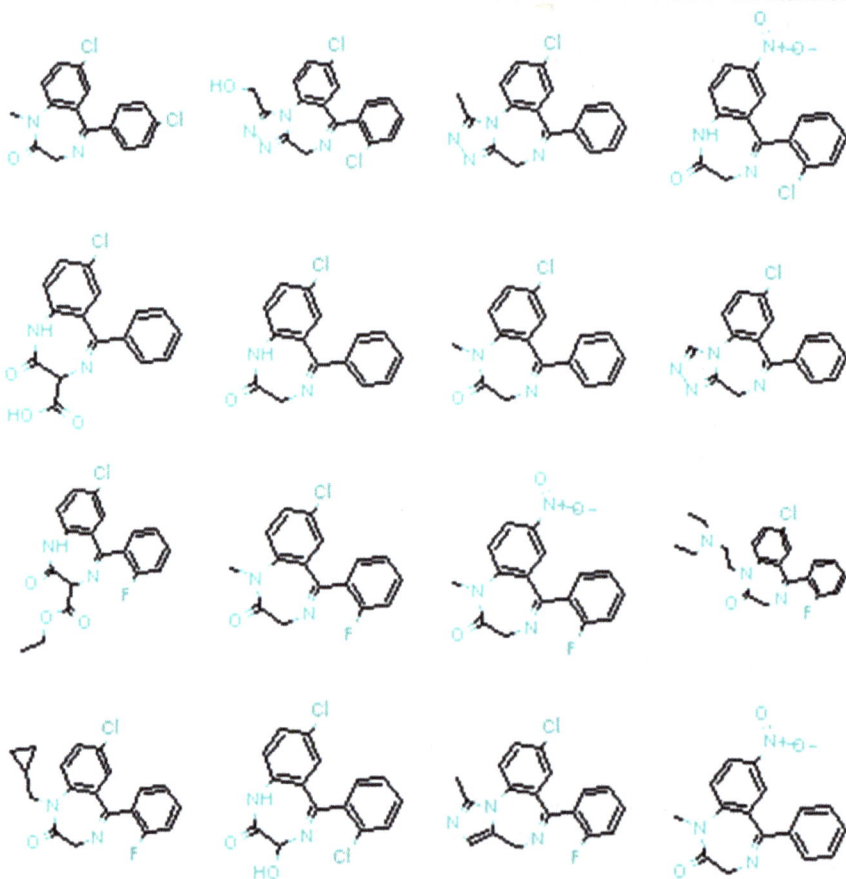

Figure 56.3: Depiction fixed substructures *(Dahlke)*.

References

1. Visualization (Wikipedia) https://en.wikipedia.org/wiki/Visualization_(graphics).
2. VRmol – Virtual Reality for Molecule https://vrmol.net/.
3. Goddard, T. D.; Brilliant, A. A.; Skillman, T. L.; Vergenz, S.; Tyrwhitt-Drake, J.; Meng, E. C.; Ferrin, T. E. Molecular Visualization on the Holodeck. *J. Mol. Biol.* **2018**, *430* (21), 3982–3996. https://doi.org/10.1016/j.jmb.2018.06.040.
4. Humphrey, W.; Dalke, A.; Schulten, K. VMD: Visual Molecular Dynamics. *J. Mol. Graph.* **1996**, *14* (1), 33–38, 27–28.
5. VMD – Visual Molecular Dynamics https://www.ks.uiuc.edu/Research/vmd/.
6. PyMOL https://pymol.org/2/.
7. PyMOL (Wikipedia) https://en.wikipedia.org/wiki/PyMOL.
8. Nguyen, H.; Case, D. A.; Rose, A. S. NGLview–interactive Molecular Graphics for Jupyter Notebooks. *Bioinformatics* **2018**, *34* (7), 1241–1242. https://doi.org/10.1093/bioinformatics/btx789.
9. Rose, A. nglview (GitHub) https://github.com/arose/nglview.
10. py3Dmol (PyPI) https://pypi.org/project/py3Dmol/.
11. Virshup, A. py3dmol (GitHub) https://github.com/avirshup/py3dmol.
12. 3Dmol.js (GitHub) https://github.com/3dmol/3Dmol.js.
13. Chemistry Toolkit Rosetta Wiki https://ctr.fandom.com/wiki/Chemistry_Toolkit_Rosetta_Wiki.

57 Molecules handling and normalization

Due to carbon's ability to catenate[1] (form chains with other carbon atoms), millions of organic compounds are known. The study of the properties, reactions, and syntheses of *organic compounds* requires manipulation and analysis of their structure representations.

RDKit

RDKit[2–4] is an open-source toolkit written in C++ and provides a bridge to use with Python via boost.[5] It includes a PostgreSQL relational database that allows molecules to be stored and retrieved via substructure and similarity searches.[2,3] RDKit has a permissive BSD license and documentation is CC BY SA 4.0. The open-source community is active.

Chapters 57–61 are based on the following literature:

Jupyter

- github.com/**rdkit**/.../Docs/Notebooks (with /data)
- github.com/rdkit/**rdkit-tutorials**/.../notebooks

Books

- www.rdkit.org/docs/...**GettingStartedInPython**.html
- www.rdkit.org/docs/...**RDKit_Book**.html
- www.rdkit.org/docs/...**Cookbook**.html

BLOGS

- **rdkit.blogspot**.com (Landrum)
- **iwatobipen.wordpress**.com (Serizawa)
- **gist.github**.com/greglandrum

The head of RDKit is mainly Greg Landrum. Major contributions to the RDKit documentation are by Takayuki Serizawa and Vincent Scalfani.

https://doi.org/10.1515/9783110629453-057

Read, write

Structure ingestion supports many data and file types: SMILES/SMARTS, SDF, TDT, SLN, Corina mol2, PDB, sequence notation, FASTA, HELM, and Canonical SMILES.

```
# Ingestion: Define, read and write structures (RDKit)
# authors: Landrum, Greg; Serizawa; Scalfani (Et al.)
# license: CC BY SA 4.0      # code: github.com/gressling/examples
# activity: active (2020)     # index: 57-1

# SMILES
m = Chem.MolFromSmiles('COc1ccc2c(c1)[nH]c(n2)[S@@](=O)Cc1ncc(c(c1C)OC)C')
m
> [img]

# Example: Lambda method transforms smiles strings to mol rdkit object
df['mol'] = df['smiles'].apply(lambda x: Chem.MolFromSmiles(x))
print(type(df['mol'][0]))
> <class 'rdkit.Chem.rdchem.Mol'>

# Reading MOLFile
m = Chem.MolFromMolFile('data/chiral.mol')
Chem.MolToSmiles(m)
> 'C[C@H](O)c1ccccc1'
Chem.MolToSmiles(m,isomericSmiles=False)
> 'CC(O)c1ccccc1'

# Mol file
molblock = """phenol
  Mrv1682210081607082D

  7  7  0  0  0  0            999 V2000
   -0.6473    1.0929    0.0000 C   0  0  0  0  0  0  0  0  0  0  0  0
   -1.3618    0.6804    0.0000 C   0  0  0  0  0  0  0  0  0  0  0  0
   -1.3618   -0.1447    0.0000 C   0  0  0  0  0  0  0  0  0  0  0  0
   -0.6473   -0.5572    0.0000 C   0  0  0  0  0  0  0  0  0  0  0  0
    0.0671   -0.1447    0.0000 C   0  0  0  0  0  0  0  0  0  0  0  0
    0.0671    0.6804    0.0000 C   0  0  0  0  0  0  0  0  0  0  0  0
    0.7816    1.0929    0.0000 O   0  0  0  0  0  0  0  0  0  0  0  0
  1  2  1  0  0  0  0
  2  3  2  0  0  0  0
  3  4  1  0  0  0  0
  4  5  2  0  0  0  0
  5  6  1  0  0  0  0
  1  6  2  0  0  0  0
  6  7  1  0  0  0  0
M  END
"""
m = Chem.MolFromMolBlock(molblock)
```

```
# File operations
m = Chem.MolFromMolFile('data/input.mol')
stringWithMolData=open('data/input.mol','r').read()
m = Chem.MolFromMolBlock(stringWithMolData)

# PDB Data
crn = Chem.MolFromPDBFile('../data/1CRN.pdb')
crn.GetNumAtoms()
> 327

# HELM
helm = 'PEPTIDE1{T.T.C.C.P.S.I.V.A.R.S.N.F.N.V.C.R.L.P.G.T.P.E.A.I.C.A.T.Y.T.
G.C.I.I.I.P.G.A.T.C.P.G.D.Y.A.N}$$$$'
m = Chem.MolFromHELM(helm)
m.GetNumAtoms()
> 327

fasta ="""
TTCCPSIVARSNFNVCRLPGTPEAICATYTGCIIIPGATCPGDYAN
"""

m = Chem.MolFromFASTA(fasta)
m.GetNumAtoms()
> 327

# Reading sets of molecules using a supplier
suppl = Chem.SDMolSupplier('data/5ht3ligs.sdf')
for mol in suppl:
    print(mol.GetNumAtoms())
```

Access atoms and their properties (0D-descriptors)

RDKit provides looping over atoms and bonds to access properties. `GetNumAtoms()` method returns a general number of all atoms in a molecule, `GetNumHeavyAtoms()` method returns a number of all atoms in a molecule with molecular weight > 1:

```
# Access atoms and their properties (RDKit)
# authors: Landrum, Greg; Serizawa; Scalfani (Et al.)
# license: CC BY SA 4.0        # code: github.com/gressling/examples
# activity: active (2020)      # index: 57-2

# CanonicalRankAtoms
mol = Chem.MolFromSmiles('C1NCN1')
list(Chem.CanonicalRankAtoms(mol, breakTies=False))
> [0,1,0,1]
```

```
# Query descriptors
df['mol'] = df['mol'].apply(lambda x: Chem.AddHs(x))
df['num_of_atoms'] = df['mol'].apply(lambda x: x.GetNumAtoms())
df['num_of_heavy_atoms'] = df['mol'].apply(lambda x: x.GetNumHeavyAtoms())


# Query descriptors
# 1 - depict molecule
m = Chem.MolFromSmiles('c1ccc2c(c1)CCC2CC=O')
for atom in m.GetAtoms():
    atom.SetProp('atomLabel',str(atom.GetIdx()))
    print(atom.GetAtomicNum())
m
```

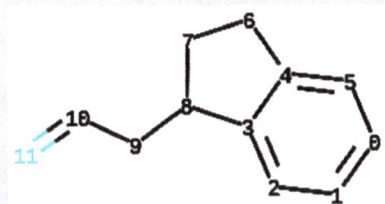

Figure 57.1: 1-Indan-Ethanone *(Gressling)*.

```
# 2 - get descriptors
i = 0
for atom in m.GetAtoms():
  print(i, ': ', atom.GetSymbol(),', Valence:',
                    atom.GetExplicitValence(), ', IsInRing:',
                    m.GetAtomWithIdx(i).IsInRing(), end=', ')
  if i<12:
    try:
      print(m.GetBondBetweenAtoms(i, i+1).GetBondType())
    except:
      print("(No bond)")
  i+=1
```

```
> 0 :  C , Valence: 3 , IsInRing: True, AROMATIC
> 1 :  C , Valence: 3 , IsInRing: True, AROMATIC
> 2 :  C , Valence: 3 , IsInRing: True, AROMATIC
> 3 :  C , Valence: 4 , IsInRing: True, AROMATIC
> 4 :  C , Valence: 4 , IsInRing: True, AROMATIC
> 5 :  C , Valence: 3 , IsInRing: True, (No bond)
> 6 :  C , Valence: 2 , IsInRing: True, SINGLE
> 7 :  C , Valence: 2 , IsInRing: True, SINGLE
> 8 :  C , Valence: 3 , IsInRing: True, SINGLE
> 9 :  C , Valence: 2 , IsInRing: False, SINGLE
> 10 :  C , Valence: 3 , IsInRing: False, DOUBLE
> 11 :  O , Valence: 2 , IsInRing: False, (No bond)
```

Substructure searching

```
# Substructure Searching (RDKit)
# authors: Landrum, Greg; Serizawa; Scalfani (Et al.)
# license: CC BY SA 4.0        # code: github.com/gressling/examples
# activity: active (2020)      # index: 57-3


# 1 - depict molecule
m = Chem.MolFromSmiles('COc1ccc2c(c1)[nH]c(n2)[S@@](=O)Cc1ncc(c(c1C)OC)C')
m
> [Fig. 57-2-a]

# 2 - find some substructures
print('OC', list(m.GetSubstructMatch(Chem.MolFromSmiles('OC'))))
print('COC', list(m.GetSubstructMatch(Chem.MolFromSmiles('COC'))))
print('S=O', list(m.GetSubstructMatch(Chem.MolFromSmiles('S=O'))))
print('CC', list(m.GetSubstructMatch(Chem.MolFromSmiles('CC'))))
> OC [1, 0]
> COC [0, 1, 2]
> S=O [11, 12]
> CC [13, 14]


tmp=m.GetSubstructMatch(Chem.MolFromSmarts('c1ccccc1'))
m
> [Fig. 57-2-b]


Chem.FastFindRings(m)
m.HasSubstructMatch(Chem.MolFromSmarts('[r]'))
> True


[list(x) for x in m.GetSubstructMatches(Chem.MolFromSmarts('[r]'))]
> [[2],
> [3],
> [4],
...
> [9],
> [10],
> [14],
...
> [19]]
m
> [Fig. 57-2-c]
```

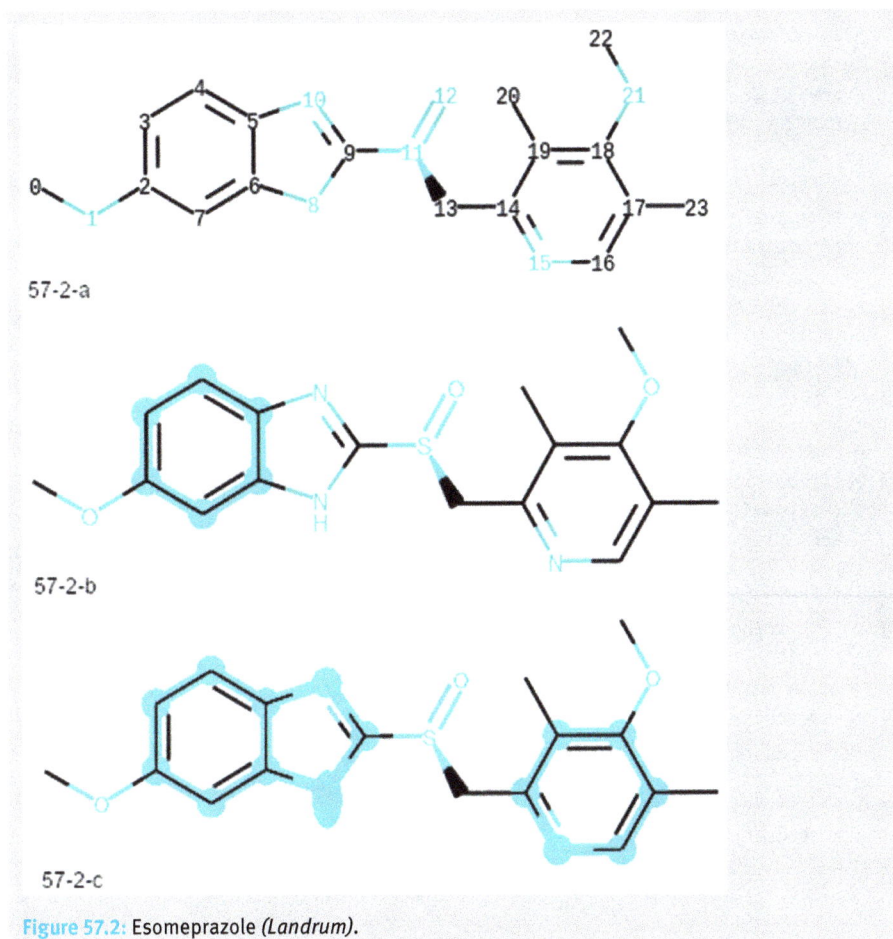

Figure 57.2: Esomeprazole *(Landrum)*.

Salt stripping and structure sanitizing

Normalization of structures includes the listing of tautomers and specifying chirality as well as adding hydrogens. It also sets protonation states and removes salt ions as well as metals. Some of the functionality provided allows molecules to be edited "in place."

Why sanitization? It ensures that the molecules are "reasonable": that they can be represented with octet-complete Lewis dot structures.

```
# Salt stripping and structure sanitizing (RDKit)
# authors: Landrum, Greg; Serizawa; Scalfani (Et al.)
# License: CC BY SA 4.0        # code: github.com/gressling/examples
# activity: active (2020)      # index: 57-4


# WITH Santitization
m = Chem.MolFromSmiles('c1ccccc1')
m.GetAtomWithIdx(0).SetAtomicNum(7)
Chem.SanitizeMol(m)
> rdkit.Chem.rdmolops.SanitizeFlags.SANITIZE_NONE
Chem.MolToSmiles(m)
> 'c1ccncc1'

# WITHOUT Sanitization
m = Chem.MolFromSmiles('c1ccccc1')
m.GetAtomWithIdx(0).SetAtomicNum(8)
Chem.MolToSmiles(m)
> 'c1ccocc1'

# Salt removing example
from rdkit.Chem.SaltRemover import SaltRemover

remover = SaltRemover(defnData="[Cl,Br]")
len(remover.salts)
> 1

mol = Chem.MolFromSmiles('CN(C)C.Cl')
res = remover.StripMol(mol)
# Draw.MolsToGridImage([mol, res])
```

References

1. Organic compound (Wikipedia) https://en.wikipedia.org/wiki/Organic_compound.
2. Landrum, G. RDKit https://www.rdkit.org/.
3. {Landrum, G.; Serizawa, T.; Scalfani, V. Getting Started with the RDKit in Python https://www.rdkit.org/docs/GettingStartedInPython.html.
4. Landrum, G. RDKit BLOG https://rdkit.blogspot.com/.
5. Boost.Python https://www.boost.org/doc/libs/1_70_0/libs/python/doc/html/index.html.

58 Features and 2D descriptors (of carbon compounds)

Descriptors were first discussed in Chapter 26. Features are selected descriptors for the solution of an ML problem. Additionally, to classic descriptors used in material science (physicochemical, 2D), other classes are *structural descriptors* in organic chemistry. In Chapter 60, vectorial representations (1D, fingerprints) will be discussed.

Structural descriptors represent the information by creating suitable concepts: bond count, atom count, and presence of particular groups of atoms. Structural descriptors can be extracted from SMILES strings, MOL representations, or calculated manually.

Descriptors in RDKit

Within structural descriptors, we have topological (3D) descriptors and geometrical (1D) descriptors. The molecular descriptor library of RDKit contains plane of best fit (PBF) and principal moments of inertia (PMI; also the normalized sum of PMIs – NPR) as well as descriptors to reflect the shape of the molecule.

RDKit contains molecular descriptor library (a subset is available via Python):
- Topological (κ3, Balaban J, etc.)
- Compositional (number of rings, number of aromatic heterocycles, etc.)
- Electrotopological state (Estate)
- clogP, MR (Wildman and Crippen approach)
- "MOE like" VSA descriptors
- MQN
- Chirality descriptor
- Multimolecule maximum common substructure
- Enumeration of molecular resonance structures
- Ring information
- Gasteiger–Marsili charges

Calculate 2D descriptors example

The following code snippet calculates:
- MaxEStateIndex
- MinEStateIndex
- MaxAbsEStateIndex
- qed
- MolWt
- HeavyAtomMolWt

https://doi.org/10.1515/9783110629453-058

- MinAbsEStateIndex
- NumValenceElectrons
- FpDensityMorgan (various)
- BalabanJ
- BertzCT
- ... and many more from ~200

- ExactMolWt
- NumRadicalElectron
- MaxPartialCharge
- MinPartialCharge
- MaxAbsPartialCharge
- MinAbsPartialCharge

```
# Calculating standard descriptors (RDKit)
# author: yamasakih
# license: MIT License      # code: gist.github.com/yamasakih/d423...965c
# activity: single example(2019)      # index: 58-1

import pandas as pd
from rdkit import Chem
from rdkit.Chem import Descriptors
from rdkit.ML.Descriptors import MoleculeDescriptors

def calculate_descriptors(mols, names=None, ipc_avg=False):
    if names is None:
        names = [d[0] for d in Descriptors._descList]
    calc = MoleculeDescriptors.MolecularDescriptorCalculator(names)
    descs = [calc.CalcDescriptors(mol) for mol in mols]
    descs = pd.DataFrame(descs, columns=names)
    if 'Ipc' in names and ipc_avg:
        descs['Ipc'] = [Descriptors.Ipc(mol, avg=True) for mol in mols]
    return descs

mols = [Chem.MolFromSmiles(s) for s in ['c1ccccc1', 'C1CCCCC1', 'CCO']]

calculate_descriptors(mols)

> MaxEStateIndex MinEStateIndex  MaxAbsEStateIndex MinAbsEStateIndex      qed
> 0              2               2                 2                 2
> 1              1.5             1.5               1.5               1.5
> 2              7.569444        0.25              7.569444          0.25

> MolWt      HeavyAtomMolWt  ExactMolWt  NumValenceElectrons NumRadicalElectrons
> 0.442628   78.114          72.066      78.04695            30
> 0.422316   84.162          72.066      84.0939             36
> 0.406808   46.069          40.021      46.041865           20
> ...
```

Descriptor: aromaticity

A property of *increased stability* (aromaticity) is given within *cyclic* (ring-shaped) *planar* structures having *resonance bonds*. Aromaticity can also be a descriptor:

```
# Structural descriptor examples (RDKit)
# authors: Landrum, Greg; Serizawa; Scalfani (Et al.)
# License: CC BY SA 4.0       # code: github.com/gressling/examples
# activity: active (2020)     # index: 58-2
# 1 - load structure and depict
m=Chem.MolFromSmiles('O=C1C=CC(=O)C2=C1OC=CO2')
for atom in m.GetAtoms():
    atom.SetProp('atomLabel',str(atom.GetIdx()))
Chem.Draw.MolToImage(m, size=(300, 300))
```

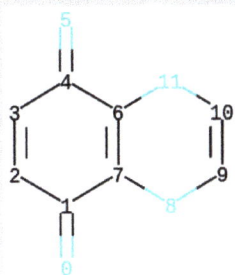

Figure 58.1: Tetrahydro-benzodioxine structure *(Landrum).*

```
# 2 - Aromaticity descriptor
print(m.GetAtomWithIdx(6).GetIsAromatic())
print(m.GetAtomWithIdx(7).GetIsAromatic())
print(m.GetBondBetweenAtoms(6,7).GetIsAromatic())

> True
> True
> False
```

Descriptor: SSSR

Finding all the rings in a structure, the SSSR is the **S**mallest **S**et of **S**mallest **R**ings.

```
# 3 - SSSR descriptor
ssr = Chem.GetSymmSSSR(mol)
len(ssr)
> 2
```

```
for ssrs in ssr:
    print(list(ssrs))
> [1, 7, 6, 4, 3, 2]
> [8, 9, 10, 11, 6, 7]
```

Descriptor: canonical rank

The canonical atom ranking for each atom describes the symmetry class for each atom. The symmetry class is used by the canonicalization routines to type each atom based on the whole chemistry of the molecular graph. Any atom with the same rank (symmetry class) is indistinguishable:

```
List(Chem.CanonicalRankAtoms(m, breakTies=False))
# O=C1C=CC(=O)C2=C1OC=CO2
> [0, 8, 4, 4, 8, 0, 10, 10, 6, 2, 2, 6]
```

Looking to the depiction of the structure in Figure 58.1 (or the SMILES), atoms 5 and 0 have the same symmetry class. From the perspective of the molecular graph, they are identical.

Features

The individual features in RDKit carry information about family (e.g., donor and acceptor), type (a more detailed description), and the atom(s) that are associated. Features in RDKit are defined using a SMARTS-based approach:

```
# Features (RDKit)
# authors: Landrum, Greg; Serizawa; Scalfani (Et al.)
# license: CC BY SA 4.0        # code: github.com/gressling/examples
# activity: active (2020)      # index: 58-3

import os
fdefName = os.path.join(RDConfig.RDDataDir,'BaseFeatures.fdef')
factory = ChemicalFeatures.BuildFeatureFactory(fdefName)

m=Chem.MolFromSmiles('OCc1ccccc1CN') # [2-(aminomethyl)phenyl]methanol
for atom in m.GetAtoms():
    atom.SetProp('atomLabel',str(atom.GetIdx()))
Chem.Draw.MolToImage(m, size=(400, 400))
```

```
feats = factory.GetFeaturesForMol(m)
display(m)
```

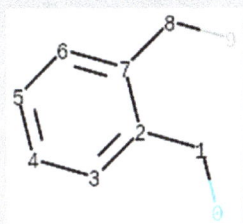

Figure 58.2: [2-(Aminomethyl)phenyl]methanol *(Landrum)*.

```
print(len(feats))
for feat in feats:
    # IDs of the atoms that participate in the feature
    print(feat.GetAtomIds(), end=", ")
    print(feat.GetFamily(), end=", ")
    print(feat.GetType())
```

```
> 8
> (0,), Donor, SingleAtomDonor
> (9,), Donor, SingleAtomDonor
> (0,), Acceptor, SingleAtomAcceptor
> (9,), PosIonizable, BasicGroup
> (2, 3, 4, 5, 6, 7), Aromatic, Arom6
> (2,), Hydrophobe, ThreeWayAttach
> (7,), Hydrophobe, ThreeWayAttach
> (2, 3, 4, 5, 6, 7), LumpedHydrophobe, RH6_6
```

Combining chemical features with topological distances between them gives a 2D pharmacophore descriptor.

Example descriptor: graph diameter

Define the distance between two atoms in the molecule as the shortest number of bonds to *get from one atom to the other*. The graph diameter is the *longest such distance* in the graph.

The graph diameter traverses the graph structure (ignore hydrogens) and finds the all-pairs shortest paths. From that list, the diameter is the largest number. For example, the graph diameter of pentane is 4 and of benzene is 3.

```
# Example descriptor: Graph diameter
# author: Gressling, T
# license: CC BY SA 4.0      # code: github.com/gressling/examples
# activity: single example   # index: 58-4


from rdkit import Chem
import numpy

# N-Methoxysuccinyl-alanyl-alanyl-prolyl-valine-4-nitroanilide
mol = Chem.MolFromSmiles('CC(C)C(C(=O)NC(=O)C1CCCN1C(=O)C(C)NC(=O)C(C)
                          NC(=O)CCC(=O)OC)NC2=CC=C(C=C2)[N+](=O)[O-]')
# GetDistanceMatrix returns the molecules 2D (topological)
# distance matrix:
dm = Chem.GetDistanceMatrix(mol)
diameter = int(numpy.max(dm))
dm
> array([[0., 1., 2., ..., 8., 9., 9.],
>        [1., 0., 1., ..., 7., 8., 8.],
>        [2., 1., 0., ..., 8., 9., 9.],
>        ...,
>        [8., 7., 8., ..., 0., 1., 1.],
>        [9., 8., 9., ..., 1., 0., 2.],
>        [9., 8., 9., ..., 1., 2., 0.]])


diameter
> 24
```

Example descriptor: Gasteiger–Marsili (PEOE) atomic partial charges

Gasteiger(–Marsili) charges are determined[1] on the iterative partial equalization of orbital electronegativity:

```
# Example descriptor: Gasteiger-Marsili atomic partial charges (RDKit)
# author: Gressling, T
# License: CC BY SA 4.0      # code: github.com/gressling/examples
# activity: single example   # index: 58-5


from rdkit.Chem.Draw import SimilarityMaps
from rdkit.Chem import AllChem
```

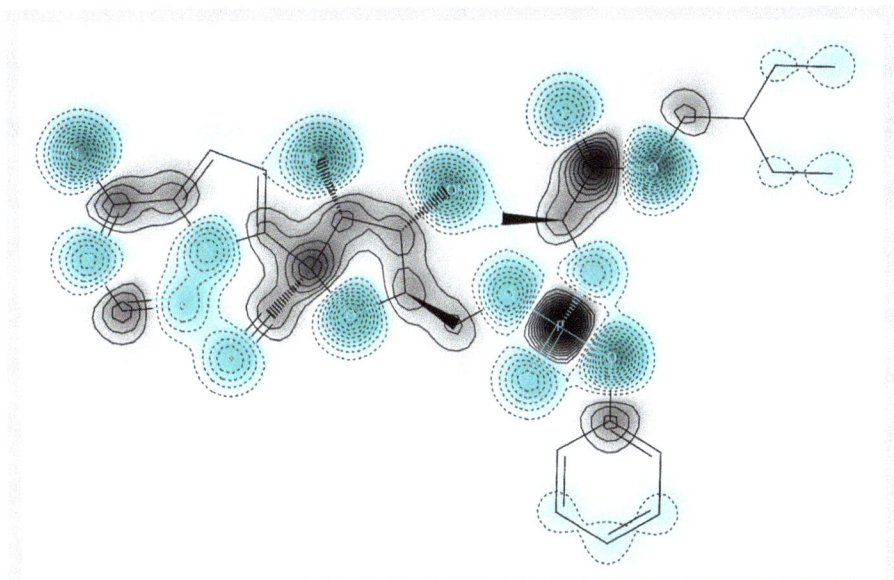

Figure 58.3: Remdesivir Gasteiger–Marsili charges *(Gressling)*.

```
# Remdesivir
mol = Chem.MolFromSmiles('CCC(CC)COC(=O)[C@H](C)N[P@](=O)(OC[C@@H]1[C@H]
                ([C@H]([C@](O1)(C#N)C2=CC=C3N2N=CN=C3N)O)O)OC4=CC=CC=C4')

AllChem.ComputeGasteigerCharges(mol)
contribs = [float(mol.GetAtomWithIdx(i).GetProp('_GasteigerCharge')) for
i in range(mol.GetNumAtoms())]
fig = SimilarityMaps.GetSimilarityMapFromWeights(mol, contribs,
                contourLines=30, size=(850, 850))
```

References

1. Gasteiger, J.; Marsili, M. Iterative Partial Equalization of Orbital Electronegativity – a Rapid Access to Atomic Charges. *Tetrahedron* **1980**, *36* (22), 3219–3228. https://doi.org/10.1016/0040-4020(80)80168-2.

More references on RDKit see Chapter 57.

59 Working with molecules and reactions

A chemical reaction is a process[1] that leads to the *transformation* of one set of chemical substances to another. Chemical reactions encompass changes that only involve the positions of electrons in the forming and breaking of chemical bonds between atoms.

Simple structure transformations

Add, delete, and replace fragments in the graph (this is not a chemical reaction):

```
# Substructure-based transformations (delete, replace) (RDKit)
# authors: Landrum, Greg; Serizawa; Scalfani (Et al.)
# License: CC BY SA 4.0        # code: github.com/gressling/examples
# activity: active (2020)      # index: 59-1

# Deleting
m = Chem.MolFromSmiles('c1(C=O)cc(OC)c(O)cc1') # Vanillin
for atom in m.GetAtoms():
    atom.SetProp('atomLabel',str(atom.GetIdx()))
patt = Chem.MolFromSmarts('cO')
rm = AllChem.DeleteSubstructs(m, patt)
Draw.MolsToGridImage([m, patt, rm], molsPerRow=3, subImgSize=(600,600))
> [Fig. 59-1-a]

# Replacing
repl = Chem.MolFromSmiles('OC')
patt = Chem.MolFromSmarts('[$(NC(=O))]')
m = Chem.MolFromSmiles('CC(=O)N') # Acetamide
rms = AllChem.ReplaceSubstructs(m, patt, repl)
Draw.MolsToGridImage([m, repl, rms[0]], ...)
> [Fig. 59-1-b]
```

https://doi.org/10.1515/9783110629453-059

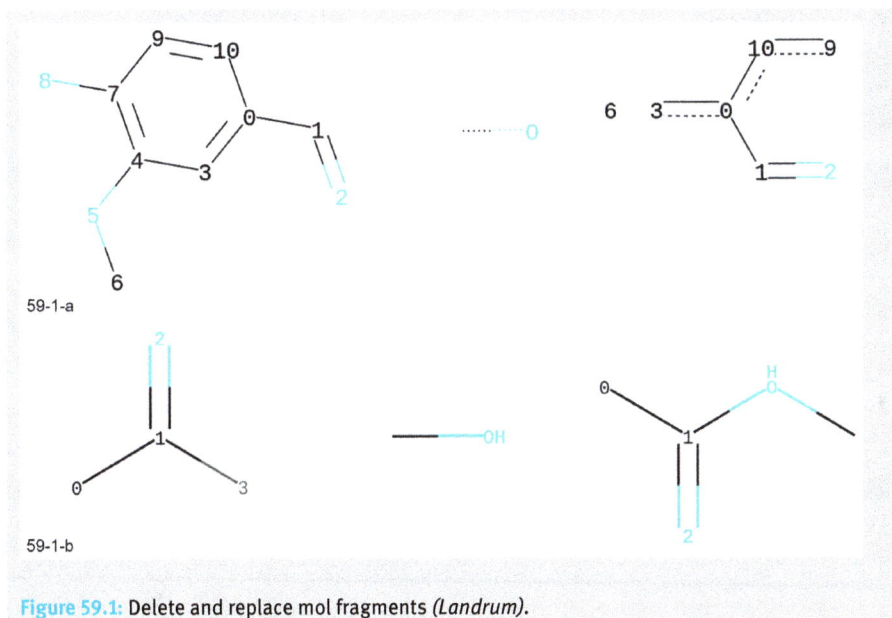

Figure 59.1: Delete and replace mol fragments *(Landrum)*.

Example: break rotatable bonds and report the fragments

Rotatable bond[2] definition in SMARTS:

```
[!$([NH]!@C(=O))&!D1&!$(*#*)]-&!@[!$([NH]!@C(=O))&!D1&!$(*#*)]
```

Use the function to write a program that takes the structure of Remdesivir and break the rotatable bonds. Then depict the fragments out.

```
# Break rotatable bonds and depict the fragments (RDKit)
# author: Gressling, T
# license: CC BY SA 4.0     # code: github.com/gressling/examples
# activity: single example  # index: 59-2

patt = Chem.MolFromSmarts(
    '[!$([NH]!@C(=O))&!D1&!$(*#*)]-&!@[!$([NH]!@C(=O))&!D1&!$(*#*)]')
mol = Chem.MolFromSmiles('CCC(CC)COC(=O)[C@H](C)N[P@](=O)(OC[C@@H]1[C@H]
    ([C@H]([C@](O1)(C#N)C2=CC=C3N2N=CN=C3N)O)O)OC4=CC=CC=C4')

# Get the rotatable bonds
bonds = mol.GetSubstructMatches(patt)

em = Chem.EditableMol(mol)
nAts = mol.GetNumAtoms()
```

```
for a,b in bonds:
    em.RemoveBond(a,b)
    em.AddAtom(Chem.Atom(0))
    em.AddBond(a, nAts, Chem.BondType.SINGLE)
    em.AddAtom(Chem.Atom(0))
    em.AddBond(b, nAts+1, Chem.BondType.SINGLE)
    nAts+=2
p = em.GetMol()
Chem.SanitizeMol(p)

mols = []
smis = [Chem.MolToSmiles(x,True) for x in
               Chem.GetMolFrags(p,asMols=True)]
for smi in smis:
    mol=Chem.MolFromSmiles(smi)
    mols.append(mol)

Draw.MolsToGridImage(mols, molsPerRow=4)
```

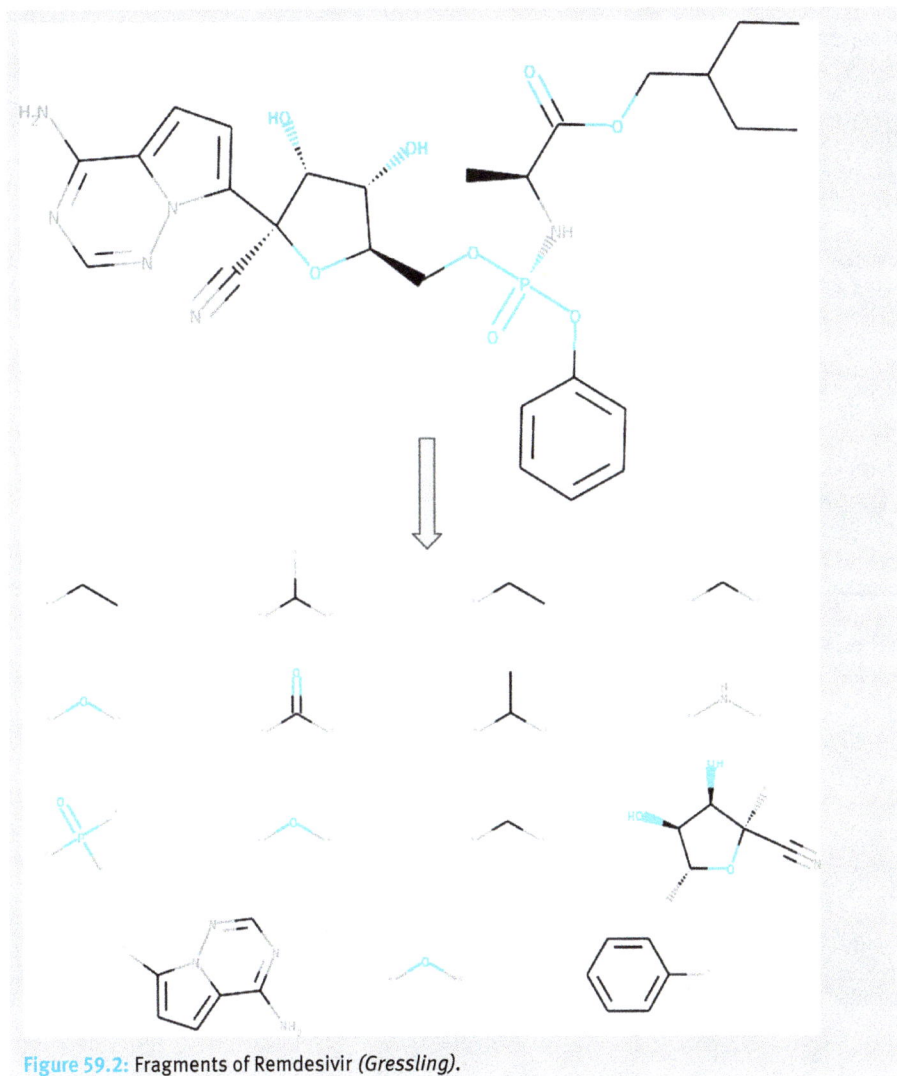

Figure 59.2: Fragments of Remdesivir *(Gressling)*.

Reactions

With RDKit, chemical reactions can be defined. These reactions can be used to create new molecules from existing reactants. A simple example[3]:

```
rxn = AllChem.ReactionFromSmarts('[#6:1]([O:2])>>[#6:1](=[O:2])')
```

1. Any carbon atom (label as :1) that ...
2. ... is bound via a single bond to an aliphatic oxygen atom (capital O) (label as :2)
3. transform the linkage between these atoms to be a double bond (=)

Examples for synthesis of thiazole

```
# Hantzsch thiazole synthesis (RDKit)
# authors:  Gao, Peng; Jiang, Shilong
# license: MIT License        # code: github.com/gressling/examples
# activity: single example (2017)   # index: 59-3


thiourea = Chem.MolFromSmiles('CN(C)C(=S)N')
haloketone = Chem.MolFromSmiles('c1ccccc1C(=O)C(C)Cl')
rxn_smarts =
'[NH2:1][C:2](=[S:3])[NH0:4].[C:5](=[O:6])[C:7][Cl:8]>>[N:4][c:2]1[s:3][c:5]
[c:7][n:1]1'


rxn = AllChem.ReactionFromSmarts(rxn_smarts)
product = rxn.RunReactants((thiourea, haloketone))[0][0]
Chem.SanitizeMol(product) # C12H14N2S


all_three = [thiourea, haloketone, product]
formulae = [rdMolDescriptors.CalcMolFormula(mol) for mol in all_three]


Draw.MolsToGridImage(all_three, legends = formulae)
```

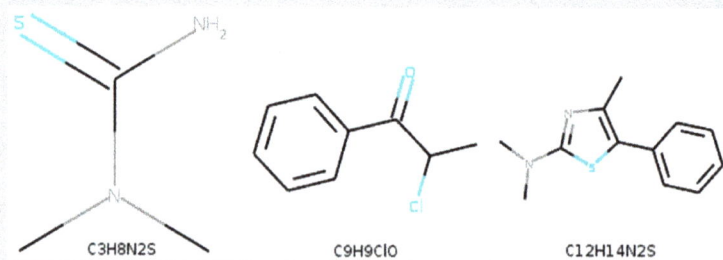

C3H8N2S   C9H9ClO   C12H14N2S

Figure 59.3: Hantzsch thiazole synthesis[4] *(Gao; Jiang)*.

References

1. Chemical reaction (Wikipedia) https://en.wikipedia.org/wiki/Chemical_reaction.
2. Chemistry Toolkit Rosetta Wiki https://ctr.fandom.com/wiki/Chemistry_Toolkit_Rosetta_Wiki.
3. Gao, P.; Jiang, S. RDKit: making new molecules using reaction SMARTS (Kesci.com) **2019**.
4. Hantzsch, A.; Weber, J. H. Ueber Verbindungen Des Thiazols (Pyridins Der Thiophenreihe). *Ber. Dtsch. Chem. Ges.* **1887**, *20* (2), 3118–3132. https://doi.org/10.1002/cber.188702002200.

More references on RDKit see Chapter 57.

60 Fingerprint descriptors (1D)

Molecular fingerprints (FPs) are a way of *encoding the structure of a molecule*.[1] **They are an essential cheminformatics tool for virtual screening and mapping chemical space.**

The most common type of FP is a series of binary digits (bits) that represent the presence or absence of particular substructures in the molecule. Comparing FPs can determine the similarity between molecules, find matches to a query substructure, and so on.

FPs are easy to implement for ML algorithms. They can be trained to an ML algorithm. The trained model then generates (for example) total energy, Mulliken charges, and other metrics for unknown molecules, as well as pharmacophores or spectrophores[TM]. Quantities can be predicted that are not available for the test set and they can be used as new input features.

Main types of chemical descriptors

FP based[2]**:**
- **ECFP 4**: atom type, Extended Connectivity FingerPrint, maximum distance = 4
- **FCFP 4**: Functional-class-based, extended conn. fingerprint, maximum distance = 4
- **MACCS**: 166 predefined MDL keys (public set)

Connectivity matrix based:
- **BCUT**: atomic charges, polarizabilities, H-bond donor and acceptor abilities, and H-bonding modes of intermolecular interaction

Shape based:
- Rapid overlay of chemical structures (**ROCS**): combo Tanimoto (shape and electrostatic score) shape-based molecular similarity method; molecules are described by smooth Gaussian function and pharmacophore points.
- **PMI**: normalized principal moment-of-inertia ratios

Pharmacophore based:
- **GpiDAPH3**: graph-based three-point pharmacophore, eight atom types computed from three atom properties (in pi system, donor, acceptor).
- **TGD**: typed graph distances; types: donor, acceptor, polar, anion, cation, hydrophobe
- **TAD**: typed atom distances

https://doi.org/10.1515/9783110629453-060

Bioactivity based:

– Bayes affinity FPs: bioactivity model based on multicategory Bayes classifier

Physicochemical property based:

– **prop2D**: physicochemical properties (such as molecular weight, atom counts, partial charges, and hydrophobicity)

Important techniques

Recap implementation

RECAP: Retrosynthetic combinatorial analysis procedure: a powerful new technique for identifying privileged molecular fragments with useful applications in combinatorial chemistry.[3]

BRICS implementation

Rules for the Breaking of Retrosynthetically Interesting Chemical Substructures (BRICS). Relative to existing methods, BRICS performs much better in retrieving compounds from various large and diverse query sets.[4]

ECFP (extended connectivity FPs)

ECFPs are circular PFs and built by applying the Morgan algorithm to a set of user-supplied atom invariants. Fragments[5] of molecules of similar size can be calculated and compared.

Implementation examples

The following two examples show a visual representation of molecular FP and an example of how to use FP as an input layer in a neural network (NN).

Depict Morgan FP bits

The rdkit.Chem.Draw package that allows direct visualization of the atoms and bonds from a molecule that set bits from Morgan and RDKit FPs display the atomic environment that sets a particular bit:

```
# Benzoic acid Morgan fingerprint visualization (RDKit)
# author: Gressling, T
# license: MIT License        # code: github.com/gressling/examples
# activity: single example    # index: 60-1

from rdkit.Chem import rdMolDescriptors

mol = Chem.MolFromSmiles('O=C(O)c1ccccc1')
bi = {}
fp = rdMolDescriptors.GetMorganFingerprintAsBitVect(mol,
                   radius=2, bitInfo=bi)

tpls = [(mol,x,bi) for x in fp.GetOnBits()]
Draw.DrawMorganBits(tpls[:14], molsPerRow=4,
                   legends=[str(x) for x in fp.GetOnBits()][:14])
```

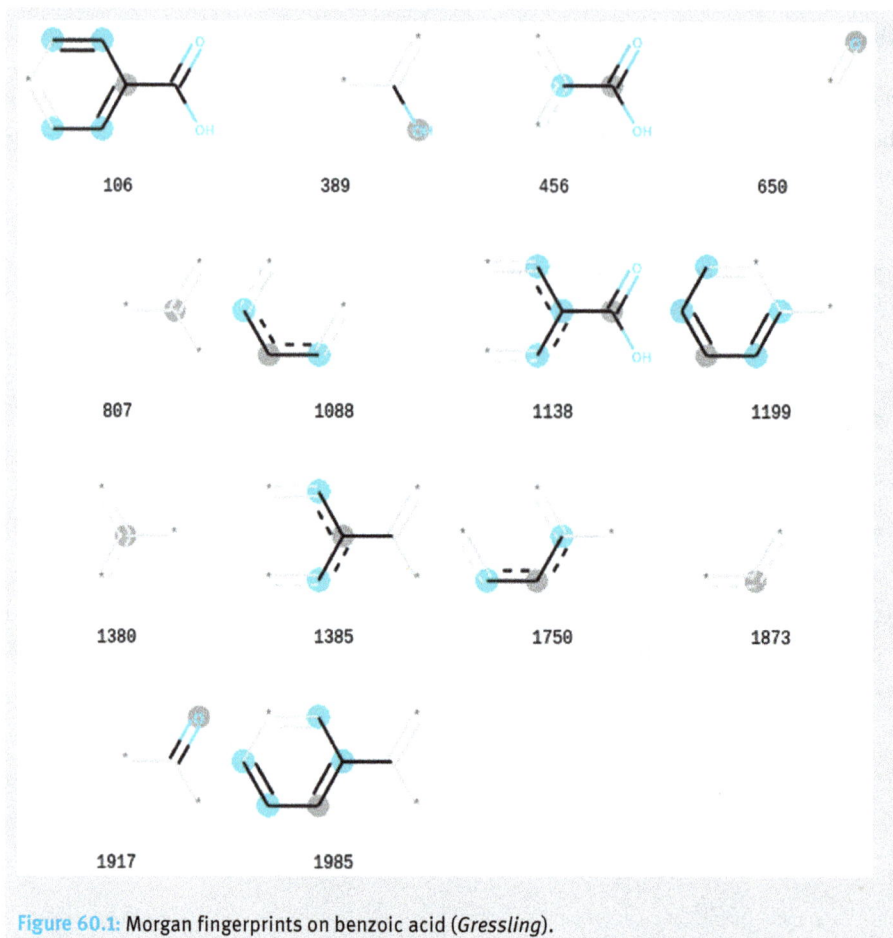

Figure 60.1: Morgan fingerprints on benzoic acid (*Gressling*).

- The central atom is highlighted in gray.
- Aromatic atoms are highlighted in cyan.
- Atoms/bonds that are drawn in light gray indicate pieces of the structure that influence the atoms' connectivity invariants but that are not directly part of the FP.

Morgan FP as input layer in NN

This code is shortened to focus on the input of the NN from the FPs:

```python
# Fingerprint as input layer in NN
# author: keiserlab.org; Evangelidis, Thomas (Et al.)
# license:  MIT License
# code: github.com/tevang/tutorials/Multilayer_Perceptron_Keras
# activity: single example (2017)   # index: 60-2
# 1. Ingest
fname = "data/cdk2.sdf"

mols = []
y = []
for mol in SDMolSupplier(fname):
    if mol is not None:
        mols.append(mol)
        y.append(float(mol.GetProp("pIC50")))

# 2. Calculate descriptors (binary Morgan fingerprint with radius 2)
# and convert them into numpy array
fp = [AllChem.GetMorganFingerprintAsBitVect(m, 2) for m in mols]

def rdkit_numpy_convert(fp):
    output = []
    for f in fp:
        arr = np.zeros((1,))
        DataStructs.ConvertToNumpyArray(f, arr)
        output.append(arr)
    return np.asarray(output)

fingerprint_input = rdkit_numpy_convert(fp)

# 3. Random seed for reproducibility
np.random.seed(seed)
mol_num, feat_num = fingerprint_input.shape
print("# molecules for training = %i, # of features = %i\n" %
        (mol_num, feat_num))

# 4. create model
def MLP(feat_num, loss):   # Multi-Layer Perceptron
    net = Sequential()
    net.add(Dense(300, input_dim=feat_num,
            kernel_initializer='normal', activation='relu'))
    net.add(Dense(1, kernel_initializer='normal'))
    net.compile(loss=loss, optimizer='adam')
    return net
```

```
estimator = KerasRegressor(build_fn=MLP,
                           feat_num=feat_num,
                           loss='mean_squared_error',
                           epochs=300,
                           batch_size=int(fingerprint_input.shape[0]/8),
                           verbose=0)
```

Layer (type)	Output Shape	Param #
dense_208 (Dense)	(None, 300)	614700
dense_209 (Dense)	(None, 1)	301

Trainable params: 615,001

```
# 5. Evaluation metrics
1) Kendall's tau (ranking correlation),
2) Pearson's R (correlation),
3) Mean Squared Error.
4) The evaluation will be done with 5-fold cross-validation.

def kendalls_tau(estimator, X, y):
    preds = estimator.predict(X)
    t = kendalltau(preds, y)[0]
    return t

def pearsons_r(estimator, X, y):
    preds = estimator.predict(X)
    r = pearsonr(preds, y)[0]
    return r

def MSE(estimator, X, y):
    preds = estimator.predict(X)
    mse = mean_squared_error(preds, y)
    return mse

# 6. Run
scoring = {'tau': kendalls_tau, 'R':pearsons_r, 'MSE':MSE}
kfold = KFold(n_splits=5, random_state=seed)
scores = cross_validate(estimator, fingerprint_input,
            y, scoring=scoring, cv=kfold, return_train_score=False)
# print(scores)
...
```

References

1. Molecular fingerprints and similarity searching – Open Babel documentation https://openbabel.org/docs/dev/Fingerprints/intro.html.
2. Wu, Y.; Wang, G. Machine Learning Based Toxicity Prediction: From Chemical Structural Description to Transcriptome Analysis. *Int. J. Mol. Sci.* **2018**, *19* (8). https://doi.org/10.3390/ijms19082358.
3. Lewell, X. Q.; Judd, D. B.; Watson, S. P.; Hann, M. M. RECAP--Retrosynthetic Combinatorial Analysis Procedure: A Powerful New Technique for Identifying Privileged Molecular Fragments with Useful Applications in Combinatorial Chemistry. *J. Chem. Inf. Comput. Sci.* **1998**, *38* (3), 511–522. https://doi.org/10.1021/ci970429i.
4. Degen, J.; Wegscheid-Gerlach, C.; Zaliani, A.; Rarey, M. On the Art of Compiling and Using "Drug-like" Chemical Fragment Spaces. *ChemMedChem* **2008**, *3* (10), 1503–1507. https://doi.org/10.1002/cmdc.200800178.
5. Evangelidis, T. Personal archive from Thomas Evangelidis (GitHub) https://github.com/tevang/tutorials.

More references on RDKit see Chapter 57.

61 Similarities

The notion of molecular similarity is one of the most important concepts in chemoinformatics. As chemical repositories of molecules continue to grow and become more open, it becomes increasingly important to develop the tools to search them efficiently.[1]

In one of the most typical settings, a query molecule is used to search millions of other compounds not only for exact matches but also frequently for approximate similarity matches.

Generating similarity maps using fingerprints

Each fingerprint bit corresponds to a fragment of a molecule. When comparing two different molecules for similarity, the result must be 0.0 (not similar) and 1.0 if they are identical. The most common approach is the Tanimoto similarity.

Lots of experience[2] shows that the best fingerprint for activities like virtual screening (finding active similar molecules in a database) depends strongly on the data set.

In this example, some fingerprints are compared[3] with the similarity of three vanillin-related structures:

```
# Fingerprint comparison for three vanillyl-derivatives (RDKit)
# author: Gressling, T
# License: MIT License          # code: github.com/gressling/examples
# activity: single example   # index: 61-1

from rdkit import Chem, DataStructs

mol1 = Chem.MolFromSmiles("CC(C)C=CCCCC(=O)NCc1ccc(c(c1)OC)O")
mol2 = Chem.MolFromSmiles("COC1=C(C=CC(=C1)C=O)O")
mol3 = Chem.MolFromSmiles("CCCCCCCCC(=O)NCC1=CC(=C(C=C1)O)OC")
display(Draw.MolsToGridImage([mol1, mol2, mol3]))

# the default RDKit fingerprint is path-based:
fp1 = Chem.RDKFingerprint(mol1)
fp2 = Chem.RDKFingerprint(mol2)
fp3 = Chem.RDKFingerprint(mol3)
print("RDKit Daylight-like; 1-2; {:.1%}".format(
                DataStructs.TanimotoSimilarity(fp1,fp2)))
print("RDKit; 2-3; {:.1%}".format(DataStructs.TanimotoSimilarity(fp2,fp3)))
print("RDKit; 1-3; {:.1%}".format(DataStructs.TanimotoSimilarity(fp1,fp3)),
end="\n\n")
```

https://doi.org/10.1515/9783110629453-061

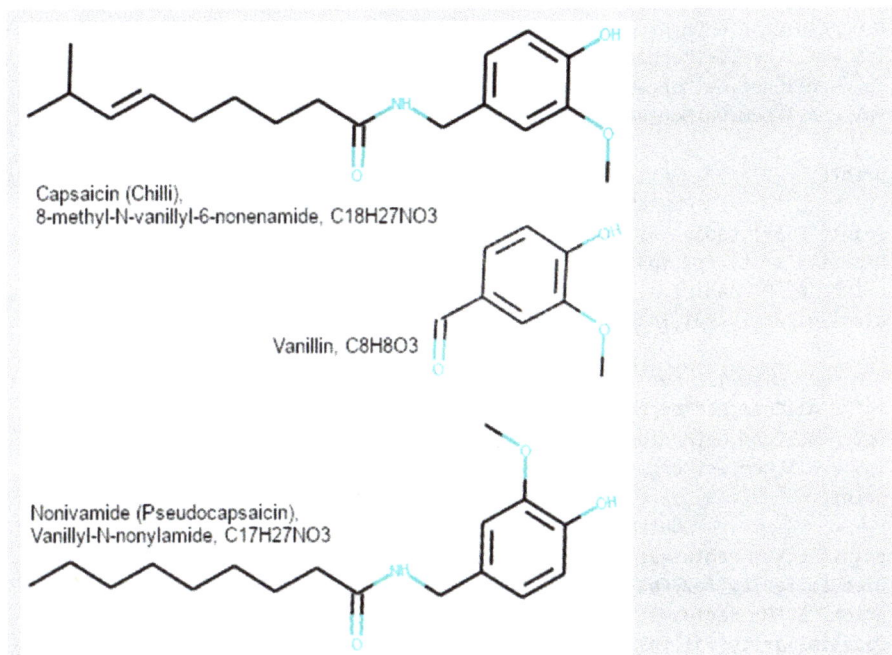

Figure 61.1: Three vanillyl derivatives *(Gressling)*.

```
# the Morgan fingerprint (similar to ECFP) is also useful:
from rdkit.Chem import rdMolDescriptors
mfp1 = rdMolDescriptors.GetMorganFingerprint(mol1,1)
mfp2 = rdMolDescriptors.GetMorganFingerprint(mol2,1)
mfp3 = rdMolDescriptors.GetMorganFingerprint(mol3,1)
print("Morgan; radius=1; 1-2; {:.1%}".format(
                DataStructs.DiceSimilarity(mfp1,mfp2)))
print ("Morgan; radius=1; 2-3; {:.1%}".format(DataStructs.DiceSimilarity
(mfp2,mfp3)))
print ("Morgan; radius=1; 1-3; {:.1%}".format(DataStructs.DiceSimilarity
(mfp1,mfp3)), end="\n\n")


# the Morgan with radius 3:
mfp1 = rdMolDescriptors.GetMorganFingerprint(mol1,3)
mfp2 = rdMolDescriptors.GetMorganFingerprint(mol2,3)
mfp3 = rdMolDescriptors.GetMorganFingerprint(mol3,3)
print("Morgan; radius=3; 1-2; {:.1%}".format(
                DataStructs.DiceSimilarity(mfp1,mfp2)))
print ("Morgan; radius=3; 2-3; {:.1%}".format(DataStructs.DiceSimilarity
(mfp2,mfp3)))
print ("Morgan; radius=3; 1-3; {:.1%}".format(DataStructs.DiceSimilarity
(mfp1,mfp3)), end="\n\n")
```

```
# Morgan with 4096 bit:
fp1 = AllChem.GetMorganFingerprintAsBitVect(mol1,radius=3,nBits=4096)
fp2 = AllChem.GetMorganFingerprintAsBitVect(mol2,radius=3,nBits=4096)
fp3 = AllChem.GetMorganFingerprintAsBitVect(mol3,radius=3,nBits=4096)

print("ECFP; radius=3; bits=4096; 1-2; {:.1%}".format(
                DataStructs.DiceSimilarity(fp1,fp2)))
print("ECFP; radius=3; bits=4096; 2-3; {:.1%}".format(DataStructs.
DiceSimilarity(fp2,fp3)))
print("ECFP; radius=3; bits=4096; 1-3; {:.1%}".format(DataStructs.
DiceSimilarity(fp1,fp3)), end="\n\n")

# Morgan with 128 bit (strange, for comparison)
fp1 = AllChem.GetMorganFingerprintAsBitVect(mol1,radius=3,nBits=128)
fp2 = AllChem.GetMorganFingerprintAsBitVect(mol2,radius=3,nBits=128)
fp3 = AllChem.GetMorganFingerprintAsBitVect(mol3,radius=3,nBits=128)
print("ECFP; radius=3; bits=128; 1-2; {:.1%}".format(
                DataStructs.DiceSimilarity(fp1,fp2)))
print("ECFP; radius=3; bits=128; 2-3; {:.1%}".format(DataStructs.
DiceSimilarity(fp2,fp3)))
print("ECFP; radius=3; bits=128; 1-3; {:.1%}".format(DataStructs.
DiceSimilarity(fp1,fp3)))
```

Results

Molecules that are similar have a lot of fragments in common,[2] there is no "right" answer for defining similarity: there is no canonical definition of "molecular similarity."

```
RDKit Daylight-like; 1-2; 42.7%    ECFP; radius=3; bits=4096; 1-2; 40.9%
RDKit Daylight-like; 2-3; 44.7%    ECFP; radius=3; bits=4096; 2-3; 41.9%
RDKit Daylight-like; 1-3; 91.3%    ECFP; radius=3; bits=4096; 1-3; 76.9%

Morgan; radius=1; 1-2; 57.6%       ECFP; radius=3; bits=128; 1-2; 51.3%
Morgan; radius=1; 2-3; 56.2%       ECFP; radius=3; bits=128; 2-3; 60.0%
Morgan; radius=1; 1-3; 79.1%       ECFP; radius=3; bits=128; 1-3; 82.6%

Morgan; radius=3; 1-2; 42.9%
Morgan; radius=3; 2-3; 41.8%
Morgan; radius=3; 1-3; 74.7%
```

References

1. Baldi, P.; Nasr, R. When Is Chemical Similarity Significant? The Statistical Distribution of Chemical Similarity Scores and Its Extreme Values. *J. Chem. Inf. Model.* **2010**, *50* (7), 1205–1222. https://doi.org/10.1021/ci100010v.
2. Landrum, G. Fingerprints in the RDKit (UGM 2012) https://www.rdkit.org/UGM/2012/Landrum_RDKit_UGM.Fingerprints.Final.pptx.pdf.
3. Chemistry Toolkit Rosetta Wiki https://ctr.fandom.com/wiki/Chemistry_Toolkit_Rosetta_Wiki.

More references on RDKit see Chapter 57.

Engineering, laboratory, and production

62 Laboratory: SILA and AnIML

A laboratory is a facility that provides *controlled conditions*[1] in which research, experiments, and measurements are performed.

Controlled data collection

Open-data formats such as AnIML and communication protocols such as SiLA are the fundamentals of the digitized laboratory. Processes like planning, execution, and documentation of experiments need to be digital. The digitalization result has to be a unified and complete data package. Up to now, there are two complementary APIs but not yet adopted as common standard[2]:
- SiLA: communication protocol, controlling of devices and software systems
- AnIML: transport, XML-based data format for describing analytical and process data

SILA

There are different repositories that can be accessed depending on the programming language of the reference implementation, all found in the GitLab group[3]:
- Java maintained by UniteLabs
- C# maintained by EQUIcon and UniteLabs
- Python maintained by University of Greifswald

Other programming languages might be added in the future, depending on the needs of the community.

SILA2 and Python

SiLA_python is a python implementation of the universal and royalty-free SiLA 2 laboratory automation standard. It provides convenient libraries, a code generator, and a collection of examples and implementations to support a fast integration of SiLA 2 into your own lab automation projects and to illustrate how SiLA 2 could be implemented.[4]

https://doi.org/10.1515/9783110629453-062

SILA feature definition

```xml
<?xml version="1.0" encoding="ISO-8859-1"?>

<Feature xsi:schemaLocation="http://www.sila-standard.org
https://gitlab.com/SiLA2/sila_base/.../FeatureDefinition.xsd"
xmlns:xsi="http://www.w3.org/2001/XMLSchema-instance"
xmlns="http://www.sila-standard.org" Category="examples"
Originator="org.silastandard" FeatureVersion="1.0" SiLA2Version="0.2">
<Identifier>GreetingProvider</Identifier>
<DisplayName>Greeting Provider</DisplayName>
<Description>
Minimum Feature Definition as example. Provides a Greeting
</Description>

<Command>
    <Identifier>SayHello</Identifier>
    <DisplayName>Say Hello</DisplayName>
    <Description>Does what it says</Description>
    <Observable>No</Observable>
    <Parameter>
        <Identifier>Name</Identifier>
        <DisplayName>Name</DisplayName>
        <Description>Your Name</Description>
        <DataType>
    </Parameter>
    <Response>
        <Identifier>Greeting</Identifier>
        <DisplayName>Greeting</DisplayName>
        <Description>The greeting coming back at you</Description>
        <DataType>
            <Basic>String</Basic>
        </DataType>
    </Response>
</Command>
</Feature>
```

Jupyter example

```
# SILA2 Notebook
# author: Gressling, Thorsten
# license: MIT License          # code: github.com/Gressling/examples
# activity: simple example      # index: 62-1
```

```
# source /home/{your_name}/py3venv/sila2_venv/bin/activate
# python HelloSiLA2_server.py
# (test: 'python HelloSiLA2_client.py')
# jupyter notebook

import sila2lib.sila_client as mySila
from sila2lib.framework import SiLAFramework_pb2 as fwpb2
import sila2lib.HelloSiLA2.GreetingProvider.gRPC.GreetingProvider_pb2 as gpr
import sila2lib.HelloSiLA2.GreetingProvider.gRPC.GreetingProvider_pb2_grpc
as gprpc
from sila2lib.framework.feature_definitions import SiLAService_pb2 as
theFeatures

# create client
self = mySila
mySila.SiLA2Client.__init__(self, name="HelloSiLA2Client",
                            server_ip='127.0.0.1', server_port=50051)
myStub = self.SiLAService_stub
response = myStub.Get_ImplementedFeatures(
           theFeatures.Get_ImplementedFeatures_Parameters())
response

> ----ImplementedFeatures {
>   FeatureIdentifier {
>     value: "GreetingProvider"
>   }
> }----

# create server GreetingProvider
self.gp = gprpc.GreetingProviderStub(self.channel)

# SayHello!
myName = fwpb2.String(value = "Linus Pauling")
response = self.gp.SayHello(gpr.SayHello_Parameters(Name = myName))
response.Greeting
> ----value: "Hello Linus Pauling!"----
```

SiLA servers

- Thermo Fisher (Cytomat, RapidStak)
- BioTek MicroFlow
- Tecan Freedom Evo 150/200
- Sartorius Balance U4600 P+
- QInstruments Bioshake
- Open Source: Camunda
- Raspberry Pi Demo (Uni-Greifswald)
- SiLA boxes (Konstanz) and others

SiLA clients

- UniteLabs: SiLA browser
- EQUIcon: niceLAB & generic client
- Camunda

labpy

labPy[5] by Mark Dörr is a collection of laboratory instrument connectors/drivers and Qt5 (PySide2)-based GUIs/instrument widget sets for fast data plotting, logging and instrument control. Its modular design is aiming at replacing LabView (R), but keeping it tiny and lean.

```
# labPy laboratory instrument connector (SILA2)
# author: Dörr, Mark
# license: "freely and openly available"   # code: gitlab.com/LARAsuite
# activity: active (2020)                   # index: 62-2

import labpy.labpyworkbench as lpwb

labPy_app = lpwb.LP_Application(sys.argv, appname=appname, description=desc)
timew.TimerWidget(LP_application=labPy_app)

# temperature toolbar widget
temp_tb_widget = tempw.TemperatureWidget(LP_application=labPy_app)
temp_tb_widget.setTemperature(42.0)
labPy_app.run()
```

AnIML: the Analytical Information Markup Language

Not to be confused with the animl Decision Tree Visualization! AnIML is
- *a core schema that defines how to store any kind of analytical data,*
- *a technique schema to describe how to store data for a given analytical technique,*
- *a set of technique definition documents.*

These XML files, one per analytical technique, apply tight constraints to the flexible core and, in turn, are defined by the Technique Schema; Extensions to Technique Definitions are possible to accommodate vendor- and institution-specific data fields.[6]

The Analytical Information Markup Language is a standardization effort of ASTM (formerly American Society for Testing and Materials) Subcommittee E13.15 on Analytical Data.[7] Other standardization formats are SpectroML, 1–3 ANDI, 4–8, and JCAMP-DX9–12 suitable for their application domains. The use of such current formats becomes difficult when multiple techniques are used or if vendor-specific data need to be stored.

Reference process for AnIML implementation

1. particular sample (acquired from a larger piece of material),
2. device (e.g., a balance),
3. dissolved in a particular solvent using a certain ratio,
4. processed using a separation technique (e.g., liquid chromatography),
5. analyzed using multiple attached detectors, for example
   a. mass spectrometer and a
   b. UV/VIS diode array detector.
6. all instruments
   a. devices have been purchased from multiple manufacturers,
   b. running with a number of instrument-specific settings,
   c. have been calibrated according to manufacturers' protocols,
   d. a blank sample has been analyzed.

This is an example of a documentation of a wide variety of laboratory workflows.

Automated generation of AniML documents by analytical instruments

The device description, created in XML, must adhere to the Device Capability Dataset schema.[8] From python, it can be used with pyxml.

AniML schema

```
# AnIML Base Schema
# author: AnIML.org
# license: GNU Library or Lesser General Public License version 2.0 (LGPLv2)
# code: sourceforge.net/projects/animl
# activity: inactive (5 yrs)

<!--Root-Node of every AnIML document-->
<xsd:element nane="AnIML" type="AnIMLType">
<xsd:compLexType name="AnIMLType">
    <xsd:annotation>
    <xsd:Sequence>
        <xsd:element ref="SampleSet" minOccurs="0"/>
        <xsd:element ref="ExperimentStepSet" minOccurs="0"/>
```

```
        <xsd:element ref="AuditTrailEntrySet" minOccurs="0"/>
        <xsd:element ref="SignatureSet" minOccurs="0"/>
    </xsd:Sequence>
    <xsd:attribute name="version" type="ShortStringType"
                   use="required" fixed="0.37">
</xsd:complexType>
```

Allotrope.org

The Allotrope™ foundation has the goal to provide ontologies and data models that are used to create linked data and standardizing experimental parameters. Up to now (2020), there is no working Python interface. The use of the protocol must be licensed, it is required to be a member of the foundation.

References

1. Laboratory (Wikipedia) https://en.wikipedia.org/wiki/Laboratory.
2. Wiley-VCH Verlag GmbH. Data Exchange in the Laboratory of the Future | Laboratory Journal https://www.laboratory-journal.com/science/information-technology-it/data-exchange-laboratory-future.
3. How to implement SiLA (GitLab) https://gitlab.com/SiLA2/sila_base/wikis/How%20to%20 implement%20SiLA.
4. SiLA2 / sila_python (GitLab) https://gitlab.com/SiLA2/sila_python.
5. Dörr, M. labpy https://pypi.org/project/labpy/.
6. Home – AnIML https://animl.org/.
7. Schäfer, B. A.; Poetz, D.; Kramer, G. W. Documenting Laboratory Workflows Using the Analytical Information Markup Language https://journals.sagepub.com/doi/abs/10.1016/j.jala.2004.10.003. https://doi.org/10.1016/j.jala.2004.10.003.
8. Roth, A.; Jopp, R.; Schäfer, R.; Kramer, G. W. Automated Generation of Animl Documents by Analytical Instruments https://journals.sagepub.com/doi/abs/10.1016/j.jala.2006.05.013. https://doi.org/10.1016/j.jala.2006.05.013.

63 Laboratory: LIMS and daily calculations

A laboratory information management system (LIMS[1]), sometimes referred to as a laboratory information system (LIS) or laboratory management system (LMS), is a software-based solution with features that support a modern laboratory's operations. Key features include – but are not limited to – workflow and data tracking support, flexible architecture, and data exchange interfaces.

LabKey

LabKey[2] Server is an open-source software platform designed to help research organizations integrate, analyze, and share complex biomedical data. Adaptable to varying research protocols, analysis tools, and data-sharing requirements, LabKey Server couples the flexibility of a custom solution with enterprise-level scalability to support scientific workflows.

The Python API[3] enables query, insert, and update data on a LabKey Server.

```python
# LIMS interaction with LabKey
# author: labkey.com
# license: Apache License 2.0    # code: github.com/LabKey
# activity: active (2020)        # index: 63-1

from labkey.utils import create_server_context
from labkey.experiment import Batch, Run, load_batch, save_batch

# credentials
labkey_server = 'localhost:8080'
project_name = 'book_example'  # Project folder name
context_path = 'labkey'
server_context = create_server_context(labkey_server, project_name, context_
path, use_ssl=False)
assay_id = 3315  # example

# example data
runTest = Run()
runTest.name = 'python upload'
runTest.data_rows = [{
    'SampleId': '08-R-56-1',
    'TimePoint': '2020/01/02 09:10:01',
    'DoubleData': 220,
    'HiddenData': 'Stirrer_1 RPM'
}}]
```

https://doi.org/10.1515/9783110629453-063

```python
# Save an Assay batch
runTest.properties['RunFieldName'] = 'Run Field Value'
# Generate the Batch object(s)
batch = Batch()
batch.runs = [runTest]
batch.name = 'python batch'
batch.properties['PropertyName'] = 'Property Value'
# Execute save api
saved_batch = save_batch(server_context, assay_id, batch)

# Load an Assay batch
batch_id = saved_batch.row_id  # provide one from your server
run_group = load_batch(server_context, assay_id, batch_id)

print("Batch Id: " + str(run_group.id))
print("Created By: " + run_group.created_by)
```

eLabFTW

It is a free and open-source electronic lab notebook[A] (ELN) designed by researchers, for researchers, with usability in mind. With eLabFTW, you get a secure, modern, and compliant system to track your experiments efficiently but also manage your lab with a powerful and flexible database.

```python
# LIMS interaction with eLabFTW
# author: elabftw.net;  Carpi, Nicolas
# License: GNU General Public License v3.0  # code: github.com/elabftw
# activity: active (2020)                    # index: 63-2

!pip install -U elabapy
import elabapy

API_KEY = "651cf...39be3"
ENDPOINT = "https://elabftw.example.com/api/v1/"
elab = elabapy.Manager(token=API_KEY, endpoint=ENDPOINT)

# create an experiment
new = elab.create_experiment()
exp_id = new["08-R-56-1"]

my_exp = elab.get_experiment(exp_id)
title, body = someFunctionThatGetsDataFromSomething()
params = {"title": title, "date": today_date, "body": body}
out = elab.post_experiment(exp_id, params)
print(f"Updating experiment with id {exp_id}: {out['result']}")
```

```
# create an item
# request available ids for the item type (as these are dynamic)
items_types = elab.get_items_types()
print(json.dumps(items_types, indent=4))
print(items_types[0]["category_id"])

# create item; the parameter here is the id of an item_type
my_item = elab.create_item(items_types[0]["category_id"])
item_id = my_item["id"]
print(f"New item created with id: {item_id}")

# now update it
out = elab.post_item(item_id, params)
# attach a file to it
my_file = {"file": open("lab_notes.wav", "rb")}
elab.upload_to_item(item_id, my_file)

# now link our experiment to that database item
params = {"link": item_id}
elab.add_link_to_experiment(exp_id, params)
# and add a tag to our experiment
params = {"tag": "automated entry"}
elab.add_tag_to_experiment(exp_id, params)
```

Daily lab calculations

pyEQL and pint

A useful Python library that provides tools for modeling aqueous electrolyte solutions is pyEQL[5]. It allows the user to manipulate solutions as Python objects, providing methods to populate them with solutes, calculate species-specific properties (such as activity and diffusion coefficients), and retrieve bulk properties (such as density, conductivity, or volume).

```
# Calculations with aqueous electrolyte solutions (pyEQL)
# author: Kingsbury, Ryan S.
# License: GNU General Public License v3.0   # code: pyeql.readthedocs.io
# activity: active (2020)            # index: 63-3

import pyEQL
# no arguments are specified, pyEQL creates a 1-L solution of water at pH 7
and 25 degC.
s1 = pyEQL.Solution()
s1
> <pyEQL.pyEQL.Solution at 0x7f9d188309b0>
```

```
print(s1.get_volume())
print(s1.get_density())

from pyEQL import unit
test_qty = pyEQL.unit('1 kg/m**3')

import pint
ureg = pint.UnitRegistry()
print([3, 4] * ureg.meter + [4, 3] * ureg.cm)
```

Buffer calculation with ionize

ionize[6] calculates the properties of individual ionic species in aqueous solution, as well as aqueous solutions containing arbitrary sets of ions.

The ionize model takes into account pH, ionic strength, and temperature effects. The object classes make these techniques directly accessible as a backend for simulations. Ionize is composed of three main components:

- **Ion Class** which represents a single ionic species. An ion contains a name, ionization states charge, a pKa, mobility, ΔH, and ΔCp, and temperature.
- **Solution Class** is used to represent an aqueous solution containing any number of ionic species.
- **Database**

```
# Calculations with aqueous electrolyte solutions (ionize)
# author: Marshall, Lewis A.
# License: GNU General Public License v2.0    # code: ionize.readthedocs.io
# activity: active (2019)                     # index: 63-4


!pip install ionize


n = 30
pKa_list = numpy.linspace(3, 13, n)
c_insult_list = numpy.linspace(-.2, .2, n)
tris = ionize.load_ion('tris')
chloride = ionize.load_ion('hydrochloric acid')
buffer = ionize.Solution([tris, chloride], [0.1, 0.05])
pH = numpy.zeros([n, n])

for i, pKa in enumerate(pKa_list):
    for j, c_insult in enumerate(c_insult_list):
        insult = ionize.Ion('insult', [int(math.copysign(1, c_insult))],
                            [pKa], [math.copysign(20e-9, c_insult)])
        new_buffer = buffer + ionize.Solution([insult], [abs(c_insult)])
        pH[j,i] = new_buffer.pH
```

```
levels = linspace(2.5, 13.5, 24)

f = contourf(pKa_list, c_insult_list, pH, cmap=plt.get_cmap('bwr'),
levels=levels)
contour(pKa_list, c_insult_list, pH, colors='k', levels=levels)
colorbar(f)
xlabel('insult pKa')
ylabel('input concentration (M) * z')
title('pH after ion insult')
show()
```

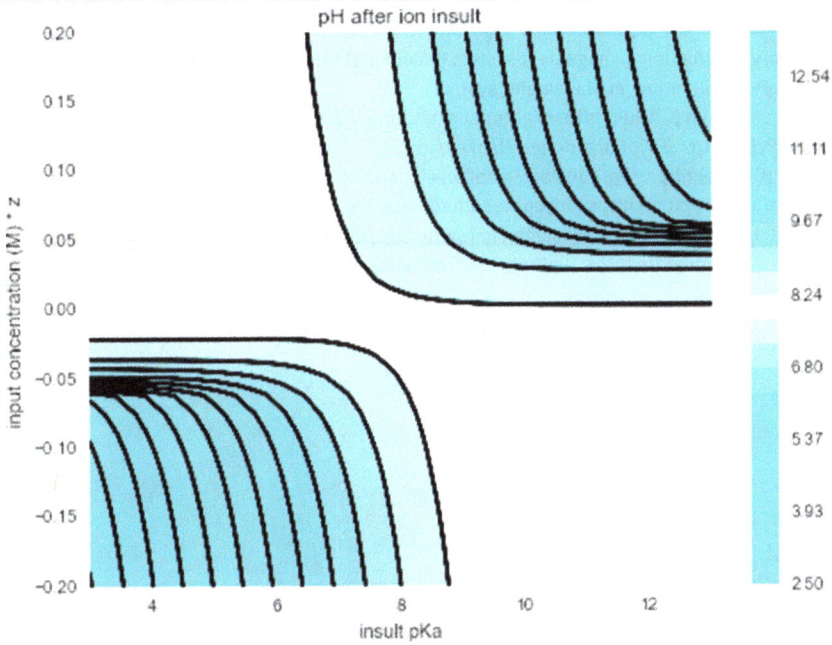

Figure 63.1: pH after ion insult *(Marshall)*.

Electrophoresis

Emigrate[7] is an electrophoresis simulator written in python. It is based on the scipy.integrate.ode solver. It is written with multiple simulation modules to allow varying solution schemes in order to allow an easy extension.

DoE – Design of Experiments

Calculations to the design of experiments (DoE) are discussed in chapter 77. DoE is used to find the minimal needed experimental resources to cover the observables of the chemical design space, Within the design space for example the Critical Process Parameters (CPP) are described.

References

1. Laboratory information management system (Wikipedia) https://en.wikipedia.org/wiki/Laboratory_information_management_system.
2. labkey-api-python (GitHub) https://github.com/LabKey/labkey-api-python.
3. labkey (PyPI) https://pypi.org/project/labkey/.
4. elabapy (GitHub) https://github.com/elabftw/elabapy.
5. pyEQL's documentation https://pyeql.readthedocs.io/en/latest/index.html.
6. Marshall, L. A. ionize http://lewisamarshall.github.io/ionize/.
7. Marshall, L. A. Emigrate https://github.com/lewisamarshall/emigrate .

64 Laboratory: robotics and cognitive assistance

A study by Richard H.R. Harper, involving two laboratories, elucidates the concept of *social organization* in laboratories.[1] The main subject of the study revolved around the *relationship* between the staff of a laboratory (researchers, administrators, receptionists, technicians, etc.) and their locator.

Laboratory design and observable coverage

About 90% of all task wet laboratories are repetitive manual tasks, short documentation fragments, and manual inventory management. No integrated IT systems, as well as seamless robotic solutions, support the worker. Data integration like Fieldbus or standardized exchange formats are not publicly available. In the last decade, several attempts for developing *Laboratory 4.0* were taken.

The implementation of cognitive chemometric[2] design principles extends the observable space in daily laboratory work. Assistance systems increase coverage and lead to better reproducibility of the experiments. It assists the worker for integration by the user's perspective.[3,4] The implementation of cognitive chemometric design principles extends the observable space in daily laboratory work. Assistance systems increase coverage and lead to better reproducibility of the experiments.

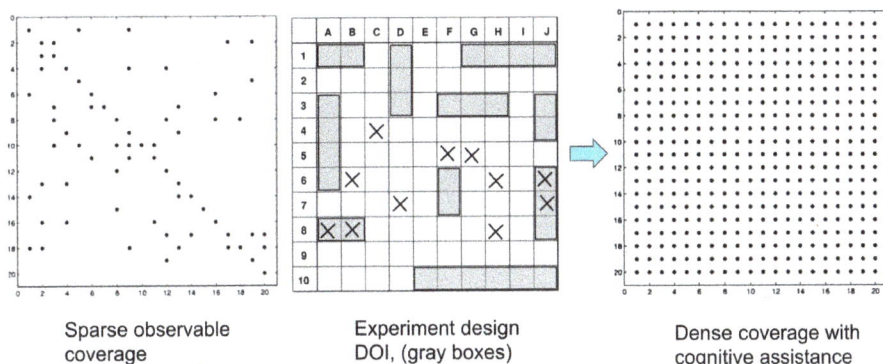

| Sparse observable coverage | Experiment design DOI, (gray boxes) | Dense coverage with cognitive assistance |

Figure 64.1: Cognitive chemometrics systems increase the observable space *(Gressling)*.

Opentrons robotic

The company[5,6] provides a common platform to share protocols and reproduce each scientist's results. The robots[7] automate experiments that would otherwise be done by hand, allowing the scientist to reduce manual workload.

https://doi.org/10.1515/9783110629453-064

```python
# Pipette roboting with opentrons
# author: Opentrons
# license: Apache License 2.0        # code: docs.opentrons.com
# activity: active (2020)            # index: 64-1

from opentrons import protocol_api

# metadata
metadata = {
    'protocolName': 'My Protocol',
    'author': 'Name <email@address.com>',
    'description': 'Simple protocol to get started using OT2'
}

# protocol run function. the part after the colon lets your editor know
# where to look for autocomplete suggestions
def run(protocol: protocol_api.ProtocolContext):

    # labware
    plate = protocol.load_labware('corning_96_wellplate_360ul_flat', '2')
    tiprack = protocol.load_labware('opentrons_96_tiprack_300ul', '1')
    # pipettes
    left_pipette = protocol.load_instrument(
        'p300_single', 'left', tip_racks=[tiprack])
    # commands
    left_pipette.pick_up_tip()
    left_pipette.aspirate(100, plate['A1'])
    left_pipette.dispense(100, plate['B2'])
    left_pipette.drop_tip()
```

Opentrons 96 Tip Rack 300 μL

TIP COUNT	96

MAX VOLUME

300 μL

FOOTPRINT (mm) ⓘ

LENGTH	127.76
WIDTH	85.48
HEIGHT	64.49

API NAME

opentrons_96_tiprack_300ul

TIP MEASUREMENTS (mm) ⓘ

TOTAL LENGTH	59.30
DIAMETER	5.23

Figure 64.2: Opentrons Tip Rack *(opentrons)*.

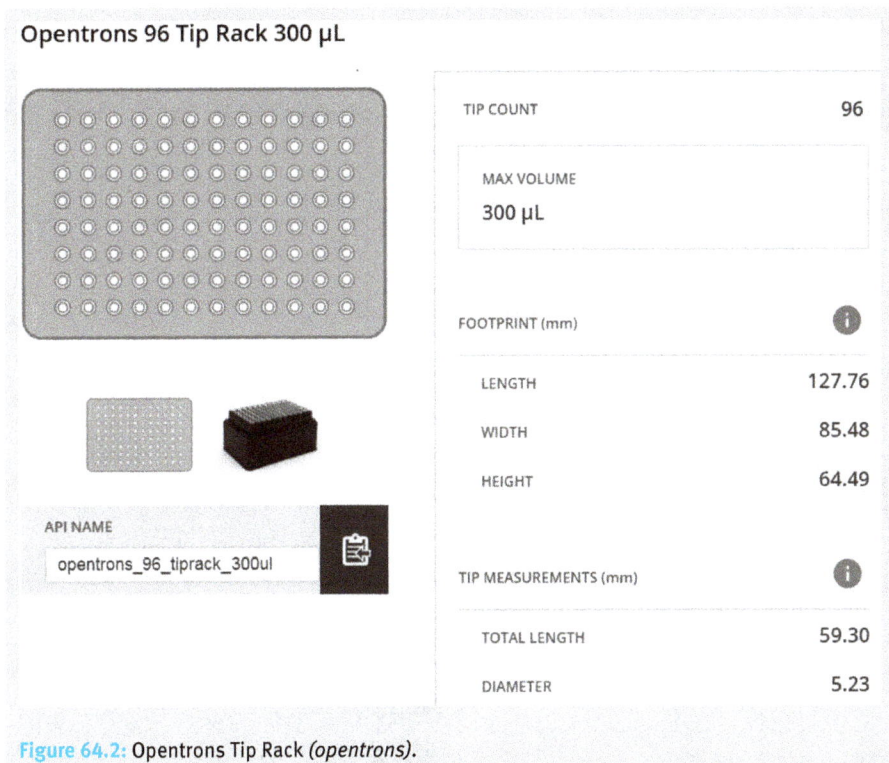

Assistance systems: labtwin and cleverlab

Labtwin[8] and cleverlab[9] are laboratory assistance systems that claim to provide *cognitive assistance* like speech-to-text, remote lab equipment control, and visual recognition. Both are in technical alpha status and proprietary but indicate the development of lab assistance systems of the future.

Python example for cognitive laboratory work

```
# Cognitive laboratory assistance (note-taking to ELN, speech)
# author: Gressling, T
# license: MIT License  # code: github.com/gressling/examples
# activity: active (2020)     # index: 64-2

from cognitiveLab import speech, ELN
```

```
ELN.connect(url, <<API Key>>)
df = speech.getWords("spokenText.wav") # obtained by device, hotword detection
ELN.parse(df).add # get meaning (simple IF/Then) and write to lab journal
ELN.listEntry.head()

> Id: 7
> Timestamp: 2020.0512.12.34.03
> Content: "Sample 267A45 contains blue crystals"
> Id: 6
> Timestamp: 2020.0512.11.43.03
> Content: "On 11:43 am stirrer stopped"
```

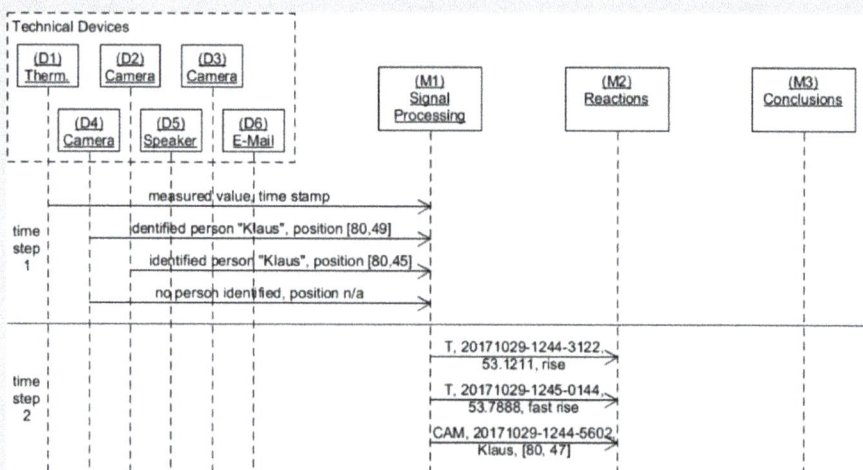

Figure 64.3: Cognitive layers: Devices layer 0, Lab situation layer 1, Cognitive layer 2 (Gressling[10]).

References

1. Laboratory (Wikipedia) https://en.wikipedia.org/wiki/Laboratory.
2. Gressling, T. Cognitive Chemometrics https://www.researchgate.net/publication/340687108_Cognitive_Chemometrics_-_how_assistant_systems_increase_observable_coverage_in_the_laboratory.
3. Gressling, T. Autochemistry: A Research Paradigm Based on Artificial Intelligence and Big Data. **2019**. https://doi.org/10.26434/chemrxiv.7264103.v1.
4. Gressling, T.; Madl, A. A New Approach to Laboratory 4.0: The Cognitive Laboratory System.
5. opentrons (GitHub) https://github.com/Opentrons/opentrons.
6. Opentrons | Open-source Pipetting Robots for Biologists https://opentrons.com/.

7. Running the Robot Using Jupyter Notebook https://support.opentrons.com/en/articles/1795367-running-the-robot-using-jupyter-notebook
8. LabTwin GmbH. LabTwin – Digital Lab Assistant https://www.labtwin.com.
9. CleverLab – cognitive laboratory assistance (ARS Computer und Consulting GmbH) https://www.cleverlab.ai/.
10. Thurner, V.; Gressling, T. A Multilayer Architecture for Cognitive Systems http://www.thinkmind.org/download.php?articleid=cognitive_2018_5_10_40030.

65 Chemical engineering

Chemical engineering[1] (CE) is a branch of engineering that uses principles of physics, mathematics, biology, and economics to efficiently use, produce, design, transport, and transform energy and materials.

Python in CE

Out from the huge list of about 85 chemical process simulators,[2] just a few mention python with none or small support:
- Comsol[TM]: currently no python[3]
- Aspen[TM] only via COM[4]
- pyANSYS scripts for automating ANSYS Fluent and SolidWorks Macro with Python[5]

REFPROP

For using the API accessing the NIST Standard Reference Database 23 (a.k.a. Refprop), there is a python package[6] for calculating thermodynamic properties using the Refprop NIST application:

```
# Boiling temperature of water (REFPROP)
# author: Bell, Ian
# license: NIST - not subject to copyright protection in the US
# code: github.com/usnistgov/REFPROP-wrappers
# activity: single example                    # index: 65-1


from ctREFPROP.ctREFPROP import REFPROPFunctionLibrary
import glob
import os
os.environ['RPPREFIX'] = r'D:/Program Files (x86)/REFPROP'


# The classical first example, calculating the boiling temperature of water
# at one atmosphere (101325 Pa) in K
p_Pa = 101325
Q = 0.0
r = RP.REFPROPdll("Water","PQ","T",MOLAR_BASE_SI,0,0,p_Pa,Q,[1.0])
r.Output[0]
```

https://doi.org/10.1515/9783110629453-065

CoolProp

coolProp[7] is a package for calculating thermodynamic properties using multiparameter equations of state. Written in C++, it implements in Python:

- Pure and pseudo-pure fluid equations of state and transport properties for 122 components
- Mixture properties using high-accuracy Helmholtz energy formulations
- Correlations of properties of incompressible fluids and brines
- High-accuracy psychrometric routines
- NIST REFPROP
- Fast IAPWS-IF97 (industrial formulation) for water/steam
- Cubic equations of state (SRK, PR)

```python
# Compute special values in SI units with CoolProp
# author: CoolProp.org
# License: MIT License        # code: coolprop.org/coolprop/wrappers/Python
# activity: active (2020)     # index: 65-2


!pip install CoolProp
import CoolProp.CoolProp as CP
fluid = 'Water'
pressure_at_critical_point = CP.PropsSI(fluid, 'pcrit')
# Massic volume (in m^3/kg) is the inverse of density
# (or volumic mass in kg/m^3). Let's compute the massic volume of liquid
# at 1bar (1e5 Pa) of pressure
vL = 1/CP.PropsSI('D','P',1e5,'Q',0,fluid)
# Same for saturated vapor
vG = 1/CP.PropsSI('D','P',1e5,'Q',1,fluid)


from CoolProp.CoolProp import Props
Rho = Props('D','T',298.15,'P',10000,'R744')
print("R744 Density at {} K and {} kPa      = {} kg/m³".format(298.15, 10000,
Rho))
```

Thermophysical data

```python
# Thermophysical Data and psychrometic chart (CoolProp)
# index: 65-3
# code: coolprop.org/dev/fluid_properties/HumidAir-1.py
# code: coolprop.org/fluid_properties/HumidAir


import CoolProp.CoolProp as CP
import matplotlib.pyplot as plt
```

```python
Tdbvec = np.linspace(-30, 55)+273.15

# Lines of constant relative humidity
for RH in np.arange(0.1, 1, 0.1):
    W = CP.HAPropsSI("W","R",RH,"P",101325,"T",Tdbvec)
    plt.plot(Tdbvec-273.15, W, color='k', lw = 0.5)

# Saturation curve
W = CP.HAPropsSI("W","R",1,"P",101325,"T",Tdbvec)
plt.plot(Tdbvec-273.15, W, color='k', lw=1.5)

# Lines of constant Vda
for Vda in np.arange(0.69, 0.961, 0.01):
    R = np.linspace(0,1)
    W = CP.HAPropsSI("W","R",R,"P",101325,"Vda",Vda)
    Tdb = CP.HAPropsSI("Tdb","R",R,"P",101325,"Vda",Vda)
    plt.plot(Tdb-273.15, W, color='b', lw=1.5
                if abs(Vda % 0.05) < 0.001 else 0.5)

# Lines of constant wetbulb
...
plt.xlabel(r'Dry bulb temperature $T_{\rm db}$ ($^{\circ}$ C)')
plt.ylabel(r'Humidity Ratio $W$ (kg/kg)')
plt.ylim(0, 0.030)
plt.xlim(-30, 55)
# plt.show()
```

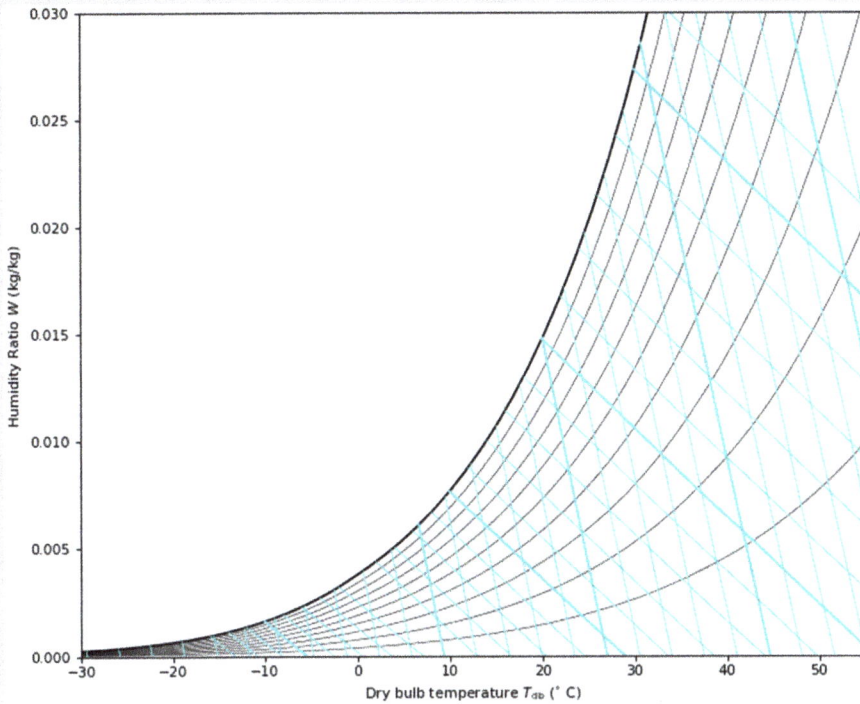

Figure 65.1: Psychrometric chart: humidity ratio versus bulb temperature.

References

1. Chemical engineering (Wikipedia) https://en.wikipedia.org/wiki/Chemical_engineering.
2. List of chemical process simulators (Wikipedia) https://en.wikipedia.org/wiki/List_of_chemical_process_simulators.
3. COMSOL + Python https://www.comsol.com/forum/thread/241931/comsol-python.
4. Kitchin, J. Running Aspen via Python https://kitchingroup.cheme.cmu.edu/blog/2013/06/14/Running-Aspen-via-Python/.
5. Horlock, A. pyANSYS (GitHub) https://github.com/Alex-Horlock/pyANSYS.
6. REFPROP-wrappers (GitHub) https://github.com/usnistgov/REFPROP-wrappers.
7. CoolProp 6.3.0 documentation http://www.coolprop.org/coolprop/wrappers/Python/index.html.

66 Reactors, process flow, and systems analysis

A chemical reactor is an enclosed volume in which a reaction takes place. In chemical engineering, it is generally understood to be a process vessel that is used to carry out a chemical reaction, which is one of the classic unit operations in chemical process analysis. Chemical engineers' design reactors to maximize the net present value for the given reaction. Normal operating expenses include energy input, energy removal, raw material costs, labor, and so on. Energy changes can come in the form of heating or cooling, pumping to increase pressure, frictional pressure loss, or agitation.[1]

Reaktoro

Reaktoro[2,3] is a computational framework developed in C++ and Python that implements numerical methods for modeling chemically reactive processes governed by either chemical equilibrium, chemical kinetics, or a combination of both.

```
# Solubility of CO2 in NaCl brines (reaktoro)
# author: Leal, Allan
# license: LGPL v2.1
# code: reaktoro.org/tutorials/equilibrium/co2-solubility
# activity: active (2020)          # index: 66-1

from reaktoro import *
# Initialize a thermodynamic database / generated from SUPCRT92 database file
db = Database('supcrt98.xml')


# Define the chemical system
editor = ChemicalEditor(db)
editor.addAqueousPhaseWithElements('H O Na Cl C')
editor.addGaseousPhase(['CO2(g)'])
# Construct the chemical system
system = ChemicalSystem(editor)


# Define the chemical equilibrium problem
problem = EquilibriumProblem(system)
problem.setTemperature(60, 'celsius')
problem.setPressure(100, 'bar')
problem.add('H2O', 1.0, 'kg')
problem.add('NaCl', 1.0, 'mol')
problem.add('CO2', 10.0, 'mol')
```

https://doi.org/10.1515/9783110629453-066

```
# Calculate the chemical equilibrium state
state = equilibrate(problem)

# The amounts of some aqueous species
print('Amount of CO2(aq):', state.speciesAmount('CO2(aq)'))
print('Amount of HCO3-:', state.speciesAmount('HCO3-'))
print('Amount of CO3--:', state.speciesAmount('CO3--'))
# The amounts of element C in both aqueous and gaseous phases
print('Amount of C in aqueous phase:', state.elementAmountInPhase
('C', 'Aqueous'))
print('Amount of C in gaseous phase:', state.elementAmountInPhase
('C', 'Gaseous'))
```

Example of reactor design

Interesting *interactive* Jupyter notebooks with simulations for chemical and process engineering courses[4] are provided by CAChemE.org. The following example shows the larger amount of code that still has to be written for the calculations. One reason is many preconditions that have to be set.

```
# Tube Diameter effect on Plug Flow Reactors
# author: Navarro, Franz (CAChemE.org)
# license:  MIT License        # code: github.com/CAChemE/Learn
# activity: active (2020)       # index: 66-2

interact_manual(diameter_PFR, diameter=(0.015, 0.3,0.001))
# diameter given in [m]
# Plug flow reactor model Gas-phase A --> B

# Reactor parameters
diameter = 0.3    # [m] Diameter of the tube reactor
diameter_0 = 0.3 # [m] Diameter initial tube reactor being studied
length = 50   # [m] Length of the tube reactor
A_cross = np.pi/4*diameter**2  # [m^2] Cross sectional aera tube reactor
A_cross_0 = np.pi/4*diameter_0**2  # [m^2] Cross sectional
number_reactors = A_cross_0 / A_cross

# Operating conditions
FA_0 =   100  # [mol m^{-3}] inlet molar flow rate
FA_0_n = 100/number_reactors # [mol m^{-3}] inlet molar flow rate within one
reactor
P_0 = 6.08e5  # [Pa] Pressure inlet
T_0 = 300  # [K] Temperature inlet
```

```python
def diffEqs(y, z, FA_0_n, T_0, P_0, A_cross):
    # y is a vector with the concentration
    # z is the indpendt variable (time)
    FA_z = y[0]  # [mol] molar flow (dependent variable)
    T_z  = y[1]  # [K] temperature (dependent variable)
    P_z  = y[2]  # [Pa] pressure (dependent variable)

    # Constants
...
    T_cool = 300  # [K] Cooling temperature inlet

    # Lenght dependent parameters
    vol_flow_0 = (FA_0_n * R_const * T_0) /  P_0  # [m^3/s] inlet gas velocity
    vol_flow = vol_flow_0 * FA_z/FA_0_n * P_0/P_z* T_z/T_0  # [m^3/s]
    volumetric flow

    # Mass balance
    Afactor = 0.0715  # [s^{-1}] pre-exponential factor
    Ea = 17000  # activation energy
...
    dFdz = -k * CA_z * A_cross # first-order exothermic reaction

    # Energy Balance
    dTdz = (Ua*(T_cool-T_z) - k*CA_z*A_cross*ΔH) / (FA_z * Cp)

    # Pressure drop
    NRe = (ρ * vol_flow/A_cross * diameter) / μ  # [-] Reynolds number
    if NRe <= 2300:
        f_factor = 64/NRe
    elif NRe > 2300:
        f_factor = 0.25 * np.log10(ε/(3.7*diameter) +  5.74/NRe**0.9)**-2
    dPdz = -ρ * f_factor/diameter * ((vol_flow/A_cross)**2/2)
    return [dFdz, dTdz, dPdz]

z  = np.linspace(0., length, 1000)
y0 = [FA_0_n, T_0, P_0] # [molar flow, temperature, pressure]
# We call the ode in order to intregate
Y = odeint(diffEqs, y0, z, args=(FA_0_n, T_0, P_0, A_cross))

# Taking back the solutions
F_A = Y[:, 0] # mol*L-1
T   = Y[:, 1] # mol*L-1
P   = Y[:, 2] # mol*L-1
F_A_total = F_A * number_reactors
T_Celsius= T - 273.15  # [K]->[ºC] Conversion of temperature
P_bar = P/1e5  # [Pa]->[bar] conversion of pressure
conversion = (1 - F_A[-1]/F_A[0]) * 100  # for the B component
```

```
# Plotting results
fig, (ax1, ax2, ax3) = plt.subplots(nrows=3, ncols=1,
                                     figsize=(4,8))
...
```

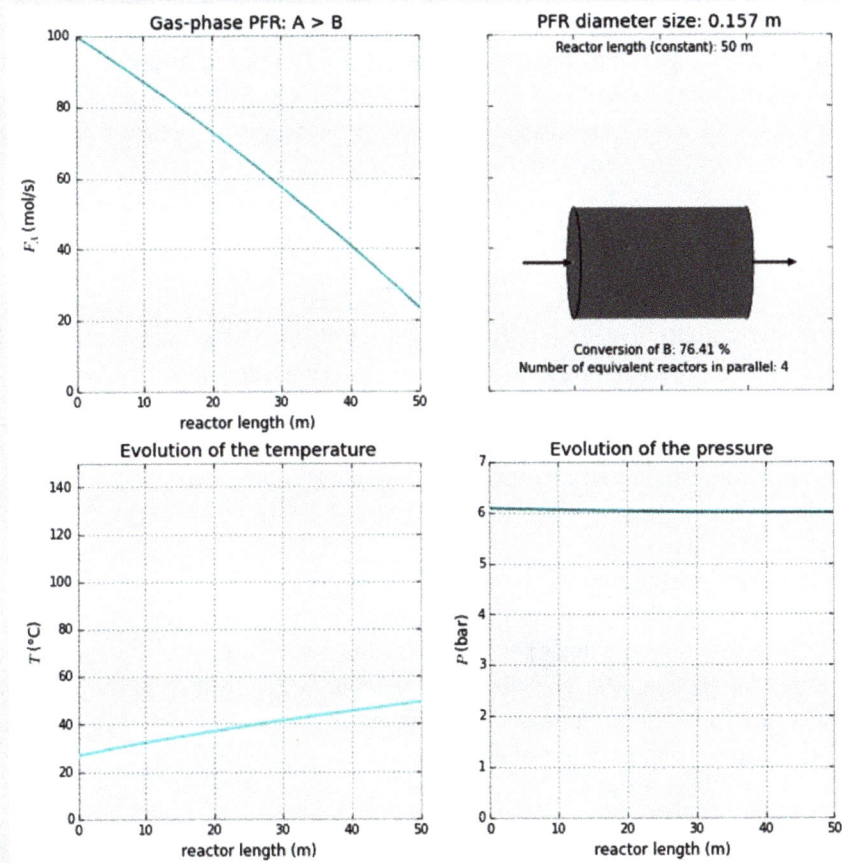

Figure 66.1: Tube diameter effect on plug flow reactors *(Navarro)*.

References

1. Chemical reactor (Wikipedia) https://en.wikipedia.org/wiki/Chemical_reactor.
2. reaktoro (GitHub) https://github.com/reaktoro/reaktoro.
3. Solubility of CO2 in NaCl brines – Reaktoro documentation https://reaktoro.org/tutorials/equilibrium/co2-solubility-nacl-brine.html.
4. Chemical and Process Engineering Interactive Simulations (CAChemE) https://github.com/CAChemE/learn.

67 Production: PLC and OPC/UA

A programmable logic controller[1] (PLC) or programmable controller is an industrial digital computer that has been ruggedized and adapted for the control of manufacturing processes, such as assembly lines, or robotic devices, or any activity that requires high reliability, ease of programming and process fault diagnosis. PLCs can range from small modular devices with tens of inputs and outputs (I/O), in a housing integral with the processor, to large rack-mounted modular devices with a count of thousands of I/O.

PLC/S7

Python is not suitable to play a role in PLC. PLCs run through their program hundreds of times per second, each time sampling inputs and computing new outputs. This is fast enough so mechanics effectively see smooth control.[2] When involving Python, it has to be considered:
–   Bus communication times to/from the PLC to the Python
–   Python wakeup time
–   Python execution time
–   Python message packing/unpacking time

All of this: several hundred times per second reliably (the OS can interrupt Python for other background processes). So Python in interaction with S7 *only acts in an advisory role*. There are a few sites proposing Python-based S7 simulators or Soft-PLC environments that can be connected to real PLC backplanes.

SNAP7

Snap7[3] is an open-source, 32/64-bit, multiplatform Ethernet communication suite for interfacing natively with Siemens S7 PLCs:

```
# Read multi vars at Siemens PLC (S7-319 CPU)
# author: Molenaar, Gijs
# license: MIT License  # code: github.com/gijzelaerr/python-snap7
# activity: active (2020)         # index: 67-1


import ctypes
import struct
```

https://doi.org/10.1515/9783110629453-067

```python
import snap7
from snap7.common import check_error
from snap7.snap7types import S7DataItem, S7AreaDB, S7WLByte

client = snap7.client.Client()
client.connect('10.100.5.2', 0, 2)

data_items = (S7DataItem * 3)()
data_items[0].Area = ctypes.c_int32(S7AreaDB)
data_items[0].WordLen = ctypes.c_int32(S7WLByte)
data_items[0].Result = ctypes.c_int32(0)
data_items[0].DBNumber = ctypes.c_int32(200)
data_items[0].Start = ctypes.c_int32(16)
data_items[0].Amount = ctypes.c_int32(4)  # reading a REAL, 4 bytes
...

# create buffers to receive the data
for di in data_items:
    buffer = ctypes.create_string_buffer(di.Amount)
    # cast the pointer to the buffer to the required type
    pBuffer = ctypes.cast(ctypes.pointer(buffer),
                          ctypes.POINTER(ctypes.c_uint8))
    di.pData = pBuffer

result, data_items = client.read_multi_vars(data_items)

for di in data_items:
    check_error(di.Result)

result_values = []
# function to cast bytes to match data_types[] above
byte_to_value = [util.get_real, util.get_real, util.get_int]

# unpack and test the result of each read
for i in range(0, len(data_items)):
    btv = byte_to_value[i]
    di = data_items[i]
    value = btv(di.pData, 0)
    result_values.append(value)
# print(result_values)

client.disconnect()
client.destroy()
```

OPC UA

Open Platform Communications[4] (OPC) is a series of standards and specifications for industrial telecommunication. An industrial automation task force developed the original standard in 1996 under the name Object Linking and Embedding for process control. OPC specifies the communication of real-time plant data between control devices from different manufacturers.

Python client example with FreeOPCUA

This software[5] is a pure Python OPC UA/IEC 62541 client and server. In this example, a connection to an OPC server is established:

```python
# OPC UA Python client example
# author: Free OPCUA Team
# license: LGPL-3  # code: github.com/FreeOpcUa/python-opcua
# activity: active (2020)          # index: 67-2

import sys
sys.path.insert(0, "..")
from opcua import Client

client = Client("opc.tcp://admin@localhost:4840/freeopcua/server/")
client.connect()
root = client.get_root_node()
print("Objects node is: ", root)
print("Children of root are: ", root.get_children())

# get a specific node knowing its node id
var = client.get_node(ua.NodeId(1002, 2))
var = client.get_node("ns=3;i=2002")

var.get_data_value() # get value of node as a DataValue object
var.get_value() # get value of node as a python builtin
var.set_value(ua.Variant([23], ua.VariantType.Int64))
#set node value using explicit data type
var.set_value(3.9) # set node value using implicit data type

client.disconnect()
```

References

1. Programmable logic controller (Wikipedia) https://en.wikipedia.org/wiki/Programmable_logic_
controller.
2. can python software take place of logic ladder program in PLC through modbus? (Stack Overflow)
https://stackoverflow.com/questions/34956823/can-python-software-take-place-of-logic-
ladder-program-in-plc-through-modbus.
3. python-snap7 (PyPI) https://pypi.org/project/python-snap7/.
4. Open Platform Communications (Wikipedia) https://en.wikipedia.org/wiki/Open_Platform_
Communications.
5. Freeopcua.github.io http://freeopcua.github.io/.

68 Production: predictive maintenance

Predictive maintenance[1] techniques are designed to help determine the condition of in-service equipment in order to estimate when maintenance should be performed. This approach promises cost savings over routine- or time-based preventive maintenance because tasks are performed only when warranted.

Thus, it is regarded as condition-based maintenance carried out as suggested by estimations of the degradation state of an item. The main promise of predictive maintenance is to allow convenient scheduling of corrective maintenance and to prevent unexpected equipment failures.

Predictive maintenance calculations are mostly done on big data architectures.

Example of a failure prediction

The turbofan engine degradation simulation data set from *Prognostics CoE* at *NASA Ames* is a famous showcase for calculating rocket engine malfunction. Four different sets were simulated under different combinations of operational conditions and fault modes. The data contain records of several sensor channels to characterize fault evolution. The *Turbofan Engine Degradation Simulation Data Sets* are free for download.[2]

Python example

This implementation uses the logistic regression (aka logit, MaxEnt) classifier. In statistics, the logistic model is used to model the probability of a certain class or event existing such as pass/fail, win/lose, alive/dead, or healthy/sick.

```
# Predictive Maintenance: Turbo-Fan Failure
# author: Gressling, T.
# license: MIT License          # code: https://github.com/gressling/examples
# activity: single example      # index: 68-1

# load data
import pandas as pd
col = ["id","cycle","setting1","setting2","setting3","s1","s2","s3","s4","s5",
    "s6","s7","s8","s9","s10","s11","s12","s13","s14","s15","s16","s17","s18",
    "s19","s20","s21"]
df_training = pd.read_csv("./CMAPSSData/train_FD001.txt",
            delim_whitespace = True, names = col, dtype = float)
df_test = pd.read_csv("./CMAPSSData/test_FD001.txt",
            delim_whitespace=True, names=col, dtype=float)
```

https://doi.org/10.1515/9783110629453-068

```python
df_max_cycles = pd.read_csv("./CMAPSSData/RL_FD001.txt",
            delim_whitespace=True, names=["malfunction"])

# prepare data
# calculate "CyclesLeft" (Residual lifespan, how long remaining)
df_max_cycles = pd.DataFrame(
        df_training.groupby(by="id")["cycle"].max().rename("max"))
df_with_max_cycles = df_training.merge(
        df_max_cycles, on="id", how='inner')
df_training["CyclesLeft"] = df_with_max_cycles["max"] -
                    df_with_max_cycles["cycle"]
df_training["malfunction"] = df_training.apply(
        lambda row: True if row["CyclesLeft"] <= 30 else False,
        axis='columns')
df_training = df_training.drop(["CyclesLeft", "id"], axis='columns')

# scale data
from sklearn.preprocessing import MinMaxScaler
scaler = MinMaxScaler()
exclude = ["malfunction"]
columns = [col for col in df_training.columns if col not in exclude]
df_training.loc[:, columns] =
        scaler.fit_transform(df_training.loc[:, columns])

# only rows with more than one value
columns = [col for col in df_training.columns
        if len(df_training.loc[:, c].unique()) > 1]
df_training = df_training.loc[:, columns]
# only last row from test data
df_test = df_test.loc[df_test.groupby('id').cycle.idxmax()]

# if engine will break -> 1, if not -> 0
df_max_cycles.index = df_max_cycles.index + 1
df_max_cycles['id'] = df_max_cycles.index
df_max_cycles['malfunction'] = df_max_cycles.apply(
            lambda row: 1 if row["malfunction"] <= 30 else 0, axis=1)

# merge malfunction and test
df_test = df_test.merge(df_max_cycles, on="id", how='inner')
df_test = df_test.drop(["id"], axis=1)
columns = [col for col in df_training.columns if col not in exclude]
df_test.loc[:, columns] = scaler.fit_transform(df_test.loc[:, columns])
df_test = df_test.loc[:, columns]

# prediction model
from sklearn.linear_model import LogisticRegression
from sklearn.metrics import confusion_matrix, accuracy_score, precision_
score, recall_score, f1_score
```

```
clf = LogisticRegression(solver='lbfgs')
clf.fit(df_training.loc[:, df_training.columns != "malfunction"],
        df_training["malfunction"])

# test
y_pred = clf.predict(df_test.loc[:, df_test.columns != "malfunction"])
y_true = df_test["malfunction"]

# metrics
scores = dict()
scores["Accuracy"] = accuracy_score(y_true, y_pred)
scores["Precision"] = precision_score(y_true, y_pred)
scores["Recall"] = recall_score(y_true, y_pred)
scores["F-Score"] = f1_score(y_true, y_pred)
scores
> 'Accuracy': 0.88,
> 'Precision': 0.6756756756756757,
> 'Recall': 1.0,
> 'F-Score': 0.8064516129032258

import seaborn as sbn
conf_matrix = confusion_matrix(y_true, y_pred)
heatmap = sbn.heatmap(conf_matrix, annot=True, cmap='Blues')
heatmap.set(ylabel='Actual')
heatmap.set(xlabel='Prediction')
```

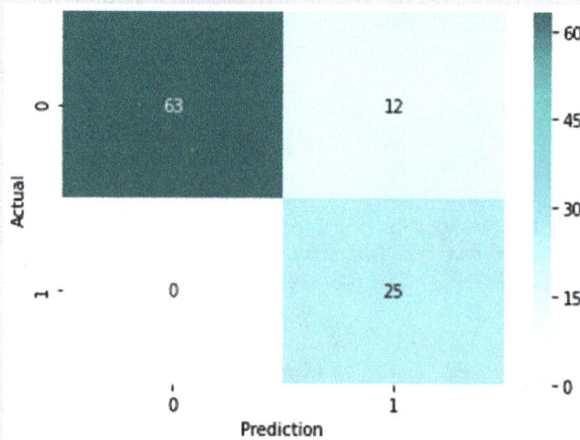

Figure 68.1: Fan degradation prediction (Gressling).

Confusion matrix interpretation:
-   OK: In the *first line*, there are **correctly** predicted 63 times **(0,0)** that the turbine would **not break** (Prediction = 0, not break, Actual = 0, not break) for the next 30 cycles.
-   Bad: But 12 times **(0,1)** there was a ***false prediction*** for failure (Prediction = 1, **will break**, Actual = 0, no break), even though it would not break.
-   OK: In the second line at **(1,1)**, the prediction = 1 (**will break**) was 25 times ***correctly*** (Actual = 1, with engine malfunction).
-   OK: And there was **correctly** no prediction for a good turbine status (**will not break** = 0) although an actual failure (= 1) occurred **(1,0).**

References

1. Predictive maintenance (Wikipedia) https://en.wikipedia.org/wiki/Predictive_maintenance.
2. Saxena, A.; Goebel, K. Turbofan Engine Degradation Simulation Data Set **2008** (NASA Ames Research Center, Moffett Field, CA,) https://ti.arc.nasa.gov/tech/dash/groups/pcoe/prognostic-data-repository/.

Part C: **Data science**
Data engineering in analytic chemistry

69 Titration and calorimetry

Titration (volumetric analysis) is a common laboratory method of quantitative chemical analysis to determine the concentration of an identified analyte (a substance to be analyzed). A reagent, called the titrant or titrator, is prepared as a standard solution of known concentration and volume. The titrant reacts with a solution of analyte (which may also be called the titrant) to determine the analyte's concentration. The volume of titrant that reacted with the analyte is called the titration volume.[1]

Calorimetry is the science or act of measuring changes in state variables of a body for the purpose of deriving the heat transfer associated with changes of its state due, for example, to chemical reactions, physical changes, or phase transitions under specified constraints. Calorimetry is performed with a calorimeter.[2,3]

pHcalc

pHcalc[4,5] is a Python library for systematically calculating solution pH, distribution diagrams, and titration curves[5]:

```
# Calculation of ph-Value, titration curve (phCalc)
# author: Ryan Nelson; (Johan Hjelm)
# license: MIT License    # code: gist.github.com/rnelsonchem
# activity: on hold (2017) # index: 69-1


from pHcalc.pHcalc import Acid, Neutral, System
import numpy as np
import matplotlib.pyplot as plt


# pH of 0.01 M; HCl is a strong acid, so we should expect complete dissociation.
cl = Neutral(charge=-1, conc=0.01)
system = System(cl)
system.pHsolve()
print(system.pH)
> 1.99999771118


# Titration of phosphoric acid
na_moles = np.linspace(1e-8, 5.e-3, 500)
sol_volume = 1. # Liter
phos = Acid(pKa=[2.148, 7.198, 12.375], charge=0, conc=1.e-3)
phs = []
```

https://doi.org/10.1515/9783110629453-069

```
for mol in na_moles:
    na = Neutral(charge=1, conc=mol/sol_volume)
    system = System(phos, na)
    system.pHsolve(guess_est=True)
    phs.append(system.pH)
plt.plot(na_moles, phs)
plt.show()
```

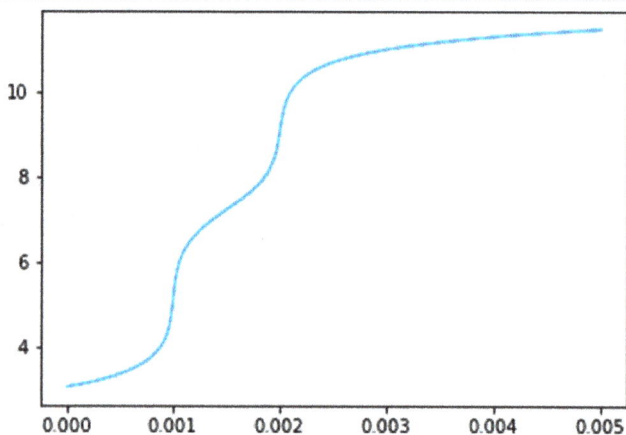

Figure 69.1: Titration curve of phosphoric acid *(Hjelm)*.

Isothermal titration and calorimetry experiments: pytc

Isothermal titration calorimetry (ITC) is a powerful technique for measuring the thermodynamics of intermolecular interactions. Pytc[6] is an open-source python package for extracting thermodynamic information from ITC experiments including Bayesian and ML fitting. Extracting thermodynamic information requires fitting models to observed heats, potentially across multiple experiments.[7,8]

```
# Calculate Ca2+/EDTA binding model (pytc)
# author: Harms lab (Harms, Mike)
# license: The Unlicense (Commercial use) # code: github.com/harmslab/pytc
# activity: active (2020)        # index: 69-2

import pytc
# Load in integrated heats from an ITC experiment
e = pytc.ITCExperiment("demos/ca-edta/tris-01.DH",
                       pytc.indiv_models.SingleSite)
# Create the global fitter, add the experiment
g = pytc.GlobalFit()
g.add_experiment(e)
```

```
# Update bounds and guess for "K".  Then fix fx_competent.
g.update_bounds("K",(1,1e12),a)
g.update_guess("K",1000,a)
g.update_fixed("fx_competent",1.0,a)

# Do the fit
g.fit()
# Print the results out
g.plot()
print(g.fit_as_csv)
```

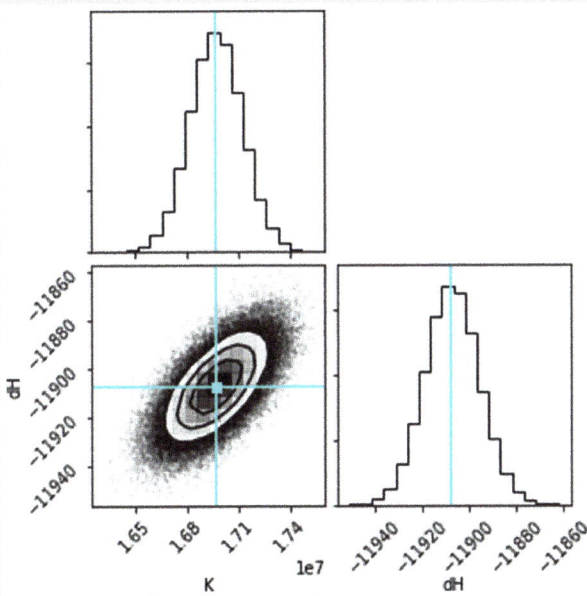

Figure 69.2: ITC: binding of calcium onto EDTA *(Harms)*.

References

1. Titration (Wikipedia) https://en.wikipedia.org/wiki/Titration.
2. Calorimetry (Wikipedia) https://en.wikipedia.org/wiki/Calorimetry.
3. Amaral Silva, D.; Löbenberg, R.; Davies, N. Are Excipients Inert? Phenytoin Pharmaceutical Investigations with New Incompatibility Insights. *J. Pharm. Pharm. Sci.* **2018**, *21* (1s), 29745. https://doi.org/10.18433/jpps29745.
4. Nelson, R. pHcalc (GitHub) https://github.com/rnelsonchem/pHcalc.
5. pHcalc https://pypi.org/project/pHcalc/.
6. Duvvuri, H.; Wheeler, L. C.; Harms, M. J. Pytc: A Python Package for Analysis of Isothermal Titration Calorimetry Experiments. *bioRxiv*, 2017, 234682. https://doi.org/10.1101/234682.
7. pytc 1.1.5 documentation https://pytc.readthedocs.io/en/latest/.
8. pytc (GitHub) https://github.com/harmslab/pytc.

70 NMR

Nuclear magnetic resonance (NMR) spectroscopy or magnetic resonance spectroscopy (MRS) is a spectroscopic technique to observe local magnetic fields around atomic nuclei. As the fields are unique or highly characteristic to individual compounds, NMR spectroscopy is the definitive method to identify monomolecular organic compounds.

NMR spectroscopy provides detailed information about the structure, dynamics, reaction state, and chemical environment of molecules.[1]

FOSS[2] for spectroscopy lists 20+ python packages[3,4] for handling NMR data and their interpretation. An example is shown here with processing data from Bruker.

nmrglue

nmrglue[5,6] is a module for working with NMR data in Python. Used with the NumPy, SciPy, and matplotlib packages, nmrglue provides an environment for rapidly developing new methods for processing, analyzing, and visualizing NMR data. nmrglue also provides a framework for connecting existing NMR software packages.

Example: processing 1D Bruker data

nmrglue has the ability to read, write, and convert between a number of NMR file formats including Agilent/Varian, Bruker, NMRPipe, Sparky, SIMPSON, and Rowland NMR Toolkit files.[7] This example of a 1D NMR spectrum of 1,3-diaminopropane is processed, and the results can be compared later with the spectrum produced from NMRPipe.

```
# Processing 1D Bruker NMR data (nmrglue)
# author: Helmus, Jonathan J.
# license: 3-Clause BSD License  # code: nmrglue.readthedocs.io/.../proc_bruker_1d
# activity: active (2020)      # index: 70-1


import nmrglue as ng
import matplotlib.pyplot as plt


# read in the bruker formatted data
dic, data = ng.bruker.read('expnmr_00001_1')
# remove the digital filter
data = ng.bruker.remove_digital_filter(dic, data)
```

https://doi.org/10.1515/9783110629453-070

```
data = ng.proc_base.zf_size(data, 32768)    # zero fill to 32768 points
data = ng.proc_base.fft(data)               # Fourier transform
data = ng.proc_base.ps(data, p0=-50.0)      # phase correction
data = ng.proc_base.di(data)                # discard the imaginaries
data = ng.proc_base.rev(data)               # reverse the data


fig = plt.figure()
ax = fig.add_subplot(111)
ax.plot(data[20000:25000])
fig.savefig('figure_nmrglue.png')
```

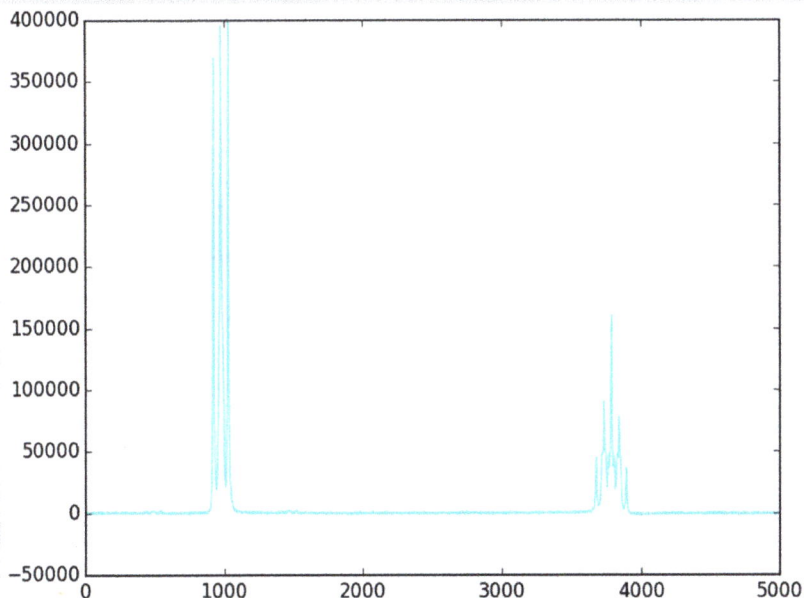

Figure 70.1: One-dimensional NMR spectrum of 1,3-diaminopropane *(Helmus)*.

References

1. Nuclear magnetic resonance spectroscopy (NMR) (Wikipedia) https://en.wikipedia.org/wiki/Nuclear_magnetic_resonance_spectroscopy.
2. Bryan A. Hanson, DePauw University. FOSS for Spectroscopy https://bryanhanson.github.io/FOSS4Spectroscopy/.
3. Sparky Manual https://www.cgl.ucsf.edu/home/sparky/manual/extensions.html.
4. peakipy 0.1.16 documentation https://j-brady.github.io/peakipy/build/index.html.
5. Helmus, J. nmrglue – A module for working with NMR data in Python https://www.nmrglue.com/.
6. Helmus, J.; Jaroniec, C. P. Nmrglue: An Open Source Python Package for the Analysis of Multidimensional NMR Data. *J. Biomol. NMR* **2013**, *55* (4), 355–367. https://doi.org/10.1007/s10858-013-9718-x.
7. nmrglue 0.7 documentation https://nmrglue.readthedocs.io/en/latest/index.html.

71 X-ray-based characterization: XAS, XRD, and EDX

X-ray absorption fine structure (XAFS) is a specific structure observed in X-ray absorption spectroscopy (XAS).[1] Their spectra contain information about an element's chemical state and local atomic coordination. It can describe any element in the local atomic coordination as well as the chemical state.

The advantage is that it works at low concentrations and with minimal sample requirements. For unknown samples, the oxidation state can thus be determined by comparison with appropriate, structurally related references.

XAS and XAFS

This code example[2] shows how to handle XAS[1] data and their plot:

```python
# X-Ray Absorption Spectroscopy, working on spectra
# author: Plews, Michael
# license: MIT License       # code: github.com/michaelplews/materials-research
# activity: active (2020)   # index: 71-1

# Supported Importers
# Object          Beamline           Facility
# IDC4   4-ID-C  APS, Argonne National Laboratory
# ALS6312        6.3.1    ALS,  Lawrence Berkeley National Laboratory
# ALS801         8.0.1    ALS,  Lawrence Berkeley National Laboratory
from . import XAS as xas

# load
sample_a = xas.IDC4(dire, base, start="248", end="250", shortname ='Sample A')
sample_b = xas.IDC4(dire, base, start="244", end="246", shortname = 'Sample B')

# analyse
# Find the peak max value between 705 eV and 710 eV and set the x value to 'x_target'
x_target, y = sample_a.max_in_range('STD', low=705,
                    high=710, plot=False, do_return=True)
# Assign the sample_b peak max value to x1 within the same range
x1, y1 = sample_b.max_in_range('STD', low=705, high=710,
                    plot=False, do_return=True)
# Align the sample_b object to sample_a
sample_b.align(x1, x_target)

# Plot (STD data)
fig = plt.figure(1, figsize=(16, 4))
sample_a.plot('STD', color='black')
sample_b.plot('STD', color='red')
plt.xlabel('Energy / eV', fontsize=15)
show()
```

https://doi.org/10.1515/9783110629453-071

XRD

X-ray crystallography (XRC) is the experimental science determining the atomic and molecular structure of a crystal, in which the crystalline structure causes a beam of incident X-rays to diffract into many specific directions.[3]

The next example shows how a BrukerBrmlFile object imports `.brml` files taken on a Bruker D8 Advance diffractometer (credit to m3wolf/scimap project). The ICDDXmlFile object imports .xml files exported from "PDF-2 2013" software to add line patterns for reference materials[2]:

```
# X-Ray diffraction (XRD) and working on spectra
# author: Plews, Michael
# license: MIT License      # code: github.com/michaelplews/materials-research
# activity: active (2020)  # index: 71-2


from . import XRD as xrd

# Load
my_sample = xrd.BrukerBrmlFile("./path/to/file.brml",
                               shortname='LiF as purchased')
ref_LiF = xrd.ICDDXmlFile("./path/to/file.xml")

# Plot
fig = plt.figure(1, figsize=(16, 5))
plt.subplot(211)
my_sample.plot(color='black')
hide_x_axis() # a custom function not included in this repo
plt.subplot(212)
ref_LiF.plot('red')
adjust_axes(10,80,0,110) # a custom function not included in this repo
plt.xlabel('2$\theta$ / $^\circ$', fontsize=20)
show()
```

XRD spectrum prediction

Python example with ipyMD

MD and MM were discussed in Chapters 47 to 55. An application of MD simulation is the calculation of XRD spectra:

```
# Calculation of XRD spectrum using MD (ipyMD)
# author: Sewell, Chris
# license: GNU GPL3         # code: ipymd.readthedocs.io/.../tutorial
# activity: on hold? (2017)  # index: 71-3
```

```
data = ipymd.data_input.crystal.Crystal()
data.setup_data(
    [[0,0,0]], ['Fe'],
    229, cellpar=[2.866, 2.866, 2.866, 90, 90, 90],
    repetitions=[5,5,5])
meta = data.get_meta_data()
atoms_df = data.get_atom_data()

wlambda = 1.542 # Angstrom (Cu K-alpha)
thetas, Is = ipymd.atom_analysis
                    .spectral.compute_xrd(atoms_df,meta,wlambda)
plot = ipymd.atom_analysis.spectral.plot_xrd_hist(thetas,Is,
        wlambda=wlambda,barwidth=1)
plot.axes.set_xlim(20,90)
plot.display_plot(True)
```

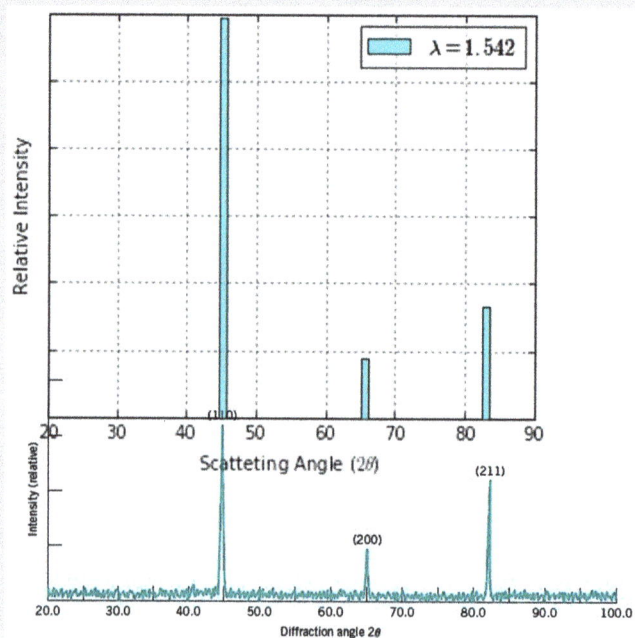

Figure 71.1: a-FE (calc.) and (obs.) *(Sevell).*

EDS, EDX

Energy-dispersive X-ray spectroscopy relies on the interaction of X-ray excitation and a sample. To stimulate the emission of characteristic X-rays from a specimen, a high-energy beam of charged particles such as electrons or protons or a beam of X-rays is focused into the sample. At rest, an atom within the sample contains ground state (or unexcited) electrons in discrete energy levels or electron shells bound to the nucleus.[4]

EDS/SEM can be analyzed with the HyperSpy[5] python package. It is an open-source Python library providing tools to interactive data analysis of multidimensional datasets. A spectrum is a 2D array of a given signal. The following example[6] shows a 4D EDS/SEM experiment using a FIB/SEM (Helios FEI).

```python
# EDS (EDX) spectrum analysis with hyperspy
# author: Burdet, Pierre⁶
# license: GPL 3.0    # code: github/hyperspy/.../electron_microscopy/EDS/SEM_EDS_4D
# activity: active (2020)    # index: 72-4

import hyperspy.api as hs

# Load data
s = hs.load("Ni_superalloy_0*.rpl", stack=True)
s.metadata.General.title = '3D EDS map'
s = s.as_signal1D(0)
s.set_signal_type('EDS_SEM')
s.plot()

# Analysis
# The position of the X-ray lines and the maps of X-ray lines intensity can
be visualized.
elements = ['Al', 'C', 'Co', 'Cr', 'Mo', 'Ni', 'Ta', 'Ti']
s.set_elements(elements)
s.sum().plot(xray_lines=True)

# PCA can be used to de-noise the spectrum
s.change_dtype('float')
s.decomposition(normalize_poissonian_noise=True)
s.plot_explained_variance_ratio()
```

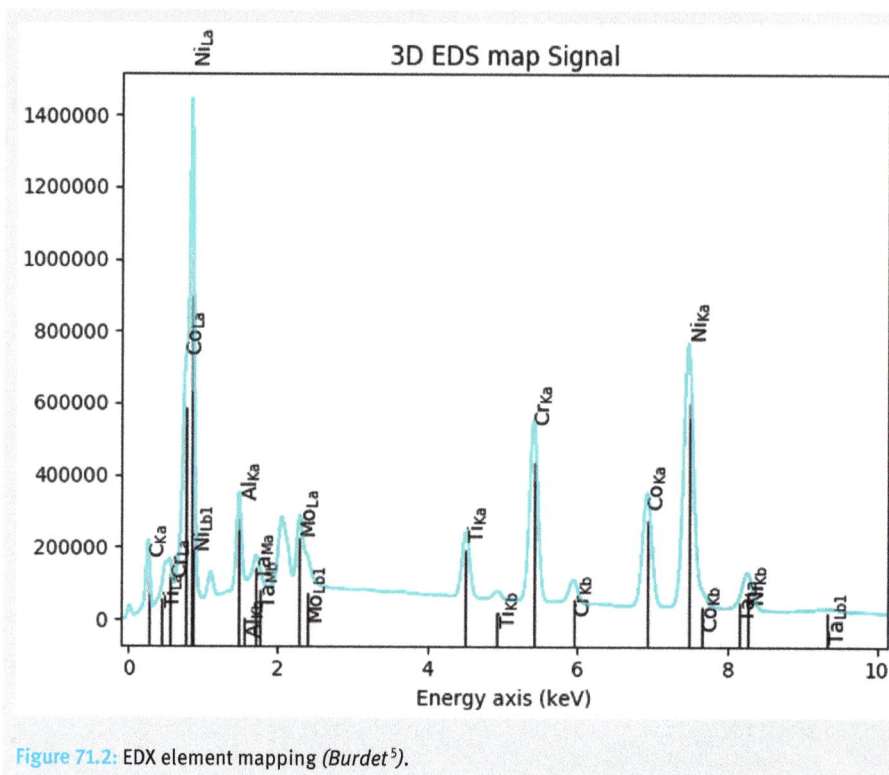

Figure 71.2: EDX element mapping (Burdet[5]).

References

1. X-ray absorption fine structure (XAS) (Wikipedia) https://en.wikipedia.org/wiki/
   X-ray_absorption_fine_structure.
2. Plews, M. materials-research (GitHub) https://github.com/michaelplews/materials-research.
3. X-ray crystallography (Wikipedia) https://en.wikipedia.org/wiki/X-ray_crystallography.
4. EDX. Energy-dispersive X-ray spectroscopy (Wikipedia) https://en.wikipedia.org/wiki/Energy-
   dispersive_X-ray_spectroscopy.
5. HyperSpy. Energy-Dispersive X-ray Spectrometry (EDS) http://hyperspy.org/hyperspy-doc/
   current/user_guide/eds.html.
6. Burdet, P. 4D EDS/SEM experiment analysis https://nbviewer.jupyter.org/github/hyperspy/
   hyperspy-demos/blob/master/electron_microscopy/EDS/SEM_EDS_4D_visualisation.ipynb.

72 Mass spectroscopy

Mass spectrometry (MS) is an analytical technique that measures the mass-to-charge ratio of ions. MS is used in many different fields and is applied to pure samples as well as complex mixtures.

A mass spectrum is used to determine the elemental or isotopic signature of a sample and to elucidate the chemical identity or structure of molecules and other chemical compounds.[1]

MS/MS spectrum processing and visualization in Python

spectrum_utils[2] is a Python package for MS data processing and visualization. It provides spectrum processing with functions like noise peak removal, intensity filtering, and peak annotations (SMILES):

```python
# MS/MS spectrum processing and visualization (spectrum_utils)
# author: Bittremieux, Wout
# license: Apache License 2.0
# code: spectrum-utils.readthedocs.io/.../quickstart
# activity: active (2020)      # index: 72-1


import matplotlib.pyplot as plt
import spectrum_utils.spectrum as sus
import spectrum_utils.plot as sup
from pyteomics import mgf

spectrum_dict = mgf.get_spectrum('spectra.mgf', 'CCMSLIB00000840351')
identifier = spectrum_dict['params']['title']
precursor_mz = spectrum_dict['params']['pepmass'][0]
precursor_charge = spectrum_dict['params']['charge'][0]
mz = spectrum_dict['m/z array']
intensity = spectrum_dict['intensity array']
spectrum = sus.MsmsSpectrum(identifier, precursor_mz,
                            precursor_charge, mz, intensity)
spectrum.filter_intensity(0.05)

charge, tol_mass, tol_mode = 1, 0.5, 'Da'
annotate_fragment_mz = [133.102, 147.080, 195.117, 237.164, 267.174, 295.170,
                        313.181, 355.192, 377.172, 391.187, 451.209, 511.231,
                        573.245, 633.269]
```

https://doi.org/10.1515/9783110629453-072

```
for fragment_mz in annotate_fragment_mz:
    spectrum.annotate_mz_fragment(fragment_mz,
                    charge, tol_mass, tol_mode)
fragment_smiles =
'[H][C@@]1([C@](C2=O)(OC3)C)[C@]4([H])[C@@]([H])(C4(C)C)'\
                    'C=C[C@@]13C=C5C2=C[C+](C5)C'
fragment_mz = 295.170
spectrum.annotate_molecule_fragment(fragment_smiles,
                    fragment_mz, charge, tol_mass, tol_mode)

fig, ax = plt.subplots(figsize=(12, 6))
sup.spectrum(spectrum, ax=ax)
plt.show()
plt.close()
```

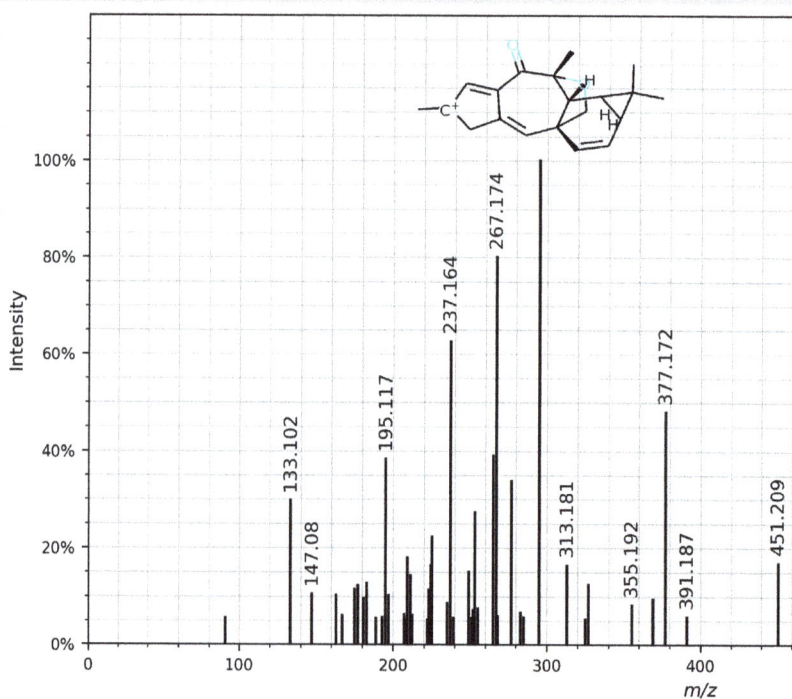

Figure 72.1: MS spectrum with annotation *(Bittremieux)*.

References

1. Mass spectrometry (Wikipedia) https://en.wikipedia.org/wiki/Mass_spectrometry.
2. Bittremieux, W. Spectrum_utils: A Python Package for Mass Spectrometry Data Processing and Visualization. *Anal. Chem.* **2020**, *92* (1), 659–661. https://doi.org/10.1021/acs.analchem.9b04884.

73 TGA, DTG

Thermogravimetric analysis or thermal gravimetric analysis (TGA) is a method of thermal analysis in which the mass of a sample is measured over time as the temperature changes.

This measurement provides information about physical phenomena, such as phase transitions, absorption, adsorption, and desorption, as well as chemical phenomena including chemisorptions, thermal decomposition, and solid–gas reactions (e.g., oxidation or reduction)[1].

Until now for TGA curve analysis, there is no library existing, so a full calculation in a notebook is required.[2,3]

Example TGA curve in Python

The TGAFile object loads data to an object from a .txt file created in Universal Analysis 2000 by TA Instruments[4]:

```
# Thermogravimetric (TGA) plot
# author: Plews, Michael
# license: MIT License      # code: github.com/michaelplews/materials-research
# activity: active (2020)   # index: 73-1


from . import TGA as tga

my_sample = tga.TGAFile("../path/TGAdata.txt", shortname=r'My Sample')
fig = plt.figure(1, figsize=(6, 4))

# plots weight percent on the y axis
my_sample.plot_percent()
# plots a point and hline for a given x value
my_sample.plot_step(550)
my_sample.plot_step(700)
adjust_axes(50, 800, 70, 110)


# a custom function not included in this repo
show()
```

https://doi.org/10.1515/9783110629453-073

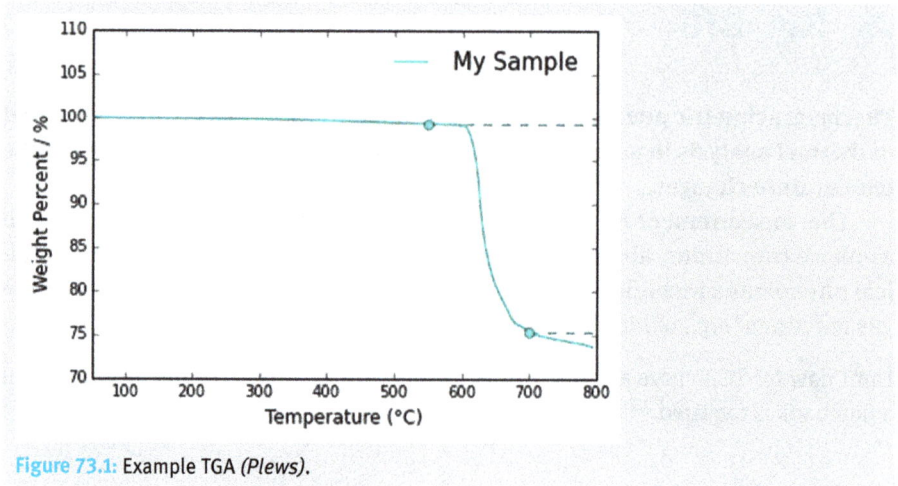

Figure 73.1: Example TGA *(Plews)*.

Analysis can be performed by calculating the regressions and then the intercepts of the curves:

```
# TGA analysis
# author: Gressling, T
# License: MIT License        # code: github.com/gressling/examples
# activity: single example    # index: 73-2

from sklearn.linear_model import LinearRegression
regressor_n = LinearRegression()
regressor_n.fit(x_n_temp, y_n_mass)
regressor_n.score(x_n_temp, y_n_mass)

m_n = regressor_n.coef_
b_n = regressor_n.intercept_

# intercept of regression lines n and m
y_mass_i = m_n*(b_m-b_n)/(m_n-m_m) + b_n
x_temp_i = (b_m-b_n)/(m_n-m_m)

# ... then calculating the mass differences for the intercept values
```

References

1. Thermogravimetric analysis TGA (Wikipedia) https://en.wikipedia.org/wiki/Thermogravimetric_analysis.
2. Cheatwood, L.; Cote, P.; Perygin, D. A Python Analysis of Mass Loss Related to Thermogravimetric Analysis of Tetrachloroethylene-Methyl Methacrylate Copolymers. *Chemical Data Collections* **2019**, *24*, 100304. https://doi.org/10.1016/j.cdc.2019.100304.
3. Perygin, D. MMA TCE DSC with Jupyter (GitHub) https://github.com/dperygin/MMA_TCE_DSC_with_Jupyter.
4. Plews, M. materials-research (GitHub) https://github.com/michaelplews/materials-research.

74 IR and Raman spectroscopy

These spectroscopic techniques used to determine vibrational modes of molecules to provide a structural fingerprint by which molecules can be identified.

Raman spectroscopy relies upon the inelastic scattering of photons, known as Raman scattering. A source of monochromatic light, usually from a laser in the visible, near-infrared, or near ultraviolet range is used, although X-rays can also be used. The laser light interacts with molecular vibrations, phonons, or other excitations in the system, resulting in the energy of the laser photons being shifted up or down. The shift in energy gives information about the vibrational modes in the system. Infrared spectroscopy typically yields similar, complementary, information.[1]

RADIS

RADIS[2,3] is a python library for calculating high-resolution infrared molecular spectra (emission/absorption, equilibrium/nonequilibrium). It also includes tools to compare experimental spectra and spectra calculated with RADIS, or with other spectral codes. It is available via anaconda. Experimental spectra can be loaded using the experimental_spectrum() function and compared with the plot_diff() function.

```
# Calculate CO infrared equilibrium spectrum (RADIS)
# author: Pannier, Erwan; Laux, Christophe O.
# License: LGPL3          # code: radis.readthedocEs.io
# activity: active (2019)  # index: 74-1


from radis import calc_spectrum
s = calc_spectrum(1900, 2300,        # cm-1
                  molecule='CO',
                  isotope='1,2,3',
                  pressure=1.01325,  # bar
                  Tgas=700,          # K
                  mole_fraction=0.1,
                  path_length=1,     # cm
                  )
s.apply_slit(0.5, 'nm')   # simulate an experimental slit
s.plot('radiance')
```

https://doi.org/10.1515/9783110629453-074

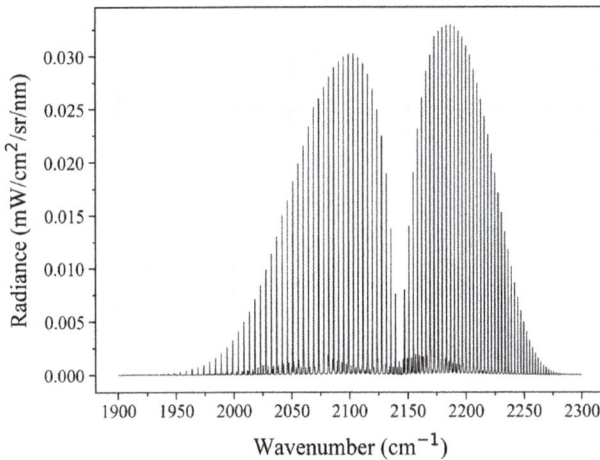

Figure 74.1: Calculated CO equilibrium spectrum *(RADIS)*.

Rampy

Rampy is a Python library[4] that helps processing spectroscopic data, such as Raman, infrared, or XAS spectra. It has functions to subtract baselines as well as to stack, resample, or smooth spectra. It integrates within a workflow that uses Numpy/Scipy as well as optimization libraries such as lmfit or emcee.

Example of Raman spectrum fitting

```
# Raman spectrum fitting with rampy
# author: Le Losq⁵, Charles
# license: GPL 2.0
# code: github.com/charlesll/rampy/Raman_spectrum_fitting
# activity: active (2020)   # index: 74-2

# First measure the centroid of a peak
import matplotlib.pyplot as plt
import numpy as np
import rampy as rp

# load data
spectrum = np.genfromtxt("myspectrum.txt")

# the frequency regions devoid of signal, used by rp.baseline()
bir = np.array([[0,100., 200., 1000]])
```

```
y_corrected, background = rp.baseline(spectrum[:,0],spectrum[:,1],bir,
"arPLS",lam=10**10)

plt.figure()
plt.plot(spectrum[:,0],spectrum[:,1],"k",label="raw data")
plt.plot(spectrum[:,0],background,"k",label="background")
plt.plot(spectrum[:,0],y_corrected,"k",label="corrected signal")
plt.show()
print("Signal centroid is %.2f" % rp.centroid(spectrum[:,0],y_corrected))
> [figure]
```

Lmfit[6] provides a high-level interface to nonlinear optimization and curve-fitting problems for Python. It builds on and extends many of the optimization methods of `scipy.optimize`. Lmfit can be used for this Raman fitting:

```
... # some data cleansing and preprocessing was done

# fit data with leastsq model from scipy
from lmfit.models import GaussianModel

result = lmfit.minimize(residual, params,
                        method = algo, args=(x_fit, y_fit[:,0]))
model = lmfit.fit_report(result.params)
yout,peak1,peak2,peak3,peak4,peak5 = residual(result.params,x_fit)
# the different peaks

# Plot
# ... plt.plot(x_fit,y_fit,'k-') ...
```

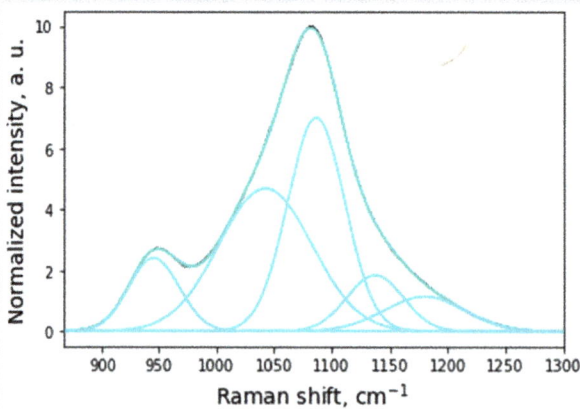

Figure 74.2: Fit of the Si-O stretch vibrationsin LS4 with the Levenberg–Marquardt (LM) algorithm (*Le Losq*).

Example: classification of NIR with PCA

This example is discussed in Chapter 78 as part of chemometrics.

References

1. Raman spectroscopy (Wikipedia) https://en.wikipedia.org/wiki/Raman_spectroscopy.
2. RADIS 0.9.25 documentation https://radis.readthedocs.io/en/latest/.
3. Pannier, E.; Laux, C. O. RADIS: A Nonequilibrium Line-by-Line Radiative Code for CO2 and HITRAN-like Database Species. *J. Quant. Spectrosc. Radiat. Transf.* **2019**, 222–223, 12–25. https://doi.org/10.1016/j.jqsrt.2018.09.027.
4. rampy (PyPI) https://pypi.org/project/rampy/.
5. Le Losq, C. rampy – Python software for spectral data processing (GitHub) https://github.com/charlesll/rampy.
6. lmfit – Built-in Fitting Models in the models module – Non-Linear Least-Squares Minimization and Curve-Fitting for Python https://lmfit.github.io/lmfit-py/builtin_models.html.

75 AFM and thermogram analysis

Pyrolysis[1] is the thermal decomposition of materials at elevated temperatures in an inert atmosphere. It involves a change in chemical composition. Pyrolysis is one of the various types of chemical degradation processes that occur at higher temperatures (above the boiling point of water or other solvents).

Piezoresponse force microscopy[2] (PFM) is a variant of atomic force microscopy (AFM) that allows imaging and manipulation of piezoelectric and ferroelectric materials. A resulting deflection of a probe cantilever is detected through a standard split photodiode detector and then demodulated. In this way, topography and ferroelectric domains can be imaged simultaneously with high resolution.

Kinetic data from thermograms

rampedpyrox[3] is a Python package for analyzing experimental kinetic data and accompanying chemical and isotope compositional information. It can read data from ramped-temperature instruments such as
– Ramped PyrOx,
– RockEval,
– pyrolysis GC (pyGC), and
– unspecified thermogravimetry (TGA)

The package converts time-series mass observations into rate/activation energy (E) distributions using a selection of reactive continuum models, including a distributed activation energy model (DAEM) for nonisothermal data. The package correlates modeled rate/E data with measured isotope compositional data and corrects isotopes for kinetic fractionation.[4]

Python example for an RPO (Ramped Pyrolysis and Oxidation system)

Getting thermogram data into the right form for importing, running the DAEM inverse model to generate an activation energy (E) probability density function [p(0,E)], determining the E range contained in each RPO fraction, correcting isotope values for blank and kinetic fractionation, and generating all necessary plots and tables for data analysis.

https://doi.org/10.1515/9783110629453-075

```python
# Thermogram analysis for activation energy
# and probability density function
# author: Hemingway, Jordon D.
# license: GNU General Public License 3
# code: github.com/FluvialSeds/rampedpyrox
# activity: on hold (2017)              # index: 75-1


import rampedpyrox as rp


# generate string to data
tg_data = '/folder_containing_data/tg_data.csv'
iso_data = '/folder_containing_data/iso_data.csv'


# make the thermogram instance tg
tg = rp.RpoThermogram.from_csv( tg_data, bl_subtract = True, nt = 250)


# generate the DAEM, assume a constant log10omega value of 10
daem = rp.Daem.from_timedata( tg, log10omega = 10,
                    E_max = 350, E_min = 50, nE = 400)


# run the inverse model to generate an energy complex,
# 'auto' calculates best-fit lambda value
ec = rp.EnergyComplex.inverse_model( daem, tg, lam = 'auto')


# forward-model back onto the thermogram
tg.forward_model(daem, ec)


# calculate isotope results
ri = rp.RpoIsotopes.from_csv( iso_data, daem, ec,
        blk_corr = True, # uses values for NOSAMS instrument
        bulk_d13C_true = [-24.9, 0.1], # true d13C value
        mass_err = 0.01,
        DE = 0.0018) # from Hemingway et al. (2017), Radiocarbon


# compare corrected isotopes and E values
print(ri.ri_corr_info)
# This displays the fractionation, mass-balance, and KIE corrected isotope
values.
```

Pycroscopy library

The library provides an analysis[5] of nanoscale materials imaging data with algorithms of k-means clustering, nonnegative matrix factorization, principal component analysis (PCA), and singular value decomposition (SVD). It supports these types of spectra:
- Micro-Raman microscope (*.wdf)
- Atomic force microscope (AFM) (*.ibw)

- AFM-IR (*.cdb)
- AFM-Raman (*.mat)
- Scanning tunneling microscope (STM) (.asc)
- Scanning transmission electron microscope (STEM) (*.dm3)

Example: singular value decomposition (SVD)

This example contains a band excitation piezoresponse force microscopy[2] (BE-PFM) imaging dataset acquired from advanced AFM. A spectrum was collected for each position in a two-dimensional grid of spatial locations that was flattened to a two-dimensional matrix in accordance with the pycroscopy data format:

```python
# PFM spectrum unmixing with Singular Value Decomposition (SVD)
# authors: Vasudevan; Giridharagopal
# credits: Suhas Somnath, Chris Smith, Numan Laanait
# License: MIT License
# code: pycroscopy.github.io/.../plot_spectral_unmixing
# activity: active (2020)      # index: 75-2

import pycroscopy as px
# Universal Spectroscopic and Imaging Data (USID)
import pyUSID as usid
# download the data file from Github:
url = 'https://raw.[...]/pycroscopy/pycroscopy/master/data/BELine_0004.h5'

# Extracting some basic parameters:
num_rows = usid.hdf_utils.get_attr(h5_meas_grp, 'grid_num_rows')
num_cols = usid.hdf_utils.get_attr(h5_meas_grp, 'grid_num_cols')

h5_main = usid.USIDataset(h5_meas_grp['Channel_000/Raw_Data'])
usid.hdf_utils.write_simple_attrs(h5_main,
            {'quantity': 'Deflection', 'units': 'V'})
h5_spec_vals = usid.hdf_utils.get_auxiliary_datasets(h5_main,
            'Spectroscopic_Values')[-1]
freq_vec = np.squeeze(h5_spec_vals.value) * 1E-3

# Example: Singular Value Decomposition (SVD)
decomposer = px.processing.svd_utils.SVD(h5_main, num_components=100)
h5_svd_group = decomposer.compute()

h5_u = h5_svd_group['U']
h5_v = h5_svd_group['V']
h5_s = h5_svd_group['S']
```

```
# Since the two spatial dimensions (x, y) have been
# collapsed to one, reshape the abundance maps:
abun_maps = np.reshape(h5_u[:, :25], (num_rows, num_cols, -1))

usid.plot_utils.plot_map_stack(abun_maps, num_comps=9,
        title='SVD Abundance Maps', reverse_dims=True,
        color_bar_mode='single', cmap='inferno', title_yoffset=0.95)
# Visualize the variance / statistical importance of each component:
usid.plot_utils.plot_scree(h5_s,
        title='Note the exponential drop of variance with number of compo-
nents')
# Visualize the eigenvectors:
_ = usid.plot_utils.plot_complex_spectra(h5_v[:9, :],
                        x_label=x_label, y_label=y_label,
                        title='SVD Eigenvectors', evenly_spaced=False)
```

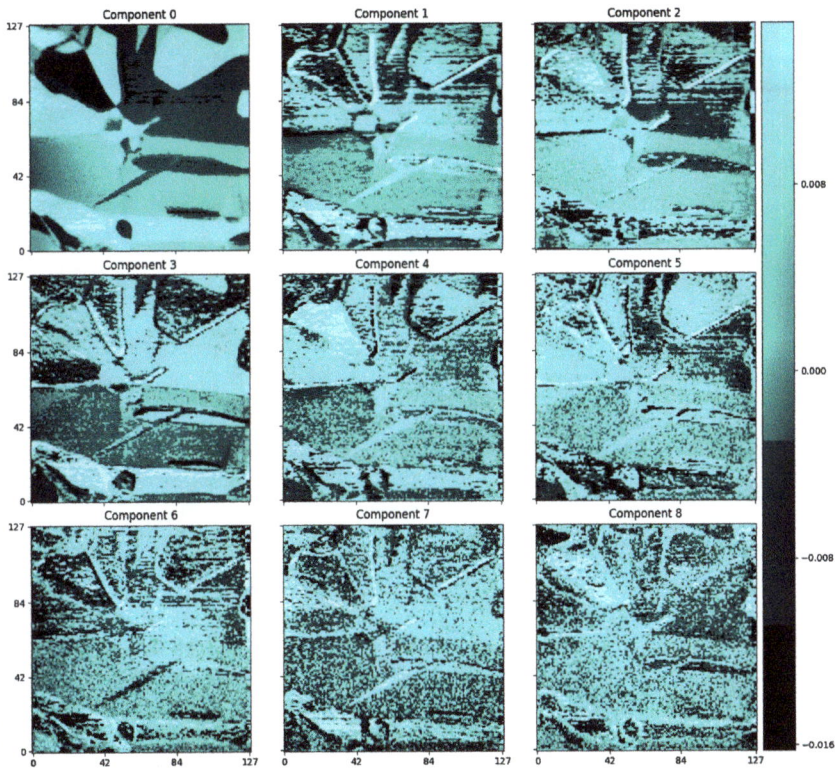

Figure 75.1: SVD abundance maps *(Somnath, S.)*.

References

1. Pyrolysis (Wikipedia) https://en.wikipedia.org/wiki/Pyrolysis.
2. Piezoresponse force microscopy (Wikipedia) https://en.wikipedia.org/wiki/Piezoresponse_force_microscopy.
3. Hemmingway. rampedpyrox Documentation https://buildmedia.readthedocs.org/media/pdf/rampedpyrox/stable/rampedpyrox.pdf.
4. Ramped PyrOx kinetic and isotope analysis (PyPI) https://pypi.org/project/rampedpyrox/.
5. pycroscopy (Github) https://github.com/pycroscopy/pycroscopy/blob/master/docs/USID_pyUSID_pycroscopy.pdf.

76 Gas chromatography-mass spectrometry (GC-MS)

Gas chromatography-mass spectrometry[1] (GC-MS) is an analytical method that combines the features of GC and MS to identify different substances within a test sample.

Applications of GC-MS include drug detection, fire investigation, environmental analysis, explosives investigation, and identification of unknown samples.

PyMassSpec[2] is based on the abandoned project pyMS[3] and is a Python package for processing GC-MS data. It provides a framework and a set of components for rapid development and testing of methods for processing of GC-MS data.

Python analysis example with PyMassSpec

```python
# Multiple GC-MS experiments peak alignment
# author: Davis-Foster, Dominic
# license: LGPL-3    # code: pymassspec.readthedocs.io/.../peak_alignment
# activity: active (2020)    # index: 76-1


# Load data
data_directory = pathlib.Path(".").resolve().parent.parent / "pyms-data"
output_directory = pathlib.Path(".").resolve() / "output"
from pyms.GCMS.IO.ANDI import ANDI_reader
andi_file = data_directory / "sample.cdf"
data = ANDI_reader(andi_file)


from pyms.IntensityMatrix import build_intensity_matrix_i
im = build_intensity_matrix_i(data)
# -> Reading netCDF file)


# Preprocess the data (Savitzky-Golay smoothing and Tophat baseline detection)
from pyms.Noise.SavitzkyGolay import savitzky_golay
from pyms.TopHat import tophat

n_scan, n_mz = im.size
for ii in range(n_mz):
    ic = im.get_ic_at_index(ii)
    ic1 = savitzky_golay(ic)
    ic_smooth = savitzky_golay(ic1)  # Why the second pass here?
    ic_bc = tophat(ic_smooth, struct="1.5m")
    im.set_ic_at_index(ii, ic_bc)


# apply Biller and Biemann based technique to detect peaks
from pyms.BillerBiemann import BillerBiemann
```

https://doi.org/10.1515/9783110629453-076

```python
pl = BillerBiemann(im, points=9, scans=2)
len(pl) # > 1191

# Trim the peak list by relative intensity
from pyms.BillerBiemann import rel_threshold, num_ions_threshold
apl = rel_threshold(pl, percent=2)
len(apl) # > 1191
# Trim the peak list by noise threshold
peak_list = num_ions_threshold(apl, n=3, cutoff=3000)
len(peak_list) # > 225

# Set the mass range, remove unwanted ions and estimate the peak area
from pyms.Peak.Function import peak_sum_area
for peak in peak_list:
    peak.crop_mass(51, 540)
    peak.null_mass(73)
    peak.null_mass(147)
    area = peak_sum_area(im, peak)
    peak.area = area

# Create an Experiment
from pyms.Experiment import Experiment

expr = Experiment("sample", peak_list)
# Set the time range for all Experiments
expr.sele_rt_range(["6.5m", "21m"])
expr.dump(output_directory / "experiments" / "sample.expr")

# LOOP in the next step a loop over the code listed above provides
# a peak alignment:
expr_codes = ["a0806_077", "a0806_078", "a0806_079"] # samples
for expr_code in expr_codes:
    print(f" -> Processing experiment '{expr_code}'")
    andi_file = data_directory / f"{expr_code}.cdf"
...

# PyMassSpec supports interactive plots on matplotlib
```

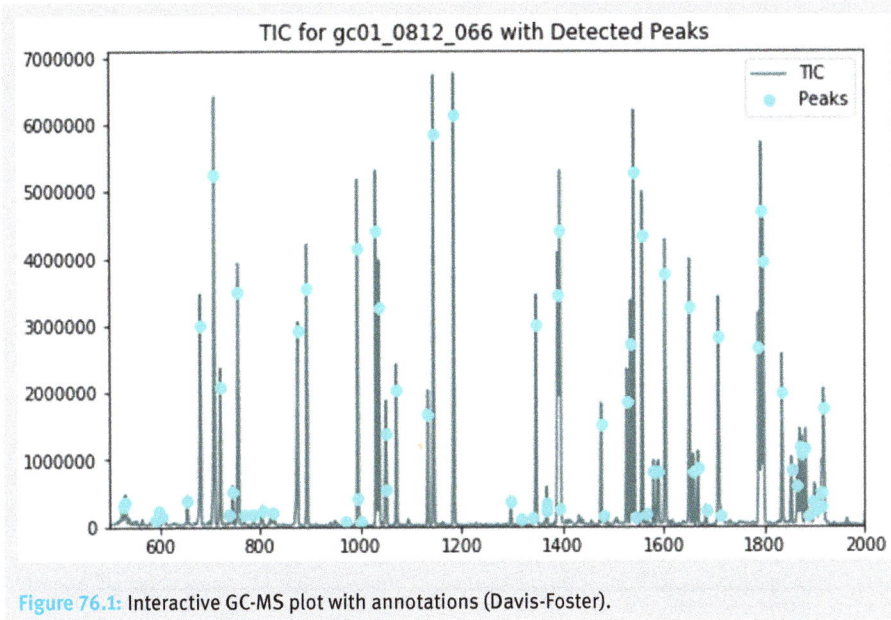

Figure 76.1: Interactive GC-MS plot with annotations (Davis-Foster).

References

1. Gas chromatography-mass spectrometry (Wikipedia) https://en.wikipedia.org/wiki/Gas_ chromatography%E2%80%93mass_spectrometry.
2. PyMassSpec 2.2.21 documentation https://pymassspec.readthedocs.io/en/master/.
3. O'Callaghan, S.; De Souza, D. P.; Isaac, A.; Wang, Q.; Hodkinson, L.; Olshansky, M.; Erwin, T.; Appelbe, B.; Tull, D. L.; Roessner, U.; Bacic, A.; McConville, M. J.; Likić, V. A. PyMS: A Python Toolkit for Processing of Gas Chromatography-Mass Spectrometry (GC-MS) Data. Application and Comparative Study of Selected Tools. *BMC Bioinformatics* **2012**, *13*, 115. https://doi.org/10.1186/1471-2105-13-115.

Applied data science and chemometrics

77 SVD chemometrics example

Chemometrics[1] is the science of extracting information from chemical systems by data-driven means. Chemometrics is inherently interdisciplinary and combines mathematics, statistics, and logic to design optimal measurement procedures and experiments.[2] It allows the extraction of relevant chemical information and helps in understanding systems.[3]

Many examples of chemometrics were already presented in previous chapters. This section intends to give an overview of the algorithms and completes the discussion with an example of the K-matrix method [singular-value decomposition (SVD)].

Use cases

Qualitative

- **Classifications** (does a sample belong to this group?)
- Errors
- Example: number of samples classified correctly

Quantitative

- **Predictions** (what is the concentration?)
- Errors
- RMSEC and RMSEP:
  - RMSEP (root mean square error of prediction)
  - RMSECV (root mean square error of cross-validation)
  - RMSEC (root mean square of calibration)
  - RMSEE (root mean square error of estimation)

Both: calibration and validations

Univariate/multivariate *calibration*

The objective of calibration[3] is to develop models that can be used to predict properties such as pressure, flow, or spectra like Raman, NMR, or mass. The models require calibration or training data sets. A reference system example is concentrations for an analyte and the corresponding infrared spectrum. Multivariate mathematical calibration techniques such as partial least squares regression, or principal component regression are used for correlating the multivariate response to the reference. For example, the resulting model can predict the concentrations of new samples.

https://doi.org/10.1515/9783110629453-077

Classification, pattern recognition, clustering

- **Supervised multivariate classification** techniques are closely related to multivariate calibration techniques. They are multivariate discriminant analysis, logistic regression, neural networks, and regression/classification trees. Examples are discriminant analysis of principal components or partial least squares scores.
- **One-class classifiers** are able to build models for an individual class of interest such as quality control and authenticity verification of products.
- **Unsupervised classification or cluster analysis** for discovering patterns in complex data sets, as already discussed in Chapter 37.

Multivariate *curve resolution*

Self-modeling mixture analysis, blind source/signal separation, and spectral unmixing are **multivariate resolution techniques** that deconstruct data sets with limited or absent reference information and system knowledge. The main problem is that many possible solutions can equivalently represent the measured data based on effects like kinetic or mass-balance constraints.

Multivariate statistical process control (MSPC)

Statistical process control modeling and optimization accounts result in in-process data which is highly amenable to chemometric modeling. Multiway modeling of batch and continuous process methods is common in the industry and is an active area of research in chemometrics and chemical engineering.

Usage in experimental design (DoE)

Design of experiments[4] and signal processing are components of almost all chemometric applications, particularly the use of signal pretreatments to condition data prior to calibration or classification.

Beyond chemometrics the estimation of experimental designs (including quasi experiments) can be performed with the pyDOE library (GitHub). This package contains statistical methods for factorial designs like Plackett-Burman, for response-surface designs like Box-Behnken or the randomized Latin Hypercube Sampling (LHS) design.

Algorithms

At least these algorithms are relevant to perform chemometric analyses. Most of them were discussed in Chapters 32–38:
- Linear algebra
- K-matrix method (SVD)
- PCA
- PLS
- PCA-R and PLS-R
- Classification

Example: K-matrix method (SVD) on Beer–Lambert–Bouguer law

Use the model (K, describes the data A_{kn} and C) or fitting parameters (slope and intercept) to calculate the concentration of unknown spectra, here applied on the Beer–Lambert–Bouguer[5] law. The matrix math involved in the K-matrix method is the same as in standard linear algebra. The command `numpy.linalg.pinv(…, M, N; array_like)` calculates the *generalized inverse of a matrix* using its SVD.

This is an example of not using whole spectra, but rather analytical wavelengths *(work of Scott Huffman[6])*:

```
# K-matrix Method (SVD) on Beer-Lambert-Bouguer law
# author: Huffman, Scott
# license: permission by author # code: agora.cs.wcu.edu/huffman/chem455
# activity: active (2020)          # index: 77-1


import numpy as np
from numpy.linalg import inv, pinv


# Data
Akn = np.matrix([
# spectra of calibration standards
    [0.157706,   0.052752 ,  0.075142],
    [0.177411 ,  0.051542  , 0.065700],
    [0.187398 ,  0.038855  , 0.066837],
    [0.200147 ,  0.047255  , 0.062150],
    [0.221439 ,  0.041524  , 0.057156],
])


# Spectra of samples of unknown concentration
Au = np.matrix([0.197056 ,  0.041629  , 0.068137])
```

```
# Concentration of calibration standards
C =  np.matrix([[60,    15 ,   25],
                [70 ,   15,    15],
                [75  ,   5 ,   20],
                [80,    10,    10],
                [90 ,    5 ,    5]])
```

```
# K-matrix regression
Compute the (Moore-Penrose) pseudo-inverse of a matrix. Calculate the
generalized inverse of a matrix using its singular-value decomposition (SVD)
and including all large singular values. Matrix or stack of matrices to be
pseudo-inverted.
```

```
K = pinv(C)*Akn  # solve for the model
Cu = Au*pinv(K)  # determine conc. of samples of unknown concentration
print('The concentrations in Cu are ', Cu)
```

```
> The concentrations in Cu are  [[78.74220601  5.71854565 18.47377805]]
```

References

1. Chemometrics (Wikipedia) https://en.wikipedia.org/wiki/Chemometrics.
2. B.y., C. PCA: The Basic Building Block of Chemometrics http://www.intechopen.com/books/
    analytical-chemistry/pca-the-basic-building-block-of-chemometrics. https://doi.org/10.5772/51429.
3. Dearing, T. Fundamentals of Chemometrics and Modeling http://depts.washington.edu/cpac/
    Activities/Meetings/documents/DearingFundamentalsofChemometrics.pdf.
4. Design of experiments (Wikipedia) https://en.wikipedia.org/wiki/Design_of_experiments.
5. Beer-Lambert law (Wikipedia) https://en.wikipedia.org/wiki/Beer%E2%80%93Lambert_law.
6. Huffman, S. W. CHEM455 Course. UNC Dataverse 2020. https://doi.org/10.15139/S3/6J9ZAU.

78 Principal component analysis (PCA)

Principal component analysis (PCA) is an unsupervised pattern recognition technique. It is the most commonly used chemometric technique. Unsupervised estimators can highlight interesting aspects of the data without reference to any known labels.

As already discussed in Chapter 37, the PCA is a *dimensionality reduction algorithm*. It uses an orthogonal transformation to convert a set of observations of possibly correlated variables (entities each of which takes on various numerical values) into a set of values of linearly uncorrelated variables called principal components.

Example: classification of NIR spectra using PCA in Python

Most of the features of NIR spectra at different wavelengths are highly correlated.[1] PCA is very efficient at performing dimensionality reduction in multidimensional data, which display a high level of correlation. PCA *get rid of correlated components* by projecting the multidimensional data set to a lower dimensionality space, for example, from 600 dimensions down to 2 *(work by Daniel Pelliccia)*:

```python
# Classification of NIR spectra using PCA
# author: Pelliccia, Daniel
# license: CC BY
# code: nirpyresearch.com/classification-nir-spectra-PCA/
# activity: active (2020)   # index: 78-1

import matplotlib.pyplot as plt
from mpl_toolkits.mplot3d import Axes3D
from scipy.signal import savgol_filter
from sklearn.decomposition import PCA as sk_pca
from sklearn.preprocessing import StandardScaler

data = pd.read_csv('milk.csv')
lab = data.values[:,1].astype('uint8') # labels
# Read the features (scans) and transform data from reflectance to absorbance
feat = np.log(1.0/(data.values[:,2:]).astype('float32'))
# Calculate first derivative applying a Savitzky-Golay filter
dfeat = savgol_filter(feat, 25, polyorder = 5, deriv=1)

skpca1 = sk_pca(n_components=10)
skpca2 = sk_pca(n_components=10)

# Scale the features to have zero mean and standard devisation of 1
# This is important when correlating data with very different variances
```

https://doi.org/10.1515/9783110629453-078

```
nfeat1 = StandardScaler().fit_transform(feat)
nfeat2 = StandardScaler().fit_transform(dfeat)

# Fit the spectral data and extract the explained variance ratio
X1 = skpca1.fit(nfeat1)
expl_var_1 = X1.explained_variance_ratio_
# Fit the first data and extract the explained variance ratio
X2 = skpca2.fit(nfeat2)
expl_var_2 = X2.explained_variance_ratio_

with plt.style.context(('ggplot')):
    ...
    plt.legend()
    plt.show()

# choose the first n principal components that account for the large
# majority of the variance, typically 90% depending on the problem:
```

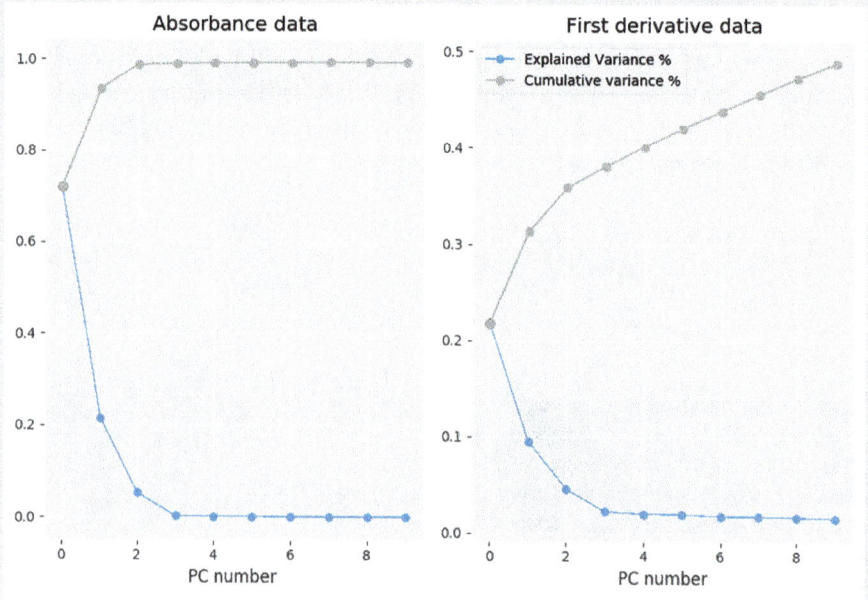

Figure 78.1: Explained variance and the cumulative variance (Pelliccia).

```
# choose n=4
skpca2 = sk_pca(n_components=4)
# Transform on the scaled features
Xt2 = skpca2.fit_transform(nfeat2)
```

```
labplot = ["0/8 Milk","1/8 Milk","2/8 Milk", "3/8 Milk", \
"4/8 Milk", "5/8 Milk","6/8 Milk","7/8 Milk", "8/8 Milk"]
unique = list(set(lab))
colors = [plt.cm.jet(float(i)/max(unique)) for i in unique]
with plt.style.context(('ggplot')):
    for i, u in enumerate(unique):
        col = np.expand_dims(np.array(colors[i]), axis=0)
        xi = [Xt2[j,0] for j in range(len(Xt2[:,0])) if lab[j] == u]
        yi = [Xt2[j,1] for j in range(len(Xt2[:,1])) if lab[j] == u]
        plt.scatter(xi, yi, c=col, s=60, edgecolors='k',label=str(u))
        ...
    plt.show()
```

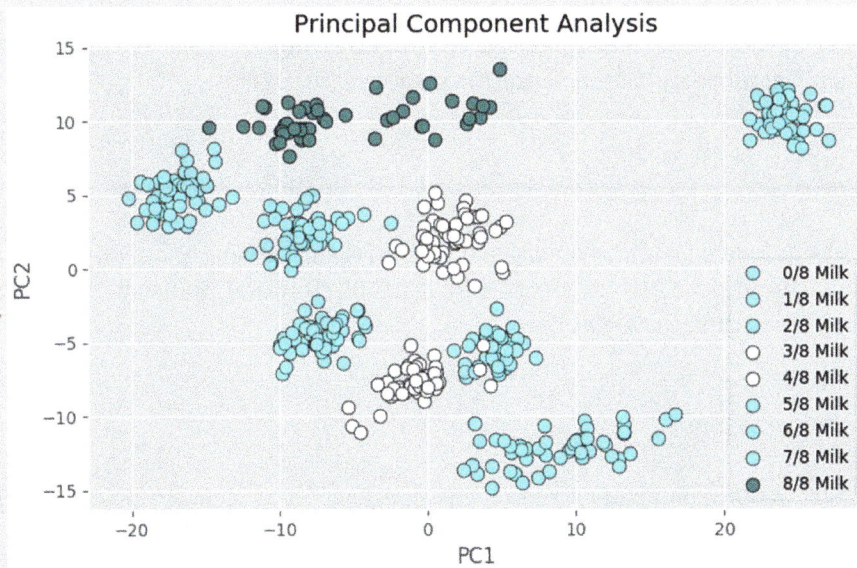

Figure 78.2: Score plot of the first two principal components *(Pelliccia)*.

Reference

1. Pelliccia, D. Statistical learning and chemometrics in Python (NIRPY Research) https://nirpyresearch.com/about/.

79 QSAR: quantitative structure–activity relationship

QSAR (quantitative structure–activity relationship) models first summarize a supposed relationship between chemical structures and biological activity in a data set of chemicals. Second, QSAR models predict the activities of new chemicals.

The models are regression or classification models. QSAR regression models relate a set of "predictor" variables (X) to the potency of the response variable (Y), while classification QSAR models relate the predictor variables to a categorical value of the response variable.

Related terms include quantitative structure–property relationships when a chemical property is modeled as the response variable.[1]

QsarDB

The QSAR DataBank provides tools that enable research groups, project teams, and institutions to share and represent QSAR data (40k structures, 500 models).[2] It also has a Python implementation.[3–5]

Python example

pyQSAR is a Python library for the QSAR Modeling.[3,4] This is an example *(work by Sinyoung Kim)*:

```
# QSAR multiple linear regression model
# author: Sinyoung, Kim
# license: MIT License      # code: github.com/crong-k/pyqsar tutorial
# activity: single example   # index: 79-1

import pyqsar
from pyqsar import data_tools as dt

csv_file_name = "qsar_input_file.csv"
sample_data = pd.read_csv(csv_file_name,sep=",")
X_data = sample_data.iloc[:,1:-1]
X_data.head()
```

https://doi.org/10.1515/9783110629453-079

```
> 5 rows × 1444 columns
> nAcid ALogP  ALogp2  AMR     apol    naAromAtom  nAromBond nAtom  n HeavyAtom
> 0      0     2.8332 8.027022 15.5429 6.481000    0         0      5 5        0
> 1      0     0.6494 0.421720 13.7114 7.993379    0         0      9 6        3


y_data = sample_data.iloc[:,-1:]
# Remove empty feature and NaN
X_data = dt.rm_empty_feature(X_data)
X_data = dt.rmNaN(X_data)


from sklearn.preprocessing import MinMaxScaler
header = list(X_data.columns.values)
scaler = MinMaxScaler()
X_data_scaled = scaler.fit_transform(X_data)
X_data = pd.DataFrame(X_data_scaled, columns=header)


# Feature selection using single core
from pyqsar import feature_selection_single as fss
select = fss.selection(X_data, y_data,
                       clust_info,
                       model='regression',
                       learning=10000,
                       bank=200,
                       component=4)
mymodel.model_corr()


>            ETA_Eta_L  TopoPSA    nRotB      nsssCH     end_point
> ETA_Eta_L  1.000000   0.076283   0.742582   -0.108054  0.266855
> TopoPSA    0.076283   1.000000   0.365734   -0.250235  0.611937
> nRotB      0.742582   0.365734   1.000000   0.021302   0.075124
> nsssCH     -0.108054  -0.250235  0.021302   1.000000   -0.655931
> end_point  0.266855   0.611937   0.075124   -0.655931  1.000000
```

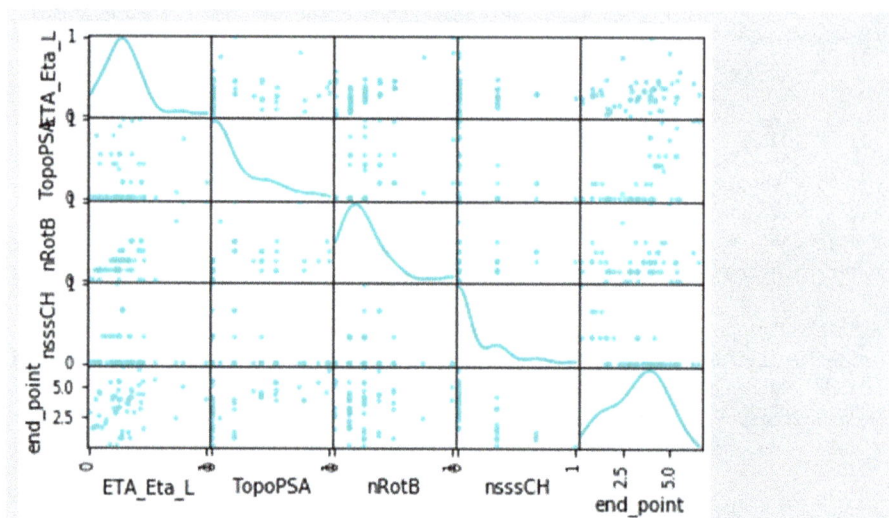

Figure 79.1: QSAR model correlations *(Sinyoung)*.

```
mymodel.mlr()
> Model features:  ['ETA_Eta_L', 'TopoPSA', 'nRotB', 'nsssCH']
> Coefficients:  [[ 5.56280362  3.45694225 -4.84833043 -2.70201862]]
> Intercept:  [2.4473551]
> RMSE: 0.647375
> R^2: 0.819407


# Draw common Substring
draw = dm.DrawMols(sdf_file_name)
commonsub=draw.common_substr(index=index)
draw.show_substr(commonsub)


> C - C - F
```

References

1. Quantitative structure–activity relationship (Wikipedia) https://en.wikipedia.org/wiki/ Quantitative_structure%E2%80%93activity_relationship.
2. Piir, G. What is QsarDB | QsarDB http://www.qsardb.org/about/qsardb.
3. Kim, S.; Cho, K. PyQSAR: A Fast QSAR Modeling Platform Using Machine Learning and Jupyter Notebook. *Bull. Korean Chem. Soc.* **2018**, *3*, 80. https://doi.org/10.1002/bkcs.11638.
4. pyqsar (PyPI) https://pypi.org/project/pyqsar/.
5. Kim, S. pyqsar tutorial (GitHub) https://github.com/crong-k/pyqsar_tutorial.

80 DeepChem: binding affinity

Binding affinity[1,2] is the strength of the binding interaction between a single biomolecule (e.g., protein or DNA) to its ligand/binding partner (e.g., drug or inhibitor). Binding affinity is typically measured and reported by the equilibrium dissociation constant (K_D), the larger the K_D value, the more weakly the target molecule and ligand are attracted to and bind to one another.

Binding affinity is influenced by noncovalent intermolecular interactions such as hydrogen bonding, electrostatic interactions, hydrophobic and Van der Waals forces between the two molecules.

Basic protein–ligand affinity models with DeepChem

Investigating models of protein–ligand affinity with machine learning can be done with DeepChem.[3-5] It was originally created by the Pande group in Stanford and is now part of Schrödinger[6] (MIT License). This example also uses mdtraj which allows to manipulate molecular dynamics (MD) trajectories and perform a variety of analyses. This was already discussed in Chapter 53.

Python example

- Loading a chemical dataset, consisting of a series of protein–ligand complexes.
- Featurizing each protein-ligand complexes with various featurization schemes.
- Fitting a model with featurized protein–ligand complexes.
- (Visualizing the results could follow)

```
# Basic Protein-Ligand Affinity Model (deepchem)
# author: Ramsundar, Bharath; Feinberg, Evan
# license: MIT License      # code: github.com/schrodingerdeepchem/notebooks
# activity: active (2020)    # index: 80-1

import deepchem as dc
from deepchem.utils import download_url
dataset_file = os.path.join(data_dir, "pdbbind_core_df.csv.gz")

# Type of dataset is: <class 'pandas.core.frame.DataFrame'>
#    pdb_id  smiles  \
# 0    2d3u          CC1CCCCC1S(O)(O)NC1CC(C2CCC(CN)CC2)SC1C(O)O
# 1    3cyx  CC(C)(C)NC(O)C1CC2CCCCC2C[NH+]1CC(O)C(CC1CCCCC...
# 2    3uo4          OC(O)C1CCC(NC2NCCC(NC3CCCCC3C3CCCCC3)N2)CC1
```

https://doi.org/10.1515/9783110629453-080

```
# 3    1p1q                              CC1ONC(O)C1CC([NH3+])C(O)O
# 4    3ag9  NC(O)C(CCC[NH2+]C([NH3+])[NH3+])NC(O)C(CCC[NH2...
import mdtraj as md

def combine_mdtraj(protein, ligand):
  chain = protein.topology.add_chain()
  residue = protein.topology.add_residue("LIG", chain, resSeq=1)
  for atom in ligand.topology.atoms:
      protein.topology.add_atom(atom.name, atom.element, residue)
  protein.xyz = np.hstack([protein.xyz, ligand.xyz])
  protein.topology.create_standard_bonds()
  return protein

def visualize_complex(complex_mdtraj):
  ligand_atoms = [a.index for a in complex_mdtraj.topology.atoms if "LIG" in
str(a.residue)]
  binding_pocket_atoms = md.compute_neighbors(complex_mdtraj, 0.5, ligand_
  atoms)[0]
  binding_pocket_residues = list(set([complex_mdtraj.topology.atom(a).
residue.resSeq for a in binding_pocket_atoms]))
  binding_pocket_residues = [str(r) for r in binding_pocket_residues]
  binding_pocket_residues = " or ".join(binding_pocket_residues)

def convert_lines_to_mdtraj(molecule_lines):
# load file from RCSB PDB
  traj = nglview.MDTrajTrajectory( complex_mdtraj )
  traj = nglview.MDTrajTrajectory( ligand_mdtraj )

protein_mdtraj = convert_lines_to_mdtraj(first_protein)
ligand_mdtraj = convert_lines_to_mdtraj(first_ligand)
complex_mdtraj = combine_mdtraj(protein_mdtraj, ligand_mdtraj)

ngltraj = visualize_complex(complex_mdtraj)
ngltraj # widget
```

ML: featurizing the dataset

The available featurizations that become standard with deepchem are
- ECFP4 fingerprints,
- RDKit descriptors,
- NNScore-style bdescriptors, and
- hybrid binding pocket descriptors.

Note how we separate the featurizers into those that featurize individual chemical compounds, compound_featurizers, and those that featurize molecular complexes, complex_featurizers:

```
grid_featurizer = dc.feat.RdkitGridFeaturizer(
    voxel_width=16.0, feature_types="voxel_combined",
    voxel_feature_types=["ecfp", "splif", "hbond", "pi_stack",
"cation_pi", "salt_bridge"],
    ecfp_power=5, splif_power=5, parallel=True, flatten=True)

compound_featurizer = dc.feat.CircularFingerprint(size=128)

# the actual featurization:
PDBBIND_tasks, (train_dataset, valid_dataset, test_dataset),
transformers = dc.molnet.load_pdbbind_grid()
```

To fit a deepchem model, instantiate one model class. In this case, a wrapper for Sci-Kit Learn can interoperate with deepchem:

```
# model and train
from sklearn.ensemble import RandomForestRegressor
sklearn_model = RandomForestRegressor(n_estimators=100)
model = dc.models.SklearnModel(sklearn_model)
model.fit(train_dataset)

# evaluate
from deepchem.utils.evaluate import Evaluator
import pandas as pd
metric = dc.metrics.Metric(dc.metrics.r2_score)

evaluator = Evaluator(model, train_dataset, transformers)
train_r2score = evaluator.compute_model_performance([metric])
print("RF Train set R^2 %f" % (train_r2score["r2_score"]))
evaluator = Evaluator(model, valid_dataset, transformers)
valid_r2score = evaluator.compute_model_performance([metric])
print("RF Valid set R^2 %f" % (valid_r2score["r2_score"]))

# computed_metrics: [0.87165637813866481]
# RF Train set R^2 0.871656
# computed_metrics: [0.1561422855082335]
# RF Valid set R^2 0.156142
```

References

1. Binding Affinity | Dissociation Constant https://www.malvernpanalytical.com/en/products/ measurement-type/binding-affinity.
2. Rezaei, M.; Li, Y.; Li, X.; Li, C. Improving the Accuracy of Protein-Ligand Binding Affinity Prediction by Deep Learning Models: Benchmark and Model, 2019. https://doi.org/10.26434/ chemrxiv.9866912.v1.
3. Ramsundar, B. Basic Protein-Ligand Affinity Models (deepchem) (GitHub) https://github.com/ schrodinger/schrodingerdeepchem/blob/master/examples/notebooks/protein_ligand_ complex_notebook.ipynb.
4. Wu, Z.; Ramsundar, B.; Feinberg, E. N.; Gomes, J.; Geniesse, C.; Pappu, A. S.; Leswing, K.; Pande, V. MoleculeNet: A Benchmark for Molecular Machine Learning http://arxiv.org/abs/1703.00564.
5. Altae-Tran, H.; Ramsundar, B.; Pappu, A. S.; Pande, V. Low Data Drug Discovery with One-Shot Learning. *ACS Cent Sci* **2017**, *3* (4), 283–293. https://doi.org/10.1021/acscentsci.6b00367.
6. Schrödingerdeepchem (GitHub) https://github.com/schrodinger/schrodingerdeepchem.

81 Stoichiometry and reaction balancing

Stoichiometry is the calculation of reactants and products in reactions.[1] Balancing means that each side of the chemical equation must represent:
– the same quantity of any particular element
– the same charge

One balances a chemical equation[2] by changing the scalar number for each chemical formula. Simple chemical equations can be balanced by inspection, that is, by trial and error. Another technique involves solving a system of linear equations: the matrix method.

Chempy

In Chapter 43, the chempy[3] library by Björn Dahlgren was introduced. It is a very powerful package including classes for representing substances, reactions, and systems of reactions. It also includes analytic solutions to differential equations. Together with sympy (see Chapter 32), it provides[4] parsing of chemical formulae, reactions, and systems:

```
# Stochiometry
# author: Dahlgren, Björn
# license: BSD 2 clause      # code: pypi.org/chempy
# activity: active (2020)    # index: 81-1

# Stoichiometry
from chempy import balance_stoichiometry
reac, prod = balance_stoichiometry(
                 {'NH4ClO4', 'Al'}, {'Al2O3', 'HCl', 'H2O', 'N2'})
from pprint import pprint
pprint(dict(reac))
pprint(dict(prod))
> {'Al': 10, 'NH4ClO4': 6}
> {'Al2O3': 5, 'H2O': 9, 'HCl': 6, 'N2': 3}

from chempy import mass_fractions
for fractions in map(mass_fractions, [reac, prod]):
    pprint({k: '{0:.3g} wt%'.format(v*100) for k, v in
               fractions.items()})
> {'Al': '27.7 wt%', 'NH4ClO4': '72.3 wt%'}
> {'Al2O3': '52.3 wt%', 'H2O': '16.6 wt%', 'HCl': '22.4 wt%', 'N2': '8.62 wt%'}

# Reactions with linear dependencies (underdetermined systems)
pprint([dict(_) for _ in
```

https://doi.org/10.1515/9783110629453-081

```
   balance_stoichiometry({'C', 'O2'}, {'CO2', 'CO'})])
> [{'C': x1 + 1, 'O2': x1 + 1/2}, {'CO': 1, 'CO2': x1}]

# get a solution with minimal (non-zero) integer coefficients: underdeter-
mined=None
pprint([dict(_) for _ in
   balance_stoichiometry({'C', 'O2'}, {'CO2', 'CO'}, underdetermined=None)])
> [{'C': 3, 'O2': 2}, {'CO': 2, 'CO2': 1}]


# Balancing reactions and chemical equilibria
# author: Dahlgren, Björn
# License: BSD 2 clause     # code: pypi.org/chempy
# activity: active (2020)   # index: 81-1

from chempy import Equilibrium
from sympy import symbols
K1, K2, Kw = symbols('K1 K2 Kw')
e1 = Equilibrium({'MnO4-': 1, 'H+': 8, 'e-': 5}, {'Mn+2': 1, 'H2O': 4}, K1)
e2 = Equilibrium({'O2': 1, 'H2O': 2, 'e-': 4}, {'OH-': 4}, K2)
coeff = Equilibrium.eliminate([e1, e2], 'e-')
coeff
> [4, -5]

redox = e1*coeff[0] + e2*coeff[1]
print(redox)
> 32 H+ + 4 MnO4- + 20 OH- = 26 H2O + 4 Mn+2 + 5 O2; K1**4/K2**5
autoprot = Equilibrium({'H2O': 1}, {'H+': 1, 'OH-': 1}, Kw)
n = redox.cancel(autoprot)
n
> 20
redox2 = redox + n*autoprot
print(redox2)
> 12 H+ + 4 MnO4- = 6 H2O + 4 Mn+2 + 5 O2; K1**4*Kw**20/K2**5


# Chemical equilibria
# Predict pH of a bicarbonate solution with pKa and pKw values
from collections import defaultdict
from chempy.equilibria import EqSystem
eqsys = EqSystem.from_string("""HCO3- = H+ + CO3-2; 10**-10.3
   H2CO3 = H+ + HCO3-; 10**-6.3
   H2O = H+ + OH-; 10**-14/55.4
   """)  # pKa1(H2CO3) = 6.3 (implicitly incl. CO2(aq)), pKa2=10.3 & pKw=14
arr, info, sane = eqsys.root(
                    defaultdict(float, {'H2O': 55.4, 'HCO3-': 1e-2}))
conc = dict(zip(eqsys.substances, arr))
from math import log10
print("pH: %.2f" % -log10(conc['H+']))
> pH: 8.30
```

References

1. Stoichiometry (Wikipedia) https://en.wikipedia.org/wiki/Stoichiometry.
2. Chemical equation (Wikipedia) https://en.wikipedia.org/wiki/Chemical_equation.
3. Dahlgren, B. ChemPy: A package useful for chemistry written in Python https://www.theoj.org/joss-papers/joss.00565/10.21105.joss.00565.pdf. https://doi.org/10.21105/joss.00565.
4. Dahlgren, B. chempy (GitHub) https://github.com/bjodah/chempy.

Applied artificial intelligence

82 ML Python libraries in chemistry

*Word*2vec[1] is a group of related models that are used to produce word embeddings. These models are shallow, two-layer neural networks that are trained to reconstruct linguistic contexts of words.

The algorithm takes as its input a large corpus of text and produces a vector space, typically of several hundred dimensions, with each unique word in the corpus being assigned a corresponding vector in the space. Word vectors are positioned in the vector space such that words that share common contexts in the corpus are located close to one another in the space.

In Chapters 28 and 29, libraries for data science and deep learning were discussed. This chapter is focusing on libraries for the domain of AI and chemistry.

mol2vec

This library integrates *molecular substructures* as information-rich *vectors*. Mol2vec[2,3] is an unsupervised machine learning approach. The substructures are obtained via the descriptors and molecular fingerprints based on the Morgan algorithm.

Mol2vec is based on the Word2vec algorithm and first encodes molecules as sentences, meaning that each substructure (represented by Morgan identifier) represents a word.

Python example

To featurize new samples using pretrained embeddings and using vector trained on uncommon samples to represent new substructures:

```
$ mol2vec featurize -i new.smi -o new.csv -m model.pkl
```

This notebook covers basic concepts of Mol2vec from the generation of "molecular sentences" to plotting embeddings obtained with Mol2vec:

```
# Plotting of an amino acid (ALA) substructure vectors (mol2vec)
# author: Turk, Samo
# License: BSD 3-Clause "New" or "Revised" License
# code: github.com/samoturk/mol2vec
# activity: active (2020)                        # index: 82-1
```

https://doi.org/10.1515/9783110629453-082

```
from rdkit import Chem
from mol2vec.features import mol2alt_sentence, MolSentence, DfVec, sentences2vec
from mol2vec.helpers import depict_identifier, IdentifierTable, mol_to_svg

# Example: Canonical SMILES of amino acids
aa_smis = ['CC(N)C(=O)O', 'N=C(N)NCCCC(N)C(=O)O', 'NC(=O)CC(N)C(=O)O',
           'NC(CC(=O)O)C(=O)O', 'NC(CS)C(=O)O', 'NC(CCC(=O)O)C(=O)O',
           'NC(=O)CCC(N)C(=O)O', 'NCC(=O)O', 'NC(Cc1cnc[nH]1)C(=O)O',
           'CCC(C)C(N)C(=O)O', 'CC(C)CC(N)C(=O)O', 'NCCCCC(N)C(=O)O',
           'CSCCC(N)C(=O)O', 'NC(Cc1ccccc1)C(=O)O', 'O=C(O)C1CCCN1',
           'NC(CO) C(=O)O', 'CC(O)C(N)C(=O)O', 'NC(Cc1c[nH]c2ccccc12)C(=O)O',
           'NC(Cc1ccc(O) cc1)C(=O)O', 'CC(C)C(N)C(=O)O']
aa_codes = ['ALA', 'ARG', 'ASN', 'ASP', 'CYS', 'GLU', 'GLN', 'GLY', 'HIS', 'ILE',
            'LEU', 'LYS', 'MET', 'PHE', 'PRO', 'SER', 'THR', 'TRP', 'TYR', 'VAL']

sentence = mol2alt_sentence(aas[0], 1)
from gensim.models import word2vec
# Pre-trained Mol2vec model trained on 20 million compounds (ZINC)
model = word2vec.Word2Vec.load('models/model_300dim.pkl')

aa_values = [get_values(x, projections) for x in aa_sentences[0]]
    plot_2D_vectors(aa_values, vector_labels=aa_sentences[0] + ['ALA']);
```

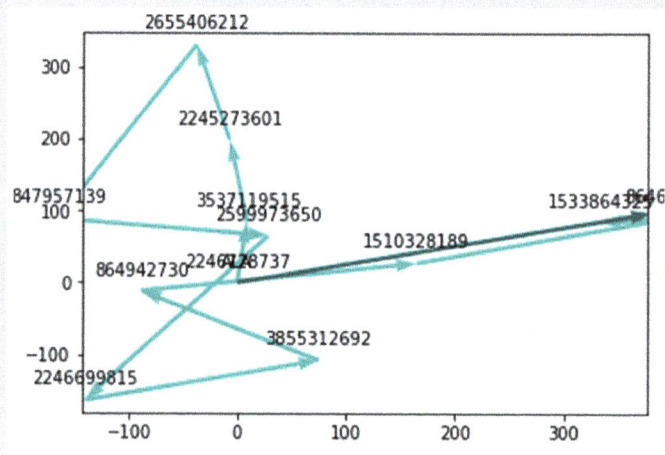

Figure 82.1: Plotting of amino acid *ALA* substructure vectors *(Turk)*.

```
# Plotting all amino acids in grid plot
for aa,name,ax in zip(aas, aa_codes, (ALA, ARG, ASN, ASP, CYS, GLU, GLN,
GLY, HIS, ILE, LEU, LYS, MET, PHE, PRO, SER, THR, TRP, TYR, VAL)):
    pca_subs = [get_values(x, projections) for x in mol2alt_sentence(aa, 1)]
    plot_2D_vectors(pca_subs, ax=ax, min_max_x=(-1000,1000), min_max_y=
    (-2000, 2000))
    ax.get_xaxis().set_visible(False)
    ax.get_yaxis().set_visible(False)
    ax.text(0.95, 0.01, u"%s" % name,
            verticalalignment='bottom', horizontalalignment='right',
            transform=ax.transAxes, weight='bold',
            fontsize=10)
```

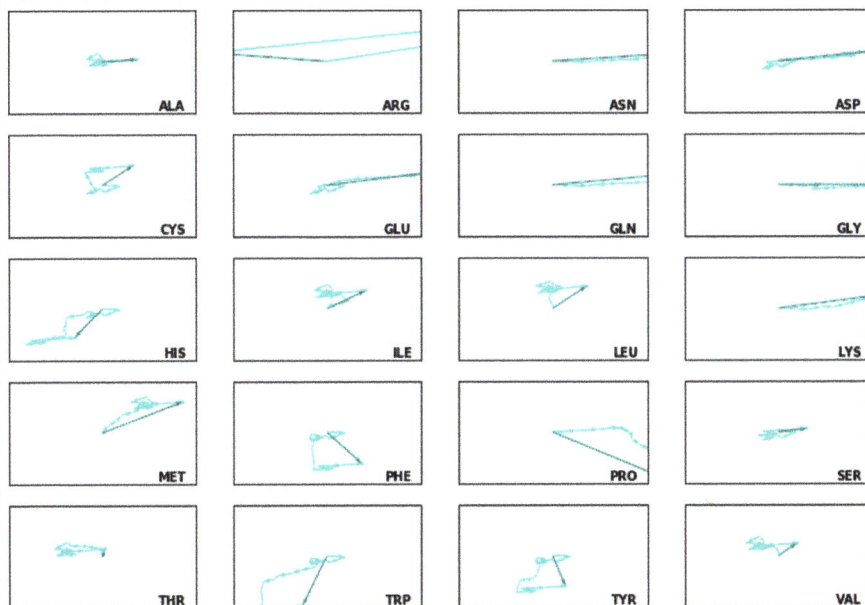

Figure 82.2: Plotting all substructure vectors *(Turk)*.

Mol2vec in chemistry

- Prediction of CYP450 enzyme–substrate selectivity based on the network-based label space division method[4]
- Deep learning-based prediction of drug-induced cardiotoxicity[5]
- Machine learning models based on molecular fingerprints and an extreme gradient boosting method lead to the discovery of JAK2 inhibitors[6]
- Rapid and accurate prediction of pKa values of C–H acids using graph convolutional neural networks[7]

- Evaluating polymer representations via quantifying structure–property relationships[8]
- Exploring tunable hyperparameters for deep neural networks with industrial ADME data sets[9]

Seq2seq

Seq2seq[10] is a *family* of machine learning approaches used for language translation, image captioning, conversational models, and text summarization. It is used in solving differential equations.

It is claimed that it could solve complex equations more rapidly and with greater accuracy than commercial solutions such as Mathematica, MATLAB, and Maple. First, the equation is parsed into a tree structure to avoid notational idiosyncrasies. An LSTM neural network then applies its standard pattern recognition facilities to process the tree.

In 2020, Google released Meena, a 2.6 billion parameter, seq2seq-based bot trained on a 341 GB data set.

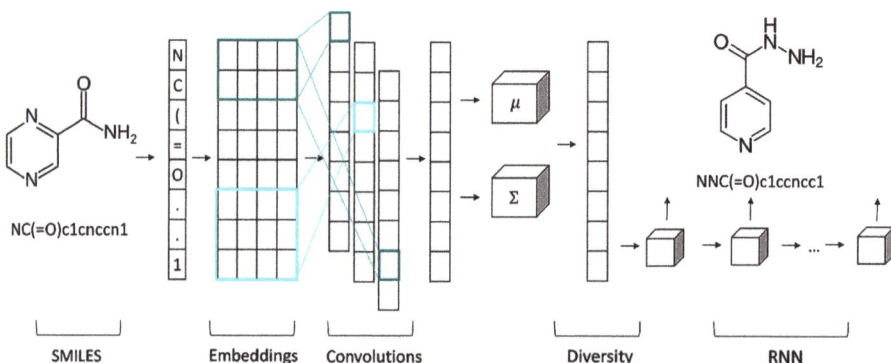

Figure 82.3: End-to-end neural net architecture for compound discovery[11] *(Harel, Radinsky)*.

seq2seq in chemistry

- Retrosynthetic reaction prediction using neural sequence-to-sequence models[12]
- Predicting retrosynthetic reactions using self-corrected transformer neural networks[13]
- De novo molecular design by combining deep autoencoder recurrent neural networks with generative topographic mapping[14]

- **Computer-assisted retrosynthesis based on molecular similarity**[15]
- **Machine learning in computer-aided synthesis planning**[16]
- **Context-aware data-driven retrosynthetic analysis**[17]
- **Enhancing retrosynthetic reaction prediction with deep learning using multiscale reaction classification**[18]
- **Prototype-based compound discovery using deep generative models**[11]

References

1. Word2vec (Wikipedia) https://en.wikipedia.org/wiki/Word2vec.
2. Jaeger, S.; Fulle, S.; Turk, S. Mol2vec: Unsupervised Machine Learning Approach with Chemical Intuition. *J. Chem. Inf. Model.* **2018**, *58* (1), 27–35. https://doi.org/10.1021/acs.jcim.7b00616.
3. Mol2vec documentation https://mol2vec.readthedocs.io/en/latest/.
4. Shan, X.; Wang, X.; Li, C.-D.; Chu, Y.; Zhang, Y.; Xiong, Y.; Wei, D.-Q. Prediction of CYP450 Enzyme-Substrate Selectivity Based on the Network-Based Label Space Division Method http://xbioinfo.sjtu.edu.cn/pdf/Shan-2019-Prediction%20of%20CYP450%20Enzyme-Substrat.pdf. https://doi.org/10.1021/acs.jcim.9b00749.
5. Cai, C.; Guo, P.; Zhou, Y.; Zhou, J.; Wang, Q.; Zhang, F.; Fang, J.; Cheng, F. Deep Learning-Based Prediction of Drug-Induced Cardiotoxicity. *J. Chem. Inf. Model.* **2019**, *59* (3), 1073–1084. https://doi.org/10.1021/acs.jcim.8b00769.
6. Yang, M.; Tao, B.; Chen, C.; Jia, W.; Sun, S.; Zhang, T.; Wang, X. Machine Learning Models Based on Molecular Fingerprints and an Extreme Gradient Boosting Method Lead to the Discovery of JAK2 Inhibitors. *J. Chem. Inf. Model.* **2019**, *59* (12), 5002–5012. https://doi.org/10.1021/acs.jcim.9b00798.
7. Roszak, R.; Beker, W.; Molga, K.; Grzybowski, B. A. Rapid and Accurate Prediction of pKa Values of C-H Acids Using Graph Convolutional Neural Networks. *J. Am. Chem. Soc.* **2019**, *141* (43), 17142–17149. https://doi.org/10.1021/jacs.9b05895.
8. Ma, R.; Liu, Z.; Zhang, Q.; Liu, Z.; Luo, T. Evaluating Polymer Representations via Quantifying Structure-Property Relationships. *J. Chem. Inf. Model.* **2019**, *59* (7), 3110–3119. https://doi.org/10.1021/acs.jcim.9b00358.
9. Zhou, Y.; Cahya, S.; Combs, S. A.; Nicolaou, C. A.; Wang, J.; Desai, P. V.; Shen, J. Exploring Tunable Hyperparameters for Deep Neural Networks with Industrial ADME Data Sets. *J. Chem. Inf. Model.* **2019**, *59* (3), 1005–1016. https://doi.org/10.1021/acs.jcim.8b00671.
10. Seq2seq (Wikipedia) https://en.wikipedia.org/wiki/Seq2seq.
11. Harel, S.; Radinsky, K. Prototype-Based Compound Discovery Using Deep Generative Models. *Mol. Pharm.* **2018**, *15* (10), 4406–4416. https://doi.org/10.1021/acs.molpharmaceut.8b00474. Reprinted with permission. Copyright (2018) American Chemical Society.
12. Liu, B.; Ramsundar, B.; Kawthekar, P.; Shi, J.; Gomes, J.; Luu Nguyen, Q.; Ho, S.; Sloane, J.; Wender, P.; Pande, V. Retrosynthetic Reaction Prediction Using Neural Sequence-to-Sequence Models. *ACS Cent Sci* **2017**, *3* (10), 1103–1113. https://doi.org/10.1021/acscentsci.7b00303.
13. Zheng, S.; Rao, J.; Zhang, Z.; Xu, J.; Yang, Y. Predicting Retrosynthetic Reactions Using Self-Corrected Transformer Neural Networks. *J. Chem. Inf. Model.* **2020**, *60* (1), 47–55. https://doi.org/10.1021/acs.jcim.9b00949.
14. Sattarov, B.; Baskin, I. I.; Horvath, D.; Marcou, G.; Bjerrum, E. J.; Varnek, A. De Novo Molecular Design by Combining Deep Autoencoder Recurrent Neural Networks with Generative Topographic Mapping. *J. Chem. Inf. Model.* **2019**, *59* (3), 1182–1196. https://doi.org/10.1021/acs.jcim.8b00751.

15. Coley, C. W.; Rogers, L.; Green, W. H.; Jensen, K. F. Computer-Assisted Retrosynthesis Based on Molecular Similarity. *ACS Cent Sci* **2017**, *3* (12), 1237–1245. https://doi.org/10.1021/acscentsci.7b00355.
16. Coley, C. W.; Green, W. H.; Jensen, K. F. Machine Learning in Computer-Aided Synthesis Planning. *Acc. Chem. Res.* **2018**, *51* (5), 1281–1289. https://doi.org/10.1021/acs.accounts.8b00087.
17. Nicolaou, C. A.; Watson, I. A.; LeMasters, M.; Masquelin, T.; Wang, J. Context Aware Data-Driven Retrosynthetic Analysis. *J. Chem. Inf. Model.* **2020**. https://doi.org/10.1021/acs.jcim.9b01141.
18. Baylon, J. L.; Cilfone, N. A.; Gulcher, J. R.; Chittenden, T. W. Enhancing Retrosynthetic Reaction Prediction with Deep Learning Using Multiscale Reaction Classification. *J. Chem. Inf. Model.* **2019**, *59* (2), 673–688. https://doi.org/10.1021/acs.jcim.8b00801.

83 AI in drug design

Scientists are adopting the use of machine learning (ML) for making potentially important decisions, such as discovery, development, optimization, synthesis, and characterization of drugs and materials.

Selected topics

"Machine learning" gives 4271 hits on ACS publications (Q1/2020). It is time to define a structure in this new field of expertise to maintain an overview and track changes.

Drug discovery: MolecularRNN

Designing new molecules with a set of predefined properties is a core challenge in drug discovery. Generating molecular graphs with optimized properties[1] can be implemented using a MolecularRNN, which is a graph-recurrent generative model for molecular structures.

Toxicity prediction

Toxicity prediction is very important to health[2] so prediction[3,4] is essential to reduce the cost and labor of a drug's trials. The prediction can benefit from ML: different ML methods that have been applied to toxicity prediction, including deep learning, random forests, k-nearest neighbors, and support vector machines are reviewed.

Deep neural networks for QSAR

Deep neural networks (DNNs) have also generated promising results for quantitative structure–activity relationship (QSAR[5]) tasks.[6] DNNs can make better predictions than traditional methods, such as random forests. The chapter discusses that during prediction a *multitask DNN* does borrow "signal" from molecules with similar structures in the training sets of the other tasks.

    The multitask DNN algorithm was implemented in Python and based on the code in the winning entry of the QSAR Kaggle competition. The codes are available at GitHub.[7]

https://doi.org/10.1515/9783110629453-083

Implementation example

Multilayer perceptron trained in Python

A multilayer perceptron model is implemented in Python[8] and trained on a CHEMBL dataset, then exported to the ONNX format. Molecular fingerprints are calculated with RDKit *(Work of Felix Eloy, shortened for readability)*:

```python
# Target prediction: multi-layer perceptron (PyTorch)
# author: Eloy, Felix
# License: MIT License       # code: github.com/eloyfelix/pistache_predictor
# activity: single example   # index: 83-1


from rdkit import Chem
import torch
import torch.nn.functional as F
import torchvision.transforms as transforms


FP_SIZE = 2048
RADIUS = 2
BATCH_SIZE = 20
N_EPOCHS = 40


def calc_morgan_fp(smiles):
    mol = Chem.MolFromSmiles(smiles)
    fp = rdMolDescriptors.GetMorganFingerprintAsBitVect(
        mol, RADIUS, nBits=FP_SIZE)
    a = np.zeros((0,), dtype=np.float32)
    Chem.DataStructs.ConvertToNumpyArray(fp, a)
    return a

class DatasetSMILES(Dataset):
    def __init__(self, file_path):
        self.data = pd.read_csv(file_path)
        self.transform = transforms.Compose([calc_morgan_fp, torch.Tensor])
    def __len__(self):
        return len(self.data)
    def __getitem__(self, index):
        fp = self.data.iloc[index, 1]
        labels = self.data.iloc[index, 0]
        fp = self.transform(fp)
        return fp, labels
```

```python
train_dataset = DatasetSMILES('CHEMBL1829.csv')
train_loader = DataLoader(dataset=train_dataset,
                          batch_size=BATCH_SIZE,
                          num_workers=0,
                          shuffle=True)

# multi layer perceptron model definition
class MLP(torch.nn.Module):
    def __init__(self, input_dim):
        super(MLP, self).__init__()
        self.fc1 = torch.nn.Linear(input_dim, 1024)
        self.fc2 = torch.nn.Linear(1024, 512)
        self.fc3 = torch.nn.Linear(512, 128)
        self.fc4 = torch.nn.Linear(128, 1)

    def forward(self, x):
        h1 = F.relu(self.fc1(x))
        h2 = F.relu(self.fc2(h1))
        h3 = F.relu(self.fc3(h2))
        out = torch.sigmoid(self.fc4(h3))
        return out

mlp = MLP(FP_SIZE)
criterion = torch.nn.BCELoss()
optimizer = torch.optim.SGD(mlp.parameters(), lr=0.05)

# model training
for epoch in range(N_EPOCHS):
    for i, (fps, labels) in enumerate(train_loader):
        optimizer.zero_grad()
        outputs = mlp(fps)
        # loss
        loss = criterion(outputs, labels.float().view(-1, 1))
        loss.backward()
        optimizer.step()
        if (i + 1) % 10 == 0:
            print('Epoch: [%d/%d], Step: [%d/%d], Loss: %.4f'
                  % (epoch + 1, N_EPOCHS,
                     i + 1, len(train_dataset)
                     BATCH_SIZE, loss.item())))

# export the model to be loaded in any ONNX runtime
torch.onnx.export(mlp, torch.ones(FP_SIZE), "target_predict.onnx",
                  export_params=True, input_names=['input'], output_
                  names=['output'])
```

```
# test
tn, fp, fn, tp = confusion_matrix(y_trues, y_preds).ravel()
sens = tp / (tp + fn)
spec = tn / (tn + fp)
prec = tp / (tp + fp)
f1 = f1_score(y_trues, y_preds)
acc = accuracy_score(y_trues, y_preds)
mcc = matthews_corrcoef(y_trues, y_preds)
auc = roc_auc_score(y_trues, y_preds_proba)

print(f"accuracy: {acc}, auc: {auc}, sens: {sens},
        spec: {spec}, prec: {prec}, mcc: {mcc}, f1: {f1}")
> accuracy: 0.8371918235997756, auc: 0.8942389411754185,
> sens: 0.7053822792666977, spec: 0.8987519347341067,
> prec: 0.7649158653846154, mcc: 0.6179805824644773,
> f1: 0.733943790291889
```

References

1. Popova, M.; Shvets, M.; Oliva, J.; Isayev, O. MolecularRNN: Generating realistic molecular graphs with optimized properties https://www.researchgate.net/publication/333564111_MolecularRNN_Generating_realistic_molecular_graphs_with_optimized_properties.
2. Wu, Y.; Wang, G. Machine Learning Based Toxicity Prediction: From Chemical Structural Description to Transcriptome Analysis. *Int. J. Mol. Sci.* **2018**, *19* (8). https://doi.org/10.3390/ijms19082358.
3. Merck Molecular Activity Challenge https://www.kaggle.com/c/MerckActivity.
4. Unterthiner, T.; Mayr, A.; Klambauer, G.; Hochreiter, S. Toxicity Prediction Using Deep Learning. *arXiv [stat.ML]*, 2015.
5. Dahl, G. E.; Jaitly, N.; Salakhutdinov, R. Multi-Task Neural Networks for QSAR Predictions. *arXiv [stat.ML]*, 2014.
6. Xu, Y.; Ma, J.; Liaw, A.; Sheridan, R. P.; Svetnik, V. Demystifying Multitask Deep Neural Networks for Quantitative Structure-Activity Relationships. *J. Chem. Inf. Model.* **2017**, *57* (10), 2490–2504. https://doi.org/10.1021/acs.jcim.7b00087.
7. DeepNeuralNet-QSAR (GitHub) https://github.com/Merck/DeepNeuralNet-QSAR.
8. Eloy, F. pistache_predictor (GitHub) https://github.com/eloyfelix/pistache_predictor.

84 Automated machine learning

AutoML[1,2] is the process of automating the process of applying machine learning to real-world problems. It covers the complete pipeline from the raw dataset to the deployable machine learning model. The high degree of automation in AutoML allows non-experts to make use of machine learning models and techniques without requiring to become an expert in this field first.

There are two major concepts[3,4]:

- **Neural architecture search (NAS)** is the process of automating the design where usually reinforcement learning or evolutionary algorithms are used.
- **Transfer learning** is a technique where one uses pretrained models to transfer what is learned when applying the model to a new but similar dataset.

Software packages

autoKeras

It is an open-source software library.[5] Both Google's AutoML and Auto-Keras are powered by an algorithm called NAS.[6,7]

auto-sklearn

It performs algorithm selection and hyperparameter tuning. It uses Bayesian optimization, meta-learning, and ensemble construction.[8]

NeuNetS/AutoAI by IBM

NeuNetS[9–12] uses AI to automatically synthesize deep neural network models. NeuNetS algorithms are designed to create new neural network models without reusing pretrained models. NeuNetS can accommodate a wide range of model synthesis scenarios.

AutoML (by Google)

A Google Cloud service charging cost provides AutoML.[13] It supports 3rd Party Packages: Deep Learning with TensorFlow & Keras, XGBoost, LightGBM, and CatBoost.

https://doi.org/10.1515/9783110629453-084

AutoDS by IBM

Automated Data Science (AutoDS)[14] by IBM is not to be confused with NeuNetS. The AutoDS package automates *data science*, that is, selects algorithms out of about 60 types.

H2O.AI

It is an open-source, in-memory, distributed, fast, and scalable machine learning and predictive analytics platform. It allows to build machine learning models on big data and provides to push the models to production in an enterprise environment. H2O's core code is written in Java and has a REST API.[15]

Examples in Python

```
# AutoKeras and auto-sklearn
# author: Gressling, T
# license: MIT License        # code: github.com/gressling/examples
# activity: single example    # index: 84-1


# !pip3 install autokeras
import autokeras as ak
cls = ak.ImageClassifier()
cls.fit(x_train, y_train)
result = clf.predict(x_test)


# auto-sklearn
import autosklearn.classification
cls = autosklearn.classification.AutoSklearnClassifier()
cls.fit(X_train, y_train)
result = cls.predict(X_test)


# NeuNetS (IBM)
#!pip install --upgrade
#git+https://github.com/pmservice/NeuNetS/#egg=neunets_
processor\&subdirectory=neunets_processor
from neunets_processor.text import text_processor
model_file = os.path.join(model_location,"keras_model.hdf5")
metadata_file = os.path.join(model_location,"metadata.json")
word_mapping_file = os.path.join(model_location,"word_mapping.json")
cls = text_processor.TextProcessor(model_file, metadata_file,
     word_mapping_file)
result = processor.predict(["Data Science in Chemistry"])
```

```
# AutoML (Google)
from auto_ml import Predictor
cls = Predictor(type_of_estimator='regressor', column_descriptions=col_desc)
cls.train(df_train)
cls.score(df_test, df_test.aCOL)
result = cls.predict(df_test)
```

References

1. Automated machine learning (Wikipedia) https://en.wikipedia.org/wiki/Automated_machine_learning.
2. Drozdal, J.; Weisz, J.; Wang, D.; Dass, G.; Yao, B.; Zhao, C.; Muller, M.; Ju, L.; Su, H. Trust in AutoML: Exploring Information Needs for Establishing Trust in Automated Machine Learning Systems. *arXiv [cs.LG]*, 2020. https://doi.org/10.1145/3377325.3377501.
3. Mwiti, D. Automated Machine Learning in Python (KDnuggets) https://www.kdnuggets.com/2019/01/automated-machine-learning-python.html.
4. automl.org https://www.automl.org/automl/.
5. AutoKeras https://autokeras.com/.
6. Zoph, B.; Vasudevan, V.; Shlens, J.; Le, Q. V. Learning Transferable Architectures for Scalable Image Recognition. *arXiv [cs.CV]* **2017**.
7. Rosebrock, A. Auto-Keras and AutoML: A Getting Started Guide – PyImageSearch https://www.pyimagesearch.com/2019/01/07/auto-keras-and-automl-a-getting-started-guide/.
8. al., F. et. Efficient and Robust Automated Machine Learning (NIPS 2015 / 5872) http://papers.nips.cc/paper/5872-efficient-and-robust-automated-machine-learning.pdf.
9. Malossi, C. NeuNetS: Automating Neural Network Model Synthesis for Broader Adoption of AI | IBM Research Blog https://www.ibm.com/blogs/research/2018/12/neunets/.
10. NeuNetS IBM (GitHub) https://github.com/pmservice/NeuNetS.
11. Sood, A.; Elder, B.; Herta, B.; Xue, C.; Bekas, C.; Malossi, A. C. I.; Saha, D.; Scheidegger, F.; Venkataraman, G.; Thomas, G.; et al. NeuNetS: An Automated Synthesis Engine for Neural Network Design. *arXiv [cs.LG]* **2019**.
12. Puri, R.; Ai, I. W. Using AI to Design AI: Game-Changing Neural Network Design with NeuNetS.
13. automl (PyPI) https://pypi.org/project/automl/.
14. Aggarwal, C.; Bouneffouf, D.; Samulowitz, H.; Buesser, B.; Hoang, T.; Khurana, U.; Liu, S.; Pedapati, T.; Ram, P.; Rawat, A.; et al. How Can AI Automate End-to-End Data Science? *arXiv [cs.AI]* **2019**.
15. AutoML: Automatic Machine Learning – H2O http://docs.h2o.ai/h2o/latest-stable/h2o-docs/automl.html.

85 Retrosynthesis and reaction prediction

Retrosynthetic analysis is a technique for solving problems in the planning of organic syntheses. This is achieved by transforming a target molecule into simpler precursor structures regardless of any potential reactivity/interaction with reagents. Each precursor material is examined using the same method.[1]

The only evaluation[2] of a model prediction is an assessment by human experts followed by validation with wet-lab experiments, which is barely scalable. Given a set of reactants/reagents, a feed-forward model or a transformer predicts the corresponding product and by-products.[3]

Implementation examples with artificial neural networks

Retrosynthesis planner .

The following example requires first to download data[4] with download.sh and the processing with extract_templates.py and process_base.py. The work by Francis Tseng is based on publications by Segler,[5-8] Schwaller,[2,3,9-11] and Coley[12-20]:

```
# Retrosynthesis planner (shortened)
# author: Tseng, Francis
# license: GNU GPL v3.0     # code: github.com/frnsys/retrosynthesis_planner
# activity: single example (2018)  # index: 85-1

import molvs
import random
import policies
from mcts import Node, mcts
import tensorflow as tf
from rdkit import Chem

def transform(mol, rule):
    # Apply transformation rule to a molecule to get reactants
    rxn = AllChem.ReactionFromSmarts(rule)
    results = rxn.RunReactants([mol])
    # Only look at first set of results
    results = results[0]
    reactants = [Chem.MolToSmiles(smi) for smi in results]
    return reactants

def expansion(node):
    # Try expanding each molecule in the current state
    # to possible reactants
```

https://doi.org/10.1515/9783110629453-085

```
    # Predict applicable rules
    preds = sess.run(expansion_net.pred_op, feed_dict={
        expansion_net.keep_prob: 1.,
        expansion_net.X: fprs,
        expansion_net.k: 5
    })
    # Generate children for reactants
    children = []
    for mol, rule_idxs in zip(mols, preds):
        # State for children will
        # not include this mol
        new_state = mols - {mol}
        mol = Chem.MolFromSmiles(mol)
        for idx in rule_idxs:
            # Extract actual rule
            rule = expansion_rules[idx]
            # Apply rule
            reactants = transform(mol, rule)
            child = Node(state=state, is_terminal=terminal,
                         parent=node, action=rule)
            children.append(child)
    return children

def rollout(node, max_depth=200):
        # Select a random mol (that's not a starting mol)
        mol = random.choice(mols)
        fprs = policies.fingerprint_mols([mol])
        # Predict applicable rules
        preds = sess.run(rollout_net.pred_op, feed_dict={
        rule = rollout_rules[preds[0][0]]
        reactants = transform(Chem.MolFromSmiles(mol), rule)
        state = cur.state | set(reactants)
        # State for children will not include this mol
        state = state - {mol}
        # Partial reward if some starting molecules are found
        reward = sum(1 for mol in cur.state if mol
                                        in starting_mols)/len(cur.state)
        # Reward of -1 if no starting molecules are found
        if reward == 0:
            return -1.
        return reward
    # Reward of 1 if solution is found
    return 1

def plan(target_mol):
    # Generate a synthesis plan for a target
    # molecule (in SMILES form).
    # If a path is found, returns a list of (action, state) tuples
    path = mcts(root, expansion, rollout, iterations=2000, max_depth=200)
```

```
    return path
# Load base compounds
smi = molvs.standardize_smiles(smi)
starting_mols.add(smi)

# Load policy networks
rules = json.load(f)
rollout_rules = rules['rollout']
expansion_rules = rules['expansion']
rollout_net = policies.RolloutPolicyNet(n_rules=len(rollout_rules))
expansion_net = policies.ExpansionPolicyNet(n_rules=len(expansion_rules))
filter_net = policies.InScopeFilterNet()

sess = tf.Session()
init = tf.global_variables_initializer()
sess.run(init)
saver = tf.train.Saver()
saver.restore(sess, 'model/model.ckpt')

# Usage Example
target_mol = 'CC(=O)NC1=CC=C(O)C=C1'
root = Node(state={target_mol})
path = plan(target_mol)
```

Code snippet: predict the outcomes of organic reactions

This code[21] by *Connor Coley* relies on Keras for its machine-learning components using the Theano background. RDKit is used for all chemistry-related parsing and processing. In the modified version, atom-mapping numbers associated with reactant molecules are preserved after calling RunReactants *(shortened to the CNN definition)*.

```
# CNN Layer in 'Predict the outcomes of organic reactions'
# (small part of the feed forward model definition)
# author: Coley, Connor (Cambridge)
# license: MIT License        # code: github.com/...coley/ochem_predict_nn
# activity: active (2020)     # index: 85-1

from keras.models import Sequential, Model, model_from_json
from keras.layers import Dense, Activation, Input, merge

FPs = Input(shape = (None, 1024),
                          name = "FPs")
FP_features = TimeDistributed(Dense(N_hf, activation = inner_act),
                          name = "FPs to features")(FPs)
unscaled_FP_score = TimeDistributed(Dense(1, activation = 'linear'),
                          name = "features to score")(FP_features)
```

```
dynamic_flattener = lambda x: T.reshape(x, (x.shape[0], x.shape[1]),
                                ndim  = x.ndim-1)
dynamic_flattener_shape = lambda x: (None, x[1])
unscaled_FP_score_flat = Lambda(dynamic_flattener,
                                output_shape = dynamic_flattener_shape,
                                name = "flatten_FP")(unscaled_FP_score)
score = Activation('softmax', name = "scores to probs")
                                (unscaled_FP_score_flat)

model = Model(input = [FPs], output = [score])
model.summary()
model.compile(loss = 'categorical_crossentropy',
              optimizer = optimizer, metrics = ['accuracy'])
```

Self-corrected retrosynthetic reaction predictor

Predicting retrosynthetic reaction using self-corrected transformer neural networks[22] is implemented in OpenNMT. The score_predictions.sh script lists the top 10 accuracies:

```
# Self-Corrected Retrosynthetic Reaction Predictor
# author: sysu-yanglab group[23] (Sun Yat-sen University)
# license: MIT License  # code: github.com/sysu-yanglab/Self-...-Predictor
```

Selected topics

Predicting organic reaction outcomes with Weisfeiler–Lehman network

This work describes the template-free prediction of organic reaction outcomes using graph convolutional neural networks. It is implemented in a graph-convolutional neural network model for the prediction of chemical reactivity[12] (Django app at http://localhost:8000/visualize)[24].

Retrosynthetic reaction prediction using neural sequence-to-sequence models

The authors describe[25] a fully data-driven model that learns to perform a retrosynthetic reaction prediction task, which is treated as a sequence-to-sequence mapping problem. The end-to-end trained model has an encoder–decoder architecture that consists of two recurrent neural networks, which has previously shown great success in solving other sequence-to-sequence prediction tasks such as machine translation.

The model is trained on 50,000 experimental reaction examples from the US patent literature, which span 10 broad reaction types that are commonly used by medicinal chemists.

The implementation uses seq2seq.[26,27] A general-purpose encoder–decoder framework for Tensorflow was discussed in Chapter 82.

Molecular transformer for chemical reaction prediction and uncertainty estimation

The molecular transformer architecture is implemented in IBM RXN[28] based on OpenNMT.[3,29]

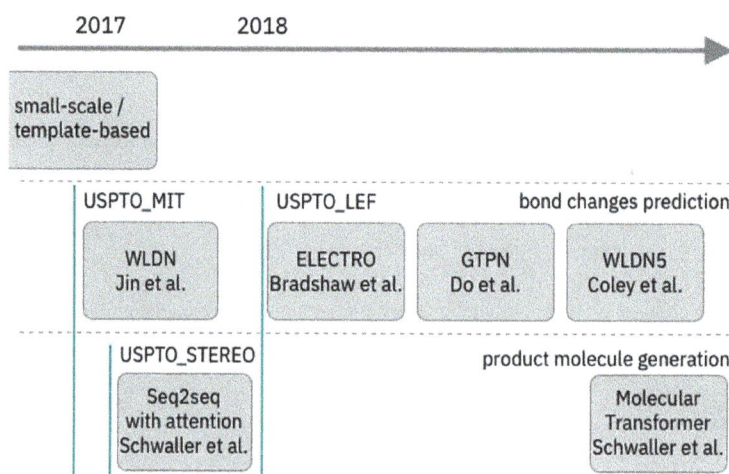

Figure 85.1: Timeline of the recent developments of large-scale data-driven reaction prediction models *(Schwaller[10])*.

Data source: USPTO applications

A famous data source for working with reactions is the USPTO application dataset by Daniel Lowe.[30,31] Reactions extracted by text-mining from US patents published between 1976 and September 2016. The reactions are available as CML or reaction SMILES. The dataset is used in many ML implementations.

References

1. Retrosynthetic analysis (Wikipedia) https://en.wikipedia.org/wiki/Retrosynthetic_analysis.
2. Schwaller, P.; Nair, V. H.; Petraglia, R.; Laino, T. Evaluation Metrics for Single-Step Retrosynthetic Models.
3. Schwaller, P.; Laino, T.; Gaudin, T.; Bolgar, P.; Hunter, C. A.; Bekas, C.; Lee, A. A. Molecular Transformer: A Model for Uncertainty-Calibrated Chemical Reaction Prediction. *ACS Cent Sci* **2019**, *5* (9), 1572–1583. https://doi.org/10.1021/acscentsci.9b00576.
4. Tseng, F. *Retrosynthesis_planner*; Github.
5. Segler, M. H. S.; Waller, M. P. Neural-Symbolic Machine Learning for Retrosynthesis and Reaction Prediction. *Chemistry* **2017**, *23* (25), 5966–5971. https://doi.org/10.1002/chem.201605499.
6. Segler, M. H. S.; Preuss, M.; Waller, M. P. Planning Chemical Syntheses with Deep Neural Networks and Symbolic AI. *Nature* **2018**, *555*, 604. https://doi.org/10.1038/nature25978.
7. Segler, M. H. S.; Preuss, M.; Waller, M. P. Learning to Plan Chemical Syntheses. *arXiv [cs.AI]*, 2017.
8. Segler, M.; Preuß, M.; Waller, M. P. Towards "AlphaChem": Chemical Synthesis Planning with Tree Search and Deep Neural Network Policies. *arXiv [cs.AI]*, 2017.
9. Schwaller, P.; Laino, T.; Gaudin, T.; Bolgar, P.; Bekas, C.; Lee, A. A. Molecular Transformer for Chemical Reaction Prediction and Uncertainty Estimation. **2018**. https://doi.org/10.26434/chemrxiv.7297379.v1.
10. Schwaller, P.; Laino, T. Data-Driven Learning Systems for Chemical Reaction Prediction: An Analysis of Recent Approaches. In *Machine Learning in Chemistry: Data-Driven Algorithms, Learning Systems, and Predictions*; ACS Symposium Series; American Chemical Society, 2019; Vol. 1326, pp 61–79. https://doi.org/10.1021/bk-2019-1326.ch004.
11. Schwaller, P.; Gaudin, T.; Lanyi, D.; Bekas, C.; Laino, T. "Found in Translation": Predicting Outcomes of Complex Organic Chemistry Reactions Using Neural Sequence-to-Sequence Models. *arXiv [cs.LG]*, 2017.
12. Coley, C. W.; Jin, W.; Rogers, L.; Jamison, T. F.; Jaakkola, T. S.; Green, W. H.; Barzilay, R.; Jensen, K. F. A Graph-Convolutional Neural Network Model for the Prediction of Chemical Reactivity. 2018. https://doi.org/10.26434/chemrxiv.7163189.v1.
13. Jin, W.; Coley, C. W.; Barzilay, R.; Jaakkola, T. Predicting Organic Reaction Outcomes with Weisfeiler-Lehman Network. *arXiv [cs.LG]*, 2017.
14. Coley, C. W.; Rogers, L.; Green, W. H.; Jensen, K. F. Computer-Assisted Retrosynthesis Based on Molecular Similarity. *ACS Cent Sci* **2017**, *3* (12), 1237–1245. https://doi.org/10.1021/acscentsci.7b00355.
15. Coley, C. W.; Barzilay, R.; Jaakkola, T. S.; Green, W. H.; Jensen, K. F. Prediction of Organic Reaction Outcomes Using Machine Learning. *ACS Cent Sci* **2017**, *3* (5), 434–443. https://doi.org/10.1021/acscentsci.7b00064.
16. Coley, C. W.; Green, W. H.; Jensen, K. F. Machine Learning in Computer-Aided Synthesis Planning. *Acc. Chem. Res.* **2018**, *51* (5), 1281–1289. https://doi.org/10.1021/acs.accounts.8b00087.
17. Schreck, J. S.; Coley, C. W.; Bishop, K. J. M. Learning Retrosynthetic Planning through Simulated Experience. *ACS Cent Sci* **2019**, *5* (6), 970–981. https://doi.org/10.1021/acscentsci.9b00055.
18. Dai, H.; Li, C.; Coley, C.; Dai, B.; Song, L. Retrosynthesis Prediction with Conditional Graph Logic Network. In *Advances in Neural Information Processing Systems 32*; Wallach, H., Larochelle, H., Beygelzimer, A., Alché-Buc, F., Fox, E., Garnett, R., Eds.; Curran Associates, Inc., 2019; pp 8872–8882.

19. Struble, T. J.; Alvarez, J. C.; Brown, S. P.; Chytil, M.; Cisar, J.; DesJarlais, R. L.; Engkvist, O.; Frank, S. A.; Greve, D. R.; Griffin, D. J.; Hou, X.; Johannes, J. W.; Kreatsoulas, C.; Lahue, B.; Mathea, M.; Mogk, G.; Nicolaou, C. A.; Palmer, A. D.; Price, D. J.; Robinson, R. I.; Salentin, S.; Xing, L.; Jaakkola, T.; Green, W. H.; Barzilay, R.; Coley, C. W.; Jensen, K. F. Current and Future Roles of Artificial Intelligence in Medicinal Chemistry Synthesis. *J. Med. Chem.* **2020**. https://doi.org/10.1021/acs.jmedchem.9b02120.
20. Coley, C. ochem_predict_nn (GitHub) https://github.com/connorcoley/ochem_predict_nn.
21. Coley, C. rexgen_direct (GitHub) https://github.com/connorcoley/rexgen_direct.
22. Zheng, S.; Rao, J.; Zhang, Z.; Xu, J.; Yang, Y. Predicting Retrosynthetic Reaction Using Self-Corrected Transformer Neural Networks. *arXiv [physics.chem-ph]*, 2019.
23. *Self-Corrected-Retrosynthetic-Reaction-Predictor*; Github.
24. Jin, W. Predicting Organic Reaction Outcomes with Weisfeiler-Lehman Network (GitHub) https://github.com/connorcoley/rexgen_direct.
25. Liu, B.; Ramsundar, B.; Kawthekar, P.; Shi, J.; Gomes, J.; Luu Nguyen, Q.; Ho, S.; Sloane, J.; Wender, P.; Pande, V. Retrosynthetic Reaction Prediction Using Neural Sequence-to-Sequence Models. *ACS Cent Sci* **2017**, *3* (10), 1103–1113. https://doi.org/10.1021/acscentsci.7b00303.
26. seq2seq (GitHub) https://github.com/google/seq2seq.
27. Ramamoorthy, S. Practical seq2seq http://suriyadeepan.github.io/2016-12-31-practical-seq2seq/.
28. IBM RXN for Chemistry https://rxn.res.ibm.com/.
29. Klein, G.; Kim, Y.; Deng, Y.; Senellart, J.; Rush, A. OpenNMT: Open-Source Toolkit for Neural Machine Translation. In *Proceedings of ACL 2017, System Demonstrations*; Association for Computational Linguistics: Stroudsburg, PA, USA, 2017; pp 67–72. https://doi.org/10.18653/v1/P17-4012.
30. Lowe, D. dan2097 / Patent Reaction Extraction / Downloads (Bitbucket) https://bitbucket.org/dan2097/patent-reaction-extraction/downloads/.
31. Lowe, D. Chemical Reactions from US Patents (1976-Sep2016), 2017. https://doi.org/10.6084/m9.figshare.5104873.v1.

86 ChemML

ChemML is a machine learning and informatics program suite for analysis, mining, and modeling of chemical and materials data.[1]

The ChemML package is built with the intension of easy use for nonprogrammers. It is a central class to construct a molecule from different chemical input formats.[2,3] This module is built on top of the RDKit and OpenBabel python API. After data ingestion either standard ML libraries like scikit or a component like "active learning" can be used.

Example of the workflow

A plot of the computation graph with the corresponding Python code:

```python
# ChemML example workflow
# author: Hachmannlab
# license: BSD 3                    # code: github.com/hachmannlab/chemml
# activity: active (2020)           # index: 86-1


# 2
from chemml.datasets import load_xyz_ polarizability
coordinates, polarizability = load_xyz_polarizability()

# 3
from chemml.chem import CoulombMatrix
model = CoulombMatrix()
df = model.represent(coordinates)

# 4, 8
import pandas as pd
from sklearn.kernel_ridge import KernelRidge
clf = KernelRidge(kernel='rbf',
                  alpha=0.1)
clf.fit(df,polarizability)
dfy_predict = clf.predict(df)
dfy_predict = pd.DataFrame(dfy_predict)
```

https://doi.org/10.1515/9783110629453-086

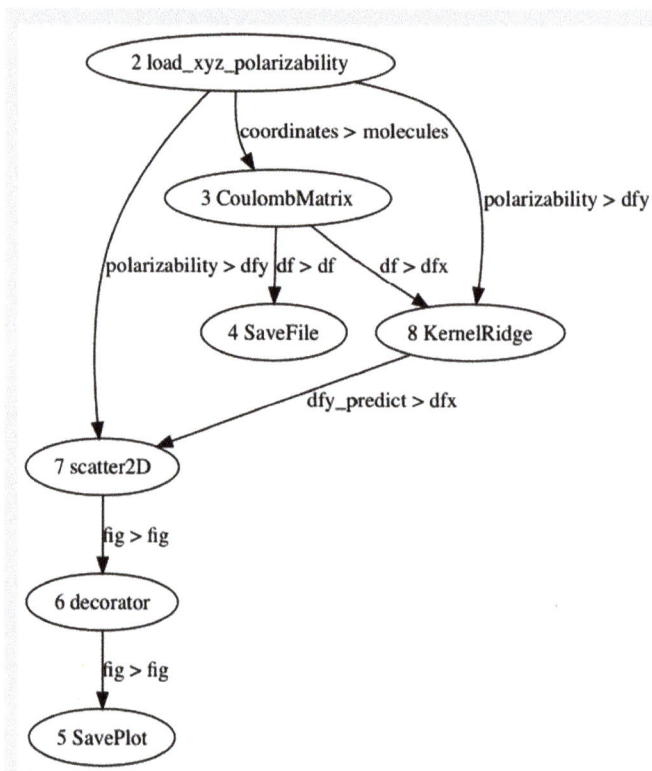

Figure 86.1: Computational graph used in ChemML *(Hachmann)*.

```
# 7
from chemml.visualization import scatter2D
sc = scatter2D(marker = 'o') (
fig = sc.plot(polarizability, dfy_predict, 0, 0)

# 6, 5
from chemml.visualization import decorator
dec = decorator(title = 'training performance',
    xlabel = 'predicted polarizability (Bohr^3)',
    ylabel = 'calculated polarizability (Bohr^3)',
    grid = True,
    grid_color = 'green'
)
dec.matplotlib font(size = 12)
fig = dec. fit(fig)
```

Example of "active learning"

The central idea behind any active learning approach is to achieve lower prediction errors (higher accuracy) in machine learning models by choosing less but more informative data points.

```python
# ChemML NN branched layer definition and 'active learning'
# author: Hachmannlab
# license: BSD 3                    # code: github.com/hachmannlab/chemml
# activity: active (2020)           # index: 86-2

def model_creator(nneurons=features.shape[1],
                  activation=['relu','tanh'], lr = 0.001):
    # branch 1
    b1_in = Input(shape=(nneurons, ), name='inp1')
    l1 = Dense(12, name='l1', activation=activation[0])(b1_in)
    b1_l1 = Dense(6, name='b1_l1', activation=activation[0])(l1)
    b1_l2 = Dense(3, name='b1_l2', activation=activation[0])(b1_l1)
    # branch 2
    b2_l1 = Dense(16, name='b2_l1', activation=activation[1])(l1)
    b2_l2 = Dense(8, name='b2_l2', activation=activation[1])(b2_l1)

    # merge branches (keas feature)
    merged = Concatenate(name='merged')([b1_l2, b2_l2])
    out = Dense(1, name='outp', activation='linear')(merged)
    model = Model(inputs = b1_in, outputs = out)
    adam = Adam(lr=lr, beta_1=0.9, beta_2=0.999, epsilon=1e-8, decay=0.0)
    model.compile(optimizer = adam, loss = 'mean_squared_error',
                  metrics=['mean_absolute_error'])
    return model

al = ActiveLearning(model_creator = model_creator, U = features,
            target_layer = ['b1_l2', 'b2_l2'],
            train_size = 50,
            test_size = 50,
            batch_size = [20,0,0]
            )
tr_ind, te_ind = al.initialize()

# Start the active search
q = al.search(n_evaluation=1, epochs=1000, verbose=0)
```

References

1. Welcome to the ChemML's documentation! https://hachmannlab.github.io/chemml/index.html.
2. Haghighatlari, M.; Vishwakarma, G.; Altarawy, D.; Subramanian, R.; Kota, B. U.; Sonpal, A.; Setlur, S.; Hachmann, J. ChemML: A Machine Learning and Informatics Program Package for the Analysis, Mining, and Modeling of Chemical and Materials Data, 2019. https://doi.org/10.26434/chemrxiv.8323271.v1.
3. chemml (GitHub) https://github.com/hachmannlab/chemml.

87 AI in material design

Molecular property prediction is a fundamental problem for materials science. Quantum-chemical simulations such as density functional theory (DFT) have been widely used for calculating the molecule properties; however, because of the heavy computational cost, it is difficult to search a huge number of potential chemical compounds.

Selected topics

Find extraordinary materials

Machine learning (ML) can extrapolate to extraordinary materials within the AFLOW[1] dataset. *Extrapolation as a classification task* can improve results.[2]

Crystallizability: how ML tools can predict novel molecules with 90% accuracy

Crystallinity can be predicted from atomic connectivity with an accuracy of 92% by classification of solid form data extracted from CSD and ZINC using ML.[3]

Data preparation was done by converting CSD and ZINC smiles to canonical smiles in RDKit to find "crystalline" molecules in ZINC.

Molecular property prediction: gated graph recursive neural networks

Shindo and Matsumoto[4] show a simple and powerful graph neural network (NN) for molecular property prediction. Experimental results show that the model achieves state-of-the-art performance on the standard benchmark dataset for molecular property prediction.

Material data discovery: reliable and explainable ML methods

Interesting in Kailkhura et al.[5] is how underrepresented/imbalanced material data break down the ML performance and lead to misleading conclusions. It is proposed that a novel pipeline employs an ensemble of simpler models to reliably predict material properties.

https://doi.org/10.1515/9783110629453-087

Implementation example

MEGNet with molecules

MatErials Graph Network[6,7] (MEGNet) is an implementation of Google's DeepMind's graph networks for universal ML in materials science. The NN topology achieves very low prediction errors in a broad array of properties in both molecules and crystals.

This example outlines how a custom model can be constructed from MEGNet-Layer, which is an implementation of a graph network using NNs.

```python
# Predict DFT trained on QM9 dataset (megnet)
# author: materialsvirtuallab/crystals.ai
# License:BSD 3-Clause "Revised" License
# code: github.com/materialsvirtuallab/megnet
# activity: active (2020)              # index: 87-1


# extract, load, transform
# convert the XYZ files (stored as strings) to an OpenBabel data object
data = pd.read_json('molecules.json').sample(1000)
from pybel import readstring
structures = [readstring('xyz', x) for x in data['xyz']]
targets = data['u0_atom'].tolist()   # Atomization energy computed with B3LYP DFT


# model setup
import tensorflow as tf
from megnet.data.molecule import MolecularGraph
from megnet.models import MEGNetModel
model = MEGNetModel(27, 2, 27, nblocks=1, lr=1e-2,
                    n1=4, n2=4, n3=4, npass=1, ntarget=1,
                graph_converter = MolecularGraph())
mg = MolecularGraph()


# train
model.train(structures, targets, epochs=24, verbose=0)


# predict
predicted_atom = [model.predict_structure(x) for x in structures]


# plot
from matplotlib import pyplot as plt
fig, ax = plt.subplots()
ax.scatter(targets, predicted_atom)
...
fig.tight_layout()
```

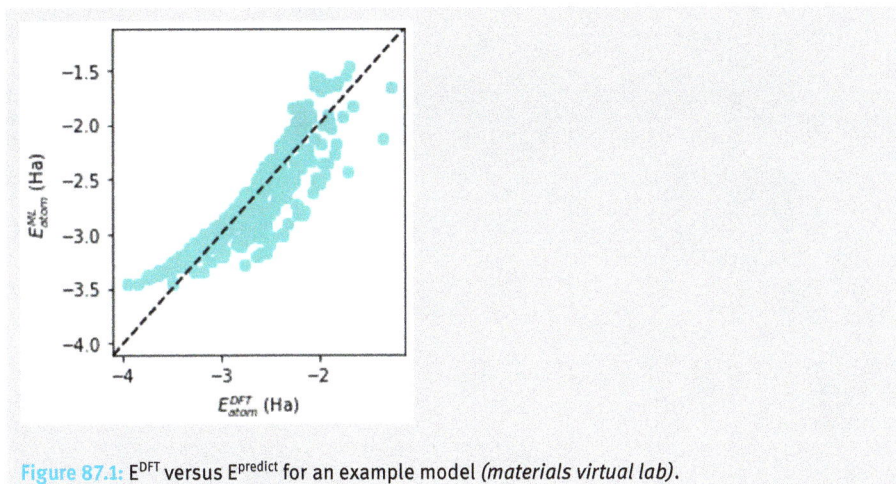

Figure 87.1: E^{DFT} versus $E^{predict}$ for an example model *(materials virtual lab)*.

References

1. Aflow – Automatic – FLOW for Materials Discovery http://www.aflowlib.org/.
2. Kauwe, S. K.; Graser, J.; Murdock, R.; Sparks, T. D. Can Machine Learning Find Extraordinary Materials? *Comput. Mater. Sci.* **2020**, *174*, 109498. https://doi.org/10.1016/j.commatsci.2019.109498.
3. Wicker, J. Will it crystallize? http://compchemkitchen.org/wp-content/uploads/2016/06/Jerome-Wicker-CCK1-May2016_final.pdf.
4. Shindo, H.; Matsumoto, Y. Gated Graph Recursive Neural Networks for Molecular Property Prediction. *arXiv [cs.LG]*, 2019.
5. Kailkhura, B.; Gallagher, B.; Kim, S.; Hiszpanski, A.; Han, Y.-J. Reliable and Explainable Machine Learning Methods for Accelerated Material Discovery https://www.researchgate.net/publication/330157800_Reliable_and_Explainable_Machine_Learning_Methods_for_Accelerated_Material_Discovery. https://doi.org/10.13140/RG.2.2.15103.25766.
6. crystals.ai https://megnet.crystals.ai/.
7. Megnet code (GitHub) https://github.com/materialsvirtuallab/megnet.

Knowledge and information

88 Ontologies and inferencing

An ontology[1] encompasses a *representation, formal naming, and definition* of the *categories, properties, and relations* between the *concepts, data, and entities* that substantiate one, many, or all domains of discourse.

More simply, an ontology is a way of showing the properties of a subject area and how they are related, by defining a set of concepts that represent the subject.

Every discipline creates ontologies to limit complexity and organize knowledge. New ontologies improve problem-solving.

An ontology defines a *common vocabulary* for researchers who need to share information in a domain. It includes machine-interpretable definitions of basic concepts and relations among them. Sharing a common understanding of the structure of information among people or software agents are one of the more common goals in developing ontologies[2]:

–   To enable reuse of domain knowledge
–   To make domain assumptions explicit
–   To separate domain knowledge from the operational knowledge
–   To analyze domain knowledge

Implementation languages

No particular upper ontology has yet gained widespread acceptance as a de facto standard. Two examples of ontology languages:

–   The **Cyc** project has its own ontology language called CycL, based on first-order predicate calculus with some higher order extensions.
–   **W3C Web Ontology Language (OWL)** is a semantic web language for making ontological statements, developed as a follow-up from RDF and RDFS, as well as earlier ontology language projects including OIL, DAML, and DAML+OIL. OWL is intended to be used over the World Wide Web, and all its elements (classes, properties, and individuals) are defined as RDF resources and identified by URIs.

https://doi.org/10.1515/9783110629453-088

OWL example

An OWL annotation list[3] has this structure:

```
<owl:Ontology rdf:about="">
  <rdfs:comment>An example OWL ontology</rdfs:comment>
  <owl:imports rdf:resource="http://www.w3.org/.../chemistry"/>
    <rdfs:label>Water (H2O)</rdfs:label>
      <owl:AnnotationProperty rdf:beilstein_reference="3587155" />
</owl:Ontology>
```

Example: Ontology Chart

Relationship extraction[4]: identification of relations between entities. example of dimethyl maleate[5]:

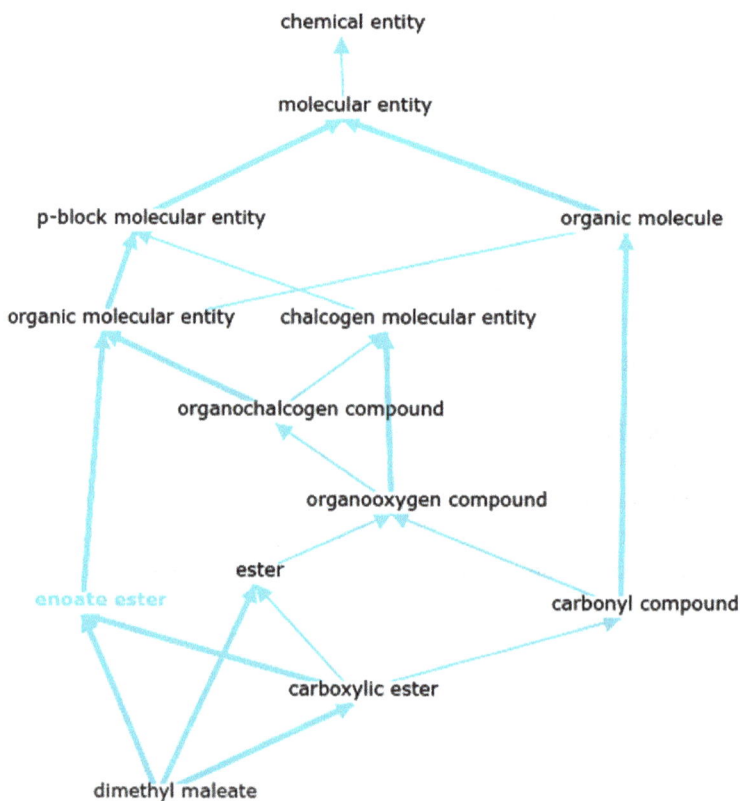

Figure 88.1: Dimethyl maleate ontology (CHEBI:*35460*).

Example of a design tool: Protégé

– WebProtégé is a web browser-based collaborative and free OWL ontology editor.[6]
– Also the original Java-based Protégé is developed by the Protégé team in the Biomedical Informatics Research Group (BMIR) at Stanford University, California, USA.

Other proprietary designs and query tools are AllegroGraph, Stardog, or PoolParty Semantic Suite.

Create an ontology

Before performing further calculations like reasoning, all classes, properties, and instances have to be created. Also, restrictions and disjointness/distinctnesses have been defined too. This can be done directly in the XML of the OWL or by Python so that the ontology is present in the program.

Example creating a "reasoning-ready" ontology with Python

Owlready2[7,8] is a package for ontology-oriented programming in Python. It can load OWL 2.0 ontologies as Python objects, modify them, save them, and perform reasoning via HermiT. Owlready2 allows transparent access to OWL ontologies.

```
# Create an ontology
# author: Lamy, Jean-Baptiste
# License:LGPL 3          # code: github.com/pwin/owlready2
# activity: active (2020)  # index: 88-1

from owlready import *

onto = Ontology("http://test.org/onto.owl")

class Drug(Thing):
    ontology = onto
    def take(self): print("I took a drug")

class ActivePrinciple(Thing):
    ontology = onto
```

```
class has_for_active_principle(Property):
    ontology = onto
    domain   = [Drug]
    range    = [ActivePrinciple]
    ANNOTATIONS[has_for_active_principle]["python_name"] = "active_principles"

class Placebo(Drug):
    equivalent_to = [Drug & NOT(restriction(
                        has_for_active_principle, SOME, ActivePrinciple))]
    def take(self): print("I took a placebo")

class SingleActivePrincipleDrug(Drug):
    equivalent_to = [Drug & restriction(
                    has_for_active_principle, EXACTLY, 1, ActivePrinciple)]
    def take(self): print("I took a drug with a single active principle")

class DrugAssociation(Drug):
    equivalent_to = [Drug & restriction(
                        has_for_active_principle, MIN, 2, ActivePrinciple)]
    def take(self): print("I took a drug with %s active principles" %
                    len(self.active_principles))

acetaminophen   = ActivePrinciple("acetaminophen")
amoxicillin     = ActivePrinciple("amoxicillin")
clavulanic_acid = ActivePrinciple("clavulanic_acid")

AllDistinct(acetaminophen, amoxicillin, clavulanic_acid)

drug1 = Drug(active_principles = [acetaminophen])
drug2 = Drug(active_principles = [amoxicillin, clavulanic_acid])
drug3 = Drug(active_principles = [])

drug1.closed_world()
drug2.closed_world()
drug3.closed_world()
```

Adding Python methods to an OWL class

Python methods can be defined in ontology classes as usual in Python.[9] In the following example, the Drug class has a Python method for computing the per-tablet cost of a drug, using two OWL properties (which have been renamed in Python, see Associating Python alias name to properties):

```
# Adding Python methods to an OWL Class
# author: Lamy, Jean-Baptiste
# license:LGPL 3           # code: github.com/pwin/owlready2
# activity: active (2020)  # index: 88-2


from owlready2 import *

onto = get_ontology("http://test.org/onto.owl")
with onto:
    class Drug(Thing):
        def get_per_tablet_cost(self):
            return self.cost / self.number_of_tablets

    class has_for_cost(Drug >> float, FunctionalProperty):
        python_name = "cost"

    class has_for_number_of_tablets(Drug >> int, FunctionalProperty):
        python_name = "number_of_tablets"

my_drug = Drug(cost = 10.0, number_of_tablets = 5)
print(my_drug.get_per_tablet_cost())
> 2.0
```

Inferencing and reasoning

In the field of artificial intelligence, the *inference* engine is a component of the system that applies logical rules to the knowledge base to deduce new information. The typical expert system consisted of a knowledge base and an inference engine.

A *reasoning* engine is a piece of software able to infer logical consequences from a set of asserted facts or axioms. The notion of a semantic reasoner *generalizes* that of an *inference engine*, by providing a richer set of mechanisms to work with. The inference rules are commonly specified by means of an ontology language, and often a description logic language. Many reasoners use first-order predicate logic to perform reasoning (see Chapter 32).

Difference between inference, reasoning, deduction, and resolution

All four terms[10] at first glance seems to be synonymous. But there are clear differences:

- **Reasoning** is the capacity for *consciously* making sense of things, applying logic, establishing and verifying facts, and changing beliefs based on new information. Reasons form
- **Inferences** which are *conclusions* drawn from propositions or assumptions that are supposed to be true (Bottom-up).
- **Deduction** is *reasoning* that goes from known truths to specific instances. It starts with a hypothesis and examines the possibilities to reach a *conclusion* (Top-down).
- **Resolution** is anything that brings the truth of the knowledge.

Reasoning with Python

Several OWL reasoners exist; Owlready includes a modified version of the HermiT reasoner, developed by the Department of Computer Science of the University of Oxford, and released under the LGPL license. The reasoner is simply run by calling the .sync_reasoner() method of the ontology:

The library gets the results of the reasoning (aka automatic classification) from HermiT and reclassifies the Classes and Instances, that is, Owlready changes the *superclasses of Classes* and the *Classes of Instances*.

```
# Reasoning example on drug intake
# author: Lamy, Jean-Baptiste
# license:LGPL 3              # code: github.com/pwin/owlready2
# activity: active (2020)  # index: 88-3


onto.sync_reasoner()

print("drug2 new Classes:", drug2.__class__)
> drug2 new Classes: onto.DrugAssociation

drug1.take()
> I took a drug with a single active principle

drug2.take()
> I took a drug with 2 active principles

drug3.take()
> I took a placebo
```

In this example, drug1, drug2, and drug3 classes have changed; the reasoner deduced that drug2 is an association drug and that drug3 is a placebo.

References

1. Ontology (information science) (Wikipedia) https://en.wikipedia.org/wiki/Ontology_ (information_science).
2. Noy, N. F.; Mc Guinness, D. L. Ontology Development 101: A Guide to Creating Your First Ontology https://protege.stanford.edu/publications/ontology_development/ontology101.pdf.
3. OWL Web Ontology Language Reference https://www.w3.org/TR/owl-ref/.
4. Kuo, K. -L.; Fuh, C.-S. A Rule-Based Clinical Decision Model to Support Interpretation of Multiple Data in Health Examinations. *J. Med. Syst.* **2011**, *35* (6), 1359–1373. https://doi.org/10.1007/s10916-009-9413-3.
5. EBI Web Team. dimethyl maleate (CHEBI:35460) https://www.ebi.ac.uk/chebi/chebiOntology. do?chebiId=35460.
6. webprotege (GitHub) https://github.com/protegeproject/webprotege.
7. Lamy, J.-B. Owlready2 0.3 documentation https://pythonhosted.org/Owlready2/index.html.
8. Lamy, J. -B. owlready2 (GitHub) https://github.com/pwin/owlready2.
9. Lamy, J. -B. Mixing Python and OWL https://owlready2.readthedocs.io/en/latest/ mixing_python_owl.html.
10. Difference between inference, reasoning, deduction, and resolution? (Quora) https://www.quora.com/What-is-the-difference-between-inference-reasoning-deduction- and-resolution.

89 Analyzing networks

Network science[1] is an academic field which studies complex networks such as telecommunication networks, computer networks, biological networks, cognitive and semantic networks, and social networks, considering distinct elements or actors represented by *nodes (or vertices)* and the connections between the elements or actors as *links (or edges)*.

In mathematics, graph theory[2] is the study of graphs, which are mathematical structures used to model pairwise relations between objects. A graph in this context is made up of vertices (also called nodes or points) which are connected by edges (also called links or lines).

In Chapter 25, an ontology with neo4j was created and the concept of a graph was discussed. Graph data can be analyzed further with the NetworkX graphs library.

Ingest graph data

Cypher query results can be coerced to NetworkX MultiDiGraphs, graphs that permit multiple edges between nodes with the get_graph method.

Neo4j Cypher cell and line magic for notebooks. With ipython-cypher, a Cypher query can be defined and then converted to a NetworkX graph. The library can be installed with `pip install ipython-cypher`[3] providing a connection to a graph database by using `neo4jrestclient` driver. This introduces a %cypher (and %%cypher) magic for Neo4j in IPython:

Python example

```
# Reading Neo4j into network
# author: Gressling, T
# license: MIT License        # code: github.com/gressling/examples
# activity: single example    # index: 89-1

!pip install networkx
import networkx as nx

%load_ext cypher
results = %cypher MATCH p = (:Labworker)-[:test]->(:Samples) RETURN p
g = results.get_graph()

%matplotlib inline
nx.draw(g)
```

https://doi.org/10.1515/9783110629453-089

Analyze networks

NetworkX[4,5] is a Python package for the creation, manipulation, and study of the structure, dynamics, and function of networks. Applications with the most important algorithms[5] are construction, shortest path, minimum spanning tree as well as page-rank and centrality measures.

Python example: laboratory workplace metric

```
# Laboratory workplace metric
# author: Gressling, T
# license: MIT License        # code: github.com/gressling/examples
# activity: single example   # index: 89-2

# data ingestion: workplaces and their distances [ft]
edgelist = [['Analytical scale workplace 2', 'HPLC', 90],\
            ['Analytical scale workplace 2', 'Chemicals storage', 80],\
            ['NMR prep', 'NMR BRUKER', 190],\
            ['Analytical scale workplace 1', 'Preparation workplace 1', 230],\
            ['Analytical scale workplace 1', 'Glassware', 68],\
            ['Preparation workplace 1', 'Preparation workplace 2', 120],\
            ['Preparation workplace 1', 'NMR BRUKER', 102],\
            ['Preparation workplace 1', 'Analytical scale workplace 1', 230],\
            ['Preparation workplace 2', 'Preparation workplace 1', 120],\
            ['Glassware', 'Analytical scale workplace 1', 68],\
            ['Glassware', 'Chemicals storage', 300],\
            ['HPLC', 'Analytical scale workplace 2', 90],\
            ['HPLC', 'NMR BRUKER', 230],\
            ['NMR BRUKER', 'Preparation workplace 1', 102],\
            ['NMR BRUKER', 'NMR prep', 190],\
            ['NMR BRUKER', 'HPLC', 230],\
            ['Chemicals storage', 'Analytical scale workplace 2', 80],\
            ['Chemicals storage', 'Glassware', 300]]

import networkx as nx
g = nx.Graph()
for edge in edgelist:
    g.add_edge(edge[0],edge[1], weight = edge[2])

nx.draw_networkx(g, node_color='cyan', node_size=5000, width=2,
font_size=20)

# Graph, draw figure a.
for i, x in enumerate(nx.connected_components(g)):
    print("cc" + str(i) + ":", x)
```

```
> cc0: {'HPLC prep', 'Chemicals storage', 'NMR prep', 'Analytical scale
workplace 2',
> 'Preparation workplace 2', 'NMR BRUKER', 'Preparation workplace 1',
'Glassware', 'HPLC', # 'Analytical scale workplace 1'}


# shortest path
print(nx.shortest_path(g, 'Analytical scale workplace 2','NMR BRUKER',
weight='weight'))
print(nx.shortest_path_length(g, 'Analytical scale workplace 2','NMR
BRUKER', weight='weight'))
> ['Analytical scale workplace 2', 'HPLC', 'NMR BRUKER']
> 320


# Minimum Spanning Tree, figure b.
nx.draw_networkx(nx.minimum_spanning_tree(g), node_color='cyan',
node_size=5000, width=2, font_size=20)
```

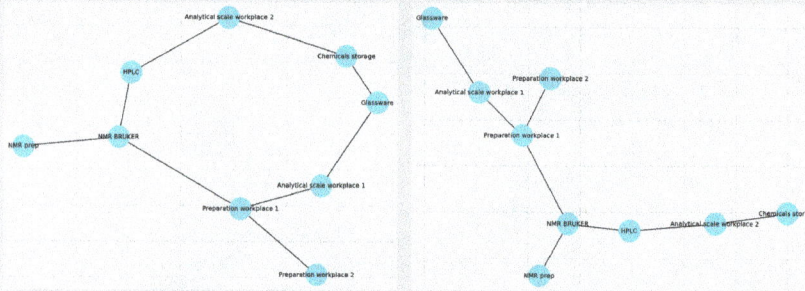

Figure 89.1: Graph (a) of lab workplaces **and** the minimum spanning tree (b) *(Gressling).*

M2M communication analysis

This example is based on a consolidated log of devices represented by ID's. Each distinct communication event is a link (or edge):

```
# M2M_Communication.prn
#  ID_from  ID_to
#     10      66
#     10      142
#     ...
```

Python example

```
# M2M Communication analysis chemical production plants
# author: Gressling, T
# license: MIT License      # code: github.com/gressling/examples
# activity: single example  # index: 89-3


comm = nx.read_edgelist('data/M2M_Communication.prn', create_using =
nx.Graph(), nodetype = int)

import matplotlib.pyplot as plt
pos = nx.spring_layout(comm)

import warnings
warnings.filterwarnings('ignore')
plt.style.use('fivethirtyeight')
plt.rcParams['figure.figsize'] = (20, 15)
plt.axis('off')
nx.draw_networkx(comm, pos, with_labels = False, node_size = 35)
plt.show()
```

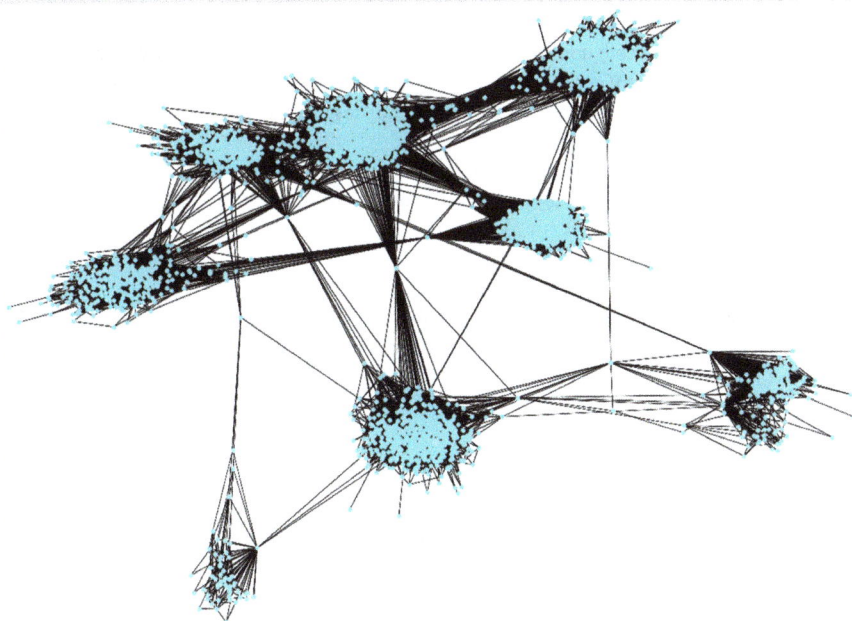

Figure 89.2: Communication pattern visualization *(Gressling).*

First- and second-degree connected nodes</H3>

Which device has a high influence capability in the data mesh? The Pagerank algorithm will give a higher score to a device contacted most by others:

```python
pageranks = nx.pagerank(comm)
print(pageranks)
# {0: 0.003787725030046891, 2: 0.0001729476557301711, 1:
0.000250061631885421, ...

import operator
sorted_pagerank = sorted(pageranks.items(), key=operator.itemgetter(1),
reverse = True)
print(sorted_pagerank) # 107 has the most connections
# [(107, 0.0038944622004718063), (1684, 0.003888183069053935), (3437,
0.0024484946788188742), ...

first_degree_connected_nodes = list(comm.neighbors(107))
second_degree_connected_nodes = []
for x in first_degree_connected_nodes:
    second_degree_connected_nodes+=list(comm.neighbors(x))
second_degree_connected_nodes.remove(107)
second_degree_connected_nodes = list(set(second_degree_connected_nodes))
subgraph = nx.subgraph(comm,
        first_degree_connected_nodes + second_degree_connected_nodes)
pos = nx.spring_layout(subgraph)
node_color = ['cyan' for v in subgraph]
node_size =  [1500 if v == 107 else 10 for v in subgraph]
plt.style.use('fivethirtyeight')
plt.rcParams['figure.figsize'] = (20, 15)
plt.axis('off')
nx.draw_networkx(subgraph, pos, with_labels = False,
                    node_color=node_color,node_size=node_size )
plt.show()
```

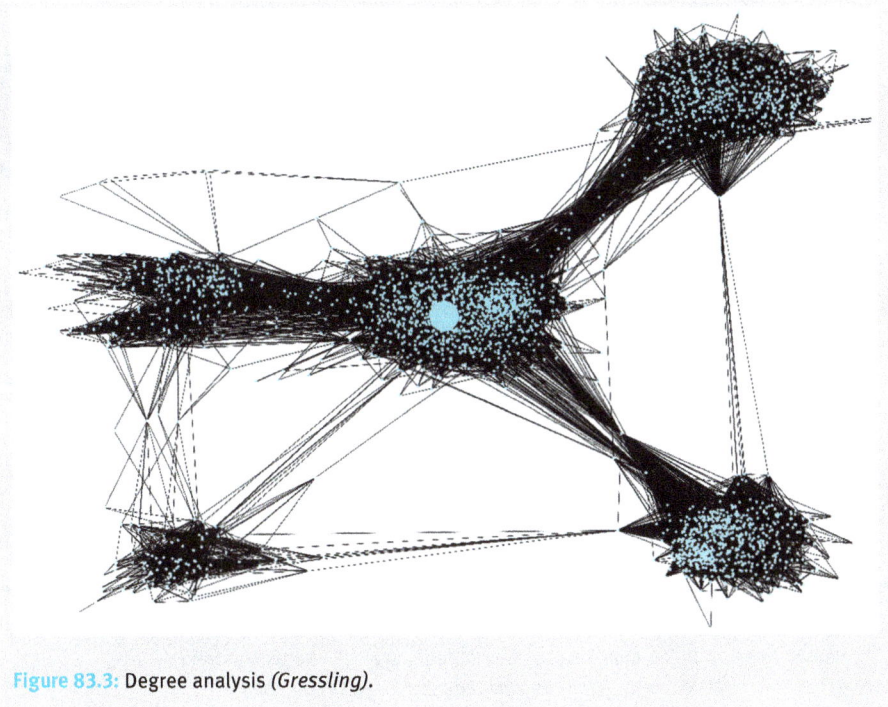

Figure 83.3: Degree analysis *(Gressling).*

Centrality measures

This type of measures can be used as features in machine learning models:
- Betweenness centrality: Not only the absolute number of connections per device may be relevant, but also devices that connect one geography to another are also important as their data contains information from *diverse geographies* (here: chemical production plants). Betweenness centrality quantifies how many times a particular node comes in the shortest chosen path between two other nodes.
- Degree centrality: It is simply the number of connections for a node.

```
pos = nx.spring_layout(subgraph)
betweennessCentrality = nx.betweenness_centrality(subgraph,
                              normalized=True, endpoints=True)
node_size =  [v * 20000 for v in betweennessCentrality.values()]
plt.figure(figsize=(20,20))
nx.draw_networkx(subgraph, pos=pos, with_labels=False,
                 node_size=node_size )
plt.axis('off')
> ...
```

Use cases in chemistry, AI, and data science

–   **Reactome**
    Explore[6] biomolecular pathways in Reactome[7] is an open-source, curated, and
    peer-reviewed pathway database. The database also contains expert-authored
    and reactions, molecular complex ingredients and structure, and protein–protein
    interactions. It also provides links from pathways and molecules to biological
    models and contains associated research papers, URLs, and books.
–   **Gated graph recursive neural networks for molecular property prediction[8]**
–   **Molecular/chemical graph[9]**
–   **MolecularRNN: generating realistic molecular graphs with optimized pro-
    perties[10]**
–   **How powerful are graph neural networks[11]?**
–   **Practical examples in Apache Spark and Neo4j[12]**

References

1.   Network science (Wikipedia) https://en.wikipedia.org/wiki/Network_science.
2.   Graph theory (Wikipedia) https://en.wikipedia.org/wiki/Graph_theory.
3.   de la Rosa, J. versae – Neo4j Cypher cell and line magic for IPython http://versae.es/.
4.   NetworkX documentation (GitHub) https://networkx.github.io/documentation/networkx-1.10/
     index.html.
5.   Agarwal, R. Data Scientists, The 5 Graph Algorithms that you should know https://
     towardsdatascience.com/data-scientists-the-five-graph-algorithms-that-you-should-know-
     30f454fa5513.
6.   Klopfenstein. ReactomePy Examples (GitHub) https://github.com/dvklopfenstein/ReactomePy;
     https://github.com/dvklopfenstein/ReactomePy/tree/master/src/ipy/tutorial.
7.   Reactome Pathway Database https://reactome.org/.
8.   Shindo, H.; Matsumoto, Y. Gated Graph Recursive Neural Networks for Molecular Property
     Prediction. *arXiv [cs.LG]*, 2019.
9.   Molecular / Chemical graph (Wikipedia) https://en.wikipedia.org/wiki/Molecular_graph.
10. Popova, M.; Shvets, M.; Oliva, J.; Isayev, O. MolecularRNN: Generating realistic molecular
     graphs with optimized properties https://www.researchgate.net/publication/333564111_
     MolecularRNN_Generating_realistic_molecular_graphs_with_optimized_properties.
11. Xu, K.; Hu, W.; Leskovec, J.; Jegelka, S. How Powerful Are Graph Neural Networks? *arXiv [cs.LG]*,
     2018.
12. Needham, M.; Hodler, A. E. Practical Examples in Apache Spark & Neo4j https://neo4j.com/
     books/free-book-graph-algorithms-practical-examples-in-apache-spark-and-neo4j/.

90 Knowledge ingestion: labeling and optical recognition

An annotation is extra information associated with a particular point in a document or other piece of information.

Labeled data[1] are a group of samples that have been tagged with one or more labels. Labeling typically takes a set of unlabeled data and augments each piece of that unlabeled data with meaningful tags that are informative.

Optical recognition for printed chemical formula

Printed chemical literature is still an important resource for chemical information.[2] Chemical named entity recognition (NER) systems can ingest information from documents, but still, there is some structural and linguistic complexity left.

Recognition engines

MOLVEC

A Java library for converting any chemical structure image into a variety of 2D molecular vectors is MOLVEC,[3,4] which achieves over 90% accuracy on the USPTO dataset.

OSRA

OSRA[5] is a utility designed to convert graphical representations of chemical structures into SMILES. OSRA can read a document in any of the 90+ graphical formats.

IMAGO OCR

Imago OCR[6] is also a toolkit for 2D chemical structure image recognition. It contains a GUI program and a command-line utility, as well as a documented API for developers. The core part of Imago is written from scratch in C++.

https://doi.org/10.1515/9783110629453-090

Python example

Test data and images are available[3] in several collections:
- TREC-CHEM (1000 binary images + molfiles)
- USPTO subset (5,710 binary images + molfiles)
- MayBridge subset (2,665 binary images + molfiles)

```
# Printed image to molecule (molvec)
# author: Gressling, T
# license: MIT License        # code: github.com/gressling/examples
# activity: single example    # index: 90-1

# {image} is the input image file and {engine} is one of the
# supported values: molvec, osra, imago, or all.
# Content-Type can be any supported image formats
# curl --data-binary @{image} -H Content-Type:image/png
# https://molvec.ncats.io/{engine}

from PIL import Image # https://pillow.readthedocs.io/
import urllib2
im = Image.open("molecule_scan.png")
url = 'https://molvec.ncats.io/{engine}'

# supported values: molvec, osra, imago, or all
req = urllib2.Request(url, data, {'Content-Type': 'image/png'})
f = urllib2.urlopen(req)
for x in f:
    print(x)
f.close()
```

Example using IMAGO with imagoPy

```
# Structure image to *.mol (imagoPy)
# author: brighteragyemang@gmail.com7 (real name unknown)
# license: GNU General Public License v3.0
# code: github.com/bbrighttaer/imagopy
# activity: active (2020)                # index: 90-2

# Imago OCR API functions calling order:
# imagoAllocSessionId()
# imagoSetSessionId(id)
# imagoLoadImageFromFile() / imagoLoadImageFromBuffer /
imagoLoadGreyscaleRawImage
# imagoSetFilter() [optional]
# imagoFilterImage()
```

```
#      imagoSetConfig() [optional]
#      imagoRecognize()
#      imagoSaveMolToFile() / imagoSaveMolToBuffer()
# imagoReleaseSessionId(id)

import rdkit as rd
import rdkit.Chem as chem
import imagopy as ocr
%matplotlib inline
x = ocr.imagoAllocSessionId()
ocr.imagoSetSessionId(x)

# Data load
> figure 90-1a (https://pubchem.ncbi.nlm.nih.gov/compound/59037449)
img = mpimg.imread('imago_test.png')
imgplot = plt.imshow(img)
ocr.imagoLoadImageFromFile('imago_test.png')

ocr.imagoRecognize()
# ocr.imagoSaveMolToFile('imago_test.mol')

from rdkit.Chem import Draw
Draw.MolToImage(mol)
> figure 90-1b

chem.MolToSmiles(mol)
> 'CC1C2CC(C3CCCC23)C1NC(=O)c1ncccc1O'
```

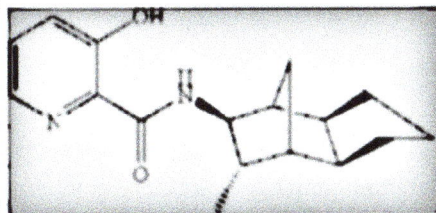

a. Original Xerox copy

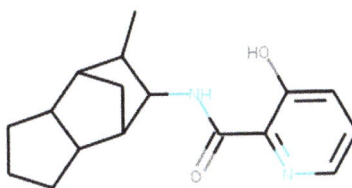

b. 3-hydroxy-N-[(2S,6S,9S)-9-methyl-8-tricyclo[5.2.1.02,6]decanyl]pyridine-2-carboxamide

Figure 90.1: Original image and extracted structure *(brighteragyemang)*.

OCR: optical character recognition

Optical character recognition[8] or optical character reader (OCR) is the electronic or mechanical conversion of images of typed, handwritten, or printed text into machine-encoded text, whether from a scanned document, a photo of a document.

Tesseract

Tesseract is Google's open-source OCR engine.[9] The program has a python wrapper.[10] Tesseract was originally developed at Hewlett-Packard. Since 2006, it has been developed by Google. The latest (LSTM based, see Chapter 38) stable version is 4.

```
# OCR with Tesseract
# author: Gressling, T
# License: MIT License        # code: github.com/gressling/examples
# activity: single example    # index: 90-3

from PIL import Image
import Image
import pytesseract

# PATH for the engine or:
 pytesseract.pytesseract.tesseract_cmd =
          r'<path_to_tesseract_executable>'

print(pytesseract.image_to_string(Image.open('gressling_handwriting.jpg')))
> Data Science in Chemistry
```

NLP

Natural language processing (NLP) is a subfield of linguistics, computer science, information engineering, and artificial intelligence concerned with the interactions between computers and human (natural) languages, in particular how to program computers to process and analyze large amounts of natural language data.

Challenges in natural language processing frequently involve speech recognition, natural language understanding, and natural language generation.

Python example with spaCy

spaCy[11,12] performs at large-scale information extraction tasks. It is written from the ground up in memory-managed Cython. The entity visualizer, ent, highlights named entities and their labels in a text.

```
# NLP of chemical literature (spaCy)
# author: Gressling, T
# license: MIT License        # code: github.com/gressling/examples
# activity: single example    # index: 90-4


# !pip install nltk
# !pip install spacy
import nltk
import spacy
from spacy import displacy
import en_core_web_sm
nlp = en_core_web_sm.load()


doc = nlp("Perovskite is composed of calcium titanate")
# create a spaCy object
doc = nlp(text)
> Perovskite --> nsubjpass --> PROPN
> is --> auxpass --> AUX
> composed --> ROOT --> VERB
> of --> prep --> ADP
> calcium --> compound --> NOUN
> titanate --> pobj --> NOUN


# Dependencies
options = {"compact": False, "distance": 130}
displacy.render(doc, style='dep',jupyter=True, options=options)
> Figure 90-2a


# Entities
doc = nlp("Perovskite's notable crystal structure was first described by
Victor Goldschmidt in 1926 in his work on tolerance factors. The crystal
structure was later published in 1945 from X-ray diffraction data.")
displacy.render(doc, style='ent',jupyter=True, options=options)
> Figure 90-2b
```

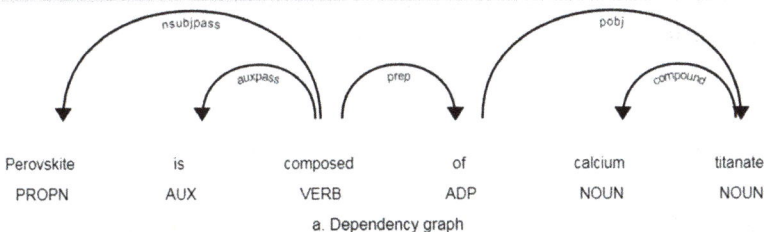

a. Dependency graph

Perovskite ORG 's notable crystal structure was first described by Victor Goldschmidt PERSON in 1926 DATE in his work on tolerance factors. The crystal structure was later published in 1945 DATE from X-ray diffraction data

b. Entity recognition

Figure 90.2: (a) and (b) Original image and extracted structure *(Gressling)*.

Labels and tags

In a supervised learning model, the algorithm learns on a labeled dataset,[13] providing an answer key that the algorithm can use to evaluate its accuracy on training data. An unsupervised model, in contrast, provides unlabeled data that the algorithm tries to make sense of by extracting features and patterns on its own. For example, labels might indicate whether a sample is rejected or has passed QC.

Doccano

Doccano[14,15] is a manual open-source text annotation tool. It provides annotation features:
– text classification,
– sequence labeling,
– sequence-to-sequence tasks.

With doccano, it is possible to create labeled data for sentiment analysis, NER, and text summarization.

Snorkel

Labeling and managing training datasets by hand is one of the biggest bottlenecks in machine learning. With Snorkel[16] training, datasets can be built *programmatically with heuristic functions*. This is called programmatic or weak supervision and the domain knowledge of humans – that is contained in a knowledge base – can be transferred into labeled data by a python script. Also, noisy sources can be cleaned with unsupervised modeling techniques to automatically clean and integrate them. This example shows how to use the labeling functions (LFs) that are one of the core operators:

```
def lf1(x):
    cid = (x.chemical_id, x.disease_id)
    return 1 if cid in KB else 0
```

```
def lf2(x):
    m = re.search(r'.*cause.*', x.between)
    return 1 if m else 0
```

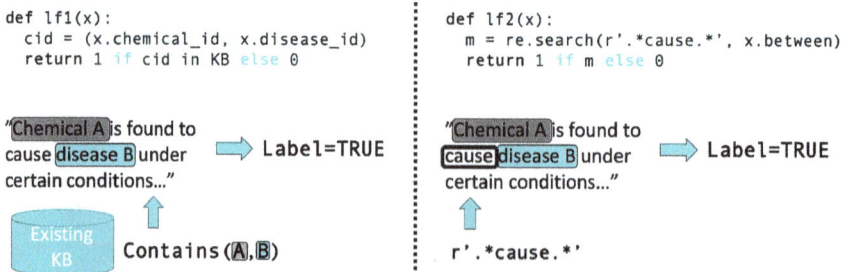

Figure 90.3: Snorkel example *(Stanford [17], Hazy Lab)*.

References

1. Labeled data (Wikipedia) https://en.wikipedia.org/wiki/Labeled_data.
2. Zhai, Z.; Nguyen, D. Q.; Akhondi, S. A.; Thorne, C.; Druckenbrodt, C.; Cohn, T.; Gregory, M.; Verspoor, K. Improving Chemical Named Entity Recognition in Patents with Contextualized Word Embeddings. *arXiv [cs.CL]*, 2019.
3. MOLVEC Demonstration https://molvec.ncats.io/.
4. molvec (GitHub) https://github.com/ncats/molvec.
5. OSRA: Optical Structure Recognition https://cactus.nci.nih.gov/osra/.
6. Imago OCR https://lifescience.opensource.epam.com/imago/index.html.
7. bbrighttaer. imagopy – A python wrapper for Imago OCR (GitHub) https://github.com/bbrighttaer/imagopy.
8. Optical character recognition OCR (Wikipedia) https://en.wikipedia.org/wiki/Optical_character_recognition.
9. Tesseract Open Source OCR Engine (GitHub) https://github.com/tesseract-ocr/tesseract.
10. Lee, M. A. pytesseract (GitHub) https://github.com/madmaze/pytesseract.
11. spaCy · Industrial-strength Natural Language Processing in Python https://spacy.io/.
12. Joshi, P. Introduction to Information Extraction Using Python and spaCy https://www.analyticsvidhya.com/blog/2019/09/introduction-information-extraction-python-spacy/.
13. Salian, I. NVIDIA Blog: What's the Difference Between Supervised & Unsupervised Learning? https://blogs.nvidia.com/blog/2018/08/02/supervised-unsupervised-learning/.
14. chakki. doccano – open source text annotation (GitHub) https://github.com/chakki-works/doccano.
15. Nakayama, H.; Kubo, T.; Kamura, J.; Taniguchi, Y.; Liang, X. doccano: Text Annotation Tool for Human https://github.com/doccano/doccano.
16. Snorkel https://www.snorkel.org/.
17. Ratner, A.; Varma, P.; Hancock, B.; Ré, C.; other members of Hazy Lab. Weak Supervision: A New Programming Paradigm for Machine Learning (Apache 2.0) http://ai.stanford.edu/blog/weak-supervision/.

91 Content mining and knowledge graphs

Text mining[1] is "the discovery by computer of new, previously unknown information, by automatically extracting information from different written resources."

Resources can be websites, books, emails, reviews, and articles. Text mining also referred to as text analytics is the process of deriving high-quality information from text.

Bring it all together: graphs and ingestion

Knowledge graphs are one of the most fascinating concepts in data science. The term is used[2] as the name for a knowledge base used by Google and its services to enhance its search engine's results with information gathered from a variety of sources.

In Chapter 25, a graph with neo4j was created and aspects of graph theory were discussed. Following Chapter 88, the ontology concept was analyzed. Finally, in this chapter, the concept is put into reality and the ingestion and usage of knowledge are discussed.

Publicly available ontologies

Semantic integration is the process of interrelating information from diverse sources. Metadata publishing within semantic integration offers the ability to link ontologies. An approach is *semantic similarity* and appropriate rules, for example, in OWL the attributes `owl:equivalentClass`, `owl:equivalentProperty` and `owl:sameAs`. Other approaches include lexical methods, as well as methodologies that rely on exploiting the structures of the ontologies.

Reaxys by Elsevier

The ontology[3,4] contains:
- substances and keywords of automatic extraction
- 150k conference papers
- 80k patent families
- 2,2m journals from 16k journals

https://doi.org/10.1515/9783110629453-091

Ontochem SciWalker open data

SciWalker is a web search engine that implements advanced information retrieval and extraction from a variety of sources such as abstracts, full-text articles, patents, web pages, and many more. It uses a set of multi-hierarchical dictionaries for annotation and ontological concept searching. SciWalker open data can be used to annotate any public or internal repository of heterogeneous documents. The search interface allows us to construct queries with logical combinations of both free text and ontological terms.[5]

Google patent search

Based on the work of Ontochem, Google[6,7] has enhanced their knowledge graph in the chemistry domain. A curated patent[8] ontology with taxonomy annotation is publicly available:

```
https://patents.google.com/?q=xylenol+orange&oq=xylenol+orange
```

It is possible to enhance the patent API with a neural network, for example, to create a patent landscape.[7] In a jupyter notebook,[9] load a seed set of patents and generate positive and negative training data. Then a deep neural network on the data can be trained and inference is performed using the model, see example 91-3.

Python example

The data model is based on the OpenSearch 1.1 specification.

```
# Patent data ingestion from Google
# author: Gressling, T
# license: MIT License        # code: github.com/gressling/examples
# activity: single example    # index: 91-1

import requests
import urllib
import time
import json

access_token = 'AIzaSyDzkpt*********MTX2pTZUXJK8'
# <<API Token>>, https://console.developers.google.com/apis/credentials
cse_id = '0059099*********S1:0twmi42d4z2'
# Search engine ID, https://cse.google.com/cse
```

```
start=1
search_text = "(~SMILES=c1(C=O)cc(OC)c(O)cc1)"
# &tbm=pts sets the patent search
url = 'https://www.googleapis.com/customsearch/v1?key='
   + access_token + '&cx=' + cse_id + '&start='
   + str(start) + '&num=10&tbm=pts&q=' + urllib.parse.quote(search_text)

response = requests.get(url)
response.json()

>   'formattedTotalResults': '7,170'},
>   'items': [{'kind': 'customsearch#result',
>     'title': 'US20040106590A1 - Methods and reagents for treating
     infections of ...',
>     'htmlTitle': 'US20040106590A1 - Methods and reagents for treating
     infections
> <b>of</b> ...',
>     'link': 'https://patents.google.com/patent/US20040106590A1/en',

f = open('Sample_patent_data'+str(int(time.time()))+'.txt', 'w')
f.write(json.dumps(response.json(), indent=4))
f.close()

# ['items'][] and the sub ['pagemap']
# get titles, thumbnails, snippets, title, link, and citations
```

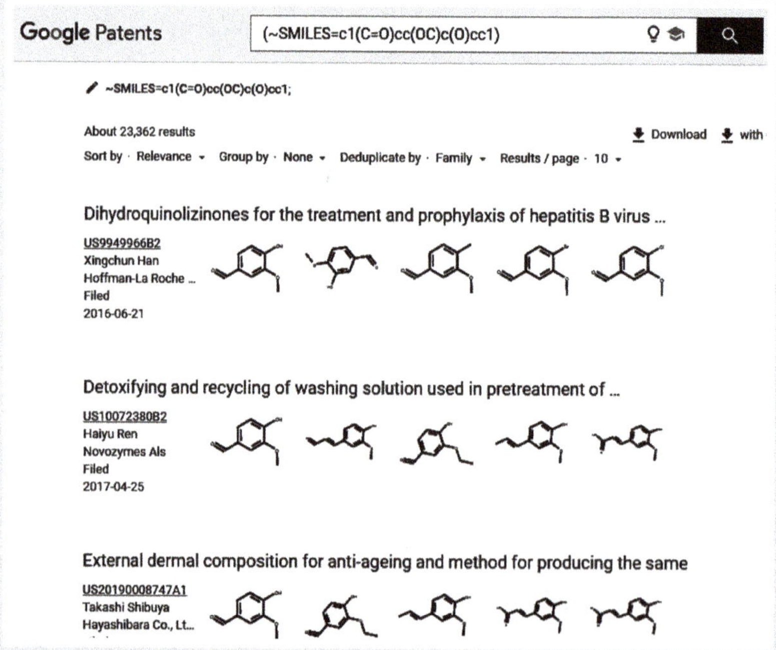

Figure 91.1: Google patents GUI example for vanillin *(Gressling)*.

CHEMBL/ChEBI

This database was discussed in Chapter 24 and also a Python example (24.4) "Query `ChEBI ontology via Python`" is listed.

SureChEMBL

SureChEMBL[10] is a publicly available large-scale resource containing compounds extracted from the full text, images, and attachments of patent documents. The data are extracted from the patent literature according to an automated text and image-mining pipeline on a daily basis.

Python example: finding key compounds in chemistry patents

Patent chemistry data extracted from SureChEMBL and after a series of filtering steps, it follows a few "traditional" chemoinformatics approaches with a set of claimed compounds.

The example identifies "key compounds" in patents using compound information[11-13]. The code is shortened to focus on the filter aspect:

```
# Finding key compounds in chemistry patents (SureChEMBL)
# author: Papadatos, George14
# license: CC BY 4.0              # code: github.com/rdkit/UGM_2014
# activity: single example (2014)  # index: 91-2

# The file was generated by extracting all chemistry from a list of patent
documents in
# SureChEMBL. The Lucene query used to retrieve the list of relevant patents
was:
# pn:"US6566360" (Bayer Levitra US patent US6566360)

df = pd.read_csv('document chemistry_20141011_114421_271.csv',sep=',')
```

The file was generated by extracting all chemistry from a list of patent documents in SureChEMBL. The Lucene query used to retrieve the list of relevant patents was: pn:"US6566360".

```
df.info()
<class 'pandas.core.frame.DataFrame'>
Int64Index: 8497 entries, 0 to 8496
Data columns (total 28 columns):
Patent ID              8497 non-null object
Annotation Reference   8497 non-null object
```

```
Chemical ID                8497 non-null int64
SMILES                     8497 non-null object
Type                       8497 non-null object
Chemical Document Count    8497 non-null int64
Annotation Document Count  3372 non-null float64
Title Count                3372 non-null float64
Abstract Count             3372 non-null float64
Claims Count               3372 non-null float64
Description Count          3372 non-null float64

# This is the base URL of the publicly accessible ChEMBL Beaker server.
# https://www.ebi.ac.uk/chembl/api/utils/docs

base_url = 'www.ebi.ac.uk/chembl/api'
pd.options.display.mpl_style = 'default'
pd.options.display.float_format = '{:.2f}'.format
rcParams['figure.figsize'] = 16, 10

# First filtering: Novel compounds appear in the description
# or claims section of the document. Alternatively,
# they are extracted from images or mol files

dff = df[(df['Claims Count'] > 0) | (df['Description Count'] > 0)
                 | (df['Type'] != "TEXT")]

# Second filtering: Simple physicochemical properties and counts
dff = dff[(dff['Rotatable Bound Count'] < 15) & (dff['Ring Count'] > 0)
                 & (df['Radical'] == 0) & (df['Singleton'] == 0)]

dff_counts
# Compounds      Link
> Patent ID
> EP-1049695-A1  2      EP-1049695-A1
> EP-1049695-B1  7      EP-1049695-B1
> EP-1174431-A2  1      EP-1174431-A2
> EP-2295436-A1  68     EP-2295436-A1
```

Knowledge Graph

How to represent knowledge in a Graph? The procedure is:
- Sentence Segmentation
- Entities Extraction
- Relations Extraction

Example: build a knowledge graph from text data

A triple represents a couple of entities and a relation between them.[15] For example, [Diels–Alder, synthesis, rings] is a triple in which "Diels–Alder" and "ring" are the related entities, and the relation between them is "synthesis."

To build a knowledge graph, the most important things are the nodes and the edges between them:

```
# Build a Knowledge Graph from Text Data
# author: Gressling, T
# license: MIT License        # code: github.com/gressling/examples
# activity: single example    # index: 91-3

import spacy
from spacy import displacy
import en_core_web_sm
nlp = en_core_web_sm.load()

from spacy.matcher import Matcher
from spacy.tokens import Span
import networkx as nx

# 0. Ingestion
# sentences by extraction from wikipedia, to CSV
candidate_sentences = pd.read_csv("sentences.csv")
candidate_sentences.shape
> (16, 1)

mydf = candidate_sentences['sentence'] # extract column
mydf
> 0    Benzoic acid is classified as a solid.
> 1    Benzoic acid has the formula C6H5CO2H.
> 2    Benzoic acid is classified as carboxylic acid.
> 3    Benzoic acid occurs naturally in many plants.
> 4    Benzoic acid is a precursor for the industrial synthesis.
> 5    Propionic acid is classified as carboxylic acid.
> 6    Propionic acid has the formula CH3CH2CO2H.
> 7    Propionic acid has a pungent smell.
> 8    Propionic acid was described in 1844.
> 9    Acetic acid is classified as a liquid.
> 10   Acetic acid has the formula CH3COOH.
> 11   Acetic acid is sometimes called glacial acid.
> 12   Acetic acid has a pungent smell.
> 13   Acetic acid is classified as weak acid.
> 14   Acetic acid is classified as carboxylic acid.
> 15   Acetic acid consists of a methyl group attached to a carboxyl group.
> Name: sentence, dtype: object
```

```python
# Tokenization test
doc = nlp("Propionic acid is a liquid with a pungent smell.")

for tok in doc:
  print(tok.text, "...", tok.dep_)

> Propionic ... amod
> acid ... nsubj
> is ... ROOT
> a ... det
> liquid ... attr
> with ... prep
> a ... det
> pungent ... amod
> smell ... pobj
> . ... punct


# 1. Entities
def get_entities(sent):
    # ... omitted, see code in GitHub
  return [ent1.strip(), ent2.strip()]

entity_pairs = []
for i in (tqdm(mydf)):
  entity_pairs.append(get_entities(i))

entity_pairs
> [['Benzoic acid', 'solid'],
> ['Benzoic acid', 'formula C6H5CO2H.'],
>  ['Benzoic acid', 'carboxylic  acid'],
> ...
>  ['Acetic  acid', 'weak  acid'],
>  ['Acetic  acid', 'carboxylic  acid'],
>  ['Acetic  acid', 'carboxyl group']]


# 2. Relations
def get_relation(sent):
  doc = nlp(sent)
  matcher = Matcher(nlp.vocab)
  pattern = [{'DEP':'ROOT'},
             {'DEP':'prep','OP':"?"},
             {'DEP':'agent','OP':"?"},
             {'POS':'ADJ','OP':"?"}]
    # ... omitted, see code in GitHub
  return(span.text)


# 3. Build knowledge graph
relations = [get_relation(i) for i in tqdm(mydf)]
```

```
# extract subject
source = [i[0] for i in entity_pairs]
# extract object
target = [i[1] for i in entity_pairs]

kg_df = pd.DataFrame({'source':source,
                      'target':target,
                      'edge':relations})

# Plot the graph
G=nx.from_pandas_edgelist(kg_df, "source", "target",
                          edge_attr=True,
                          create_using=nx.MultiDiGraph())
plt.figure(figsize=(12,12))
pos = nx.spring_layout(G)
nx.draw(G, with_labels=True, node_color='skyblue', edge_cmap=plt.cm.Blues,
pos = pos)
plt.show()
```

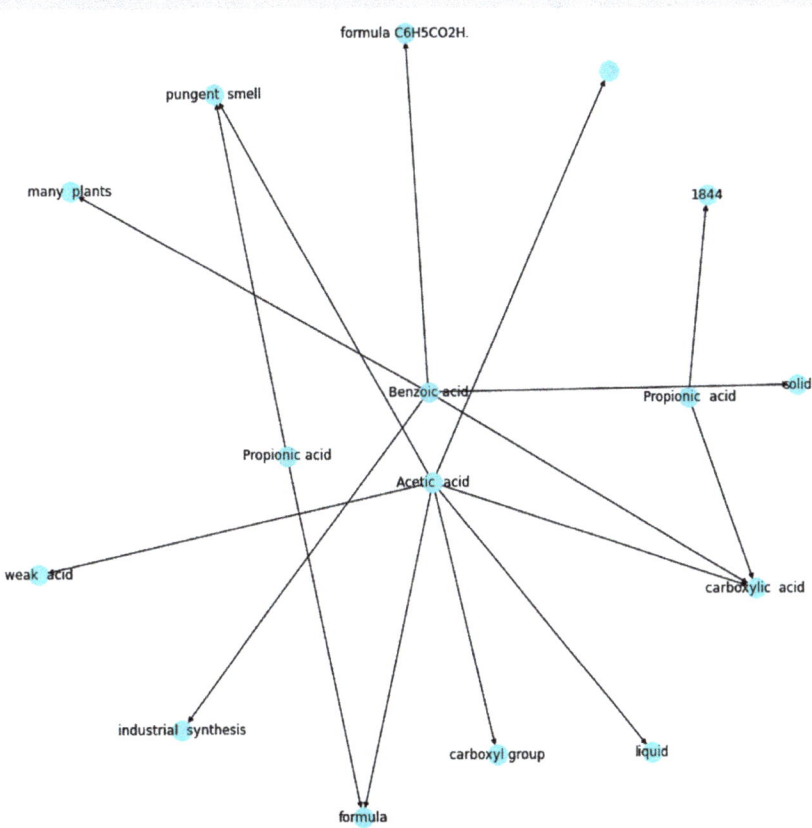

Figure 91.2: Knowledge graph (Gressling).

References

1. Text mining (Wikipedia) https://en.wikipedia.org/wiki/Text_mining.
2. Knowledge Graph (Wikipedia) https://en.wikipedia.org/wiki/Knowledge_Graph.
3. Elsevier. Reaxys API: Integrating Reaxys with other chemistry research systems – Reaxys | Elsevier Solutions https://www.elsevier.com/solutions/reaxys/how-reaxys-works/api.
4. Lawson, A. J.; Swienty-Busch, J.; Géoui, T.; Evans, D. The Making of Reaxys – Towards Unobstructed Access to Relevant Chemistry Information. In *The Future of the History of Chemical Information*; ACS Symposium Series; American Chemical Society, 2014; Vol. 1164, pp 127–148. https://doi.org/10.1021/bk-2014-1164.ch008.
5. Weber, L. SciWalker Open Data https://sciwalker.com/sciwalker/faces/search.xhtml.
6. Deka, M. Google-Patents (GitHub) https://github.com/minz27/Google-Patents.
7. patents-public-data (GitHub) https://github.com/google/patents-public-data.
8. GreyB. The Definitive Guide To Google Patents Search https://www.greyb.com/google-patents-search-guide/.
9. Automated Patent Landscaping (GitHub, Google) https://github.com/google/patents-public-data/blob/master/models/landscaping/LandscapeNotebook.ipynb.
10. Papadatos, G.; Davies, M.; Dedman, N.; Chambers, J.; Gaulton, A.; Siddle, J.; Koks, R.; Irvine, S. A.; Pettersson, J.; Goncharoff, N.; Hersey, A.; Overington, J. P. SureChEMBL: A Large-Scale, Chemically Annotated Patent Document Database. *Nucleic Acids Res.* **2016**, *44* (D1), D1220–D1228. https://doi.org/10.1093/nar/gkv1253.
11. Papadatos, G. ChEMBL blog http://chembl.github.io/page3/.
12. Hattori, K.; Wakabayashi, H.; Tamaki, K. Predicting Key Example Compounds in Competitors' Patent Applications Using Structural Information Alone. *J. Chem. Inf. Model.* **2008**, *48* (1), 135–142. https://doi.org/10.1021/ci7002686.
13. Tyrchan, C.; Boström, J.; Giordanetto, F.; Winter, J.; Muresan, S. Exploiting Structural Information in Patent Specifications for Key Compound Prediction. *J. Chem. Inf. Model.* **2012**, *52* (6), 1480–1489. https://doi.org/10.1021/ci3001293.
14. Papadatos, G. SureChEMBL iPython Notebook Tutorial https://nbviewer.jupyter.org/github/rdkit/UGM_2014/blob/master/Notebooks/Vardenafil.ipynb.
15. Joshi, P. Introduction to Information Extraction Using Python and spaCy https://www.analyticsvidhya.com/blog/2019/09/introduction-information-extraction-python-spacy/.

Part D: **Quantum computing and chemistry**
Introduction

92 Quantum concepts

Q-concepts — Q-computing — Q-Algorithms ⊤ Q-Chemistry on Q-computers
QChemistry software ⌙

Quantum *computing*[1] is the use of quantum-mechanical phenomena (such as superposition and entanglement) to perform *computation*.

Quantum *chemistry*[2] (molecular quantum mechanics) is a branch of chemistry focused on the application of quantum mechanics on chemical systems. Understanding electronic structure and molecular dynamics using the Schrödinger equations as well as studies of the ground state of molecules and the transition states that occur during chemical reactions are central topics in quantum chemistry. (Theoretical quantum chemistry tends to fall under the category of computational chemistry.)

General procedure

Performing quantum chemistry + calculations on quantum computers → means using *algorithms* that have *supremacy* on quantum computers (which means they are hard to calculate on classic computers).

All current *quantum chemical calculations* on quantum computers → use standard QM software that has special plugins → to use these quantum algorithms → that run best on quantum processors.

Selected concepts

Uncertainty principle

In quantum mechanics, the uncertainty principle[3] (also known as Heisenberg's uncertainty principle) is the fundamental limit to the precision with which the values for certain pairs of physical quantities can be predicted.
    The uncertainty principle implies that it is in general not possible to predict the value of a quantity with arbitrary certainty, even if all initial conditions are specified.

https://doi.org/10.1515/9783110629453-092

Energy states

A quantum mechanical system or particle that is bound can only take on certain discrete values of energy,[4] called energy levels. This contrasts with classical particles, which can have any amount of energy.

The term is commonly used for the energy levels of electrons in atoms, ions, or molecules. The energy spectrum of a system with such discrete energy levels is said to be quantized.

Wave function

A wave function[5] in quantum physics is a mathematical description of the quantum state of an isolated quantum system. The wave function is a complex-valued probability amplitude, and the probabilities for the possible results of measurements made on the system can be derived from it. The most common symbols for a wave function are the Greek letters ψ and Ψ (lower-case and capital psi, respectively).

The wave function is a function of the degrees of freedom corresponding to some maximal set of commuting observables. Once such a representation is chosen, the wave function can be derived from the quantum state.

Hartree–Fock calculation

The simplest type of ab initio electronic structure calculation is the Hartree–Fock[6] (HF) method, an extension of molecular orbital theory.[7]

The concept of Slater-type orbitals are exact solutions for the hydrogen atom and provide an accurate basis set for many-electron molecules however the calculations of the integrals are expensive as there is no simple exact solution for the integrals.

Schrödinger's equation

The Schrödinger equation is a linear partial differential equation that describes the wave function or state function of a quantum-mechanical system. This example animates a simple quantum mechanical system with python:

```
# General Numerical Solver 1D Time-Dependent Schrodinger's equation
# author: Vanderplas, Jake
# license: BSD
# code: jakevdp.github.io/blog/2012/09/05/quantum-python
# activity: single Example (2012)    # index: 92-1
```

```python
from scipy.fftpack import fft,ifft
class Schrodinger(object):
def time_step(self, dt, Nsteps = 1):
    for i in xrange(Nsteps - 1):
        self.compute_k_from_x()
        self.psi_mod_k *= self.k_evolve
        self.compute_x_from_k()
        self.psi_mod_x *= self.x_evolve
    self.t += dt * Nsteps

def theta(x):
    x = np.asarray(x)
    y = np.zeros(x.shape)
    y[x > 0] = 1.0
    return y

def square_barrier(x, width, height):
    return height * (theta(x) - theta(x - width))

# specify time steps and duration
dt = 0.01
N_steps = 50
t_max = 120
frames = int(t_max / float(N_steps * dt))

# specify constants
hbar = 1.0    # planck's constant
m = 1.9       # particle mass

# specify a range in x coordinate
N = 2 ** 11
dx = 0.1
x = dx * (np.arange(N) - 0.5 * N)

# specify potential
V0 = 1.5
L = hbar / np.sqrt(2 * m * V0)
a = 3 * L
x0 = -60 * L
V_x = square_barrier(x, a, V0)
V_x[x < -98] = 1E6
V_x[x > 98] = 1E6

# specify initial momentum and quantities derived from it
p0 = np.sqrt(2 * m * 0.2 * V0)
dp2 = p0 * p0 * 1./80
d = hbar / np.sqrt(2 * dp2)
```

```
k0 = p0 / hbar
v0 = p0 / m
psi_x0 = gauss_x(x, d, x0, k0)

# define the Schrodinger object which performs the calculations
S = Schrodinger(x=x,
                psi_x0=psi_x0,
                V_x=V_x,
                hbar=hbar,
                m=m,
                k0=-28)


# Set up plot
fig = pl.figure()

xlim = (-100, 100)
klim = (-5, 5)
# top axes show the x-space data
ymin = 0
ymax = V0
ax1 = fig.add_subplot(211, xlim=xlim,
                      ylim=(ymin - 0.2 * (ymax - ymin),
                            ymax + 0.2 * (ymax - ymin)))
psi_x_line, = ax1.plot([], [], c='r', label=r'$|\psi(x)|$')
V_x_line, = ax1.plot([], [], c='k', label=r'$V(x)$')
center_line = ax1.axvline(0, c='k', ls=':',
                          label = r"$x_0 + v_0t$")

title = ax1.set_title("")
ax1.legend(prop=dict(size=12))
ax1.set_xlabel('$x$')
ax1.set_ylabel(r'$|\psi(x)|$')
ax2.legend(prop=dict(size=12))
ax2.set_xlabel('$k$')
ax2.set_ylabel(r'$|\psi(k)|$')
V_x_line.set_data(S.x, S.V_x)

pl.show()
```

$$t = 65.00$$

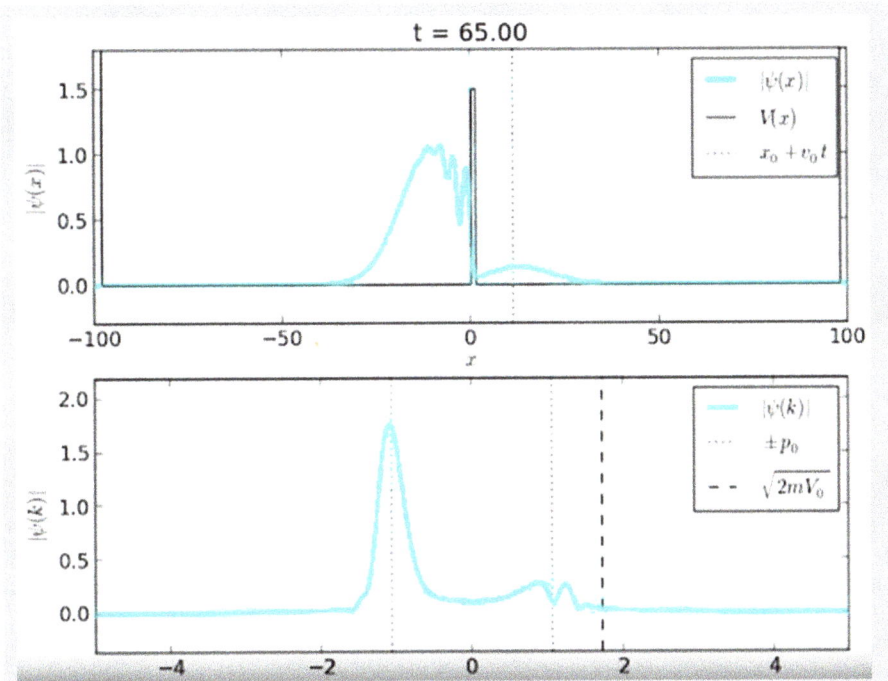

Figure 92.1: Animation of Schrödinger's equation *(Vanderplas).*

Pauli's exclusion

The Pauli exclusion principle[8] is the quantum mechanical principle which states that two or more identical fermions (particles with half-integer spin) cannot occupy the same quantum state within a quantum system simultaneously. It is fundamental to use qubits.

Dirac (Bra-ket) notation

Dirac notation, also known as Bra-ket notation, is a symbol set for depict quantum states, that is, vectors in the Hilbert space (where the algebra of the observables acts is located, for example, all possible wave functions).

Dirac notation solves these issues by presenting a new language to fit the precise needs of quantum mechanics. While column vector notation is ubiquitous in linear algebra, it is often ponderous in quantum computing, especially when dealing with qubits.

```
# Dirac notation (sympy)
# author: SymPy Development Team
# License: New BSD          # code: github.com/gressling/examples
# activity: active (2020)   # index: 92-2

from sympy.physics.quantum import Ket, Bra
from sympy import symbols, I

k = Ket('psi')
k
>  |psi>

k.hilbert_space
>  H
>  k.is_commutative
>  False

# Take a linear combination of two kets:
k0 = Ket(0)
k1 = Ket(1)
2*I*k0 - 4*k1
>    2*I*|0> - 4*|1>
```

Quantum operations

In quantum computing and specifically the quantum circuit model of computation, a quantum logic gate[9] (or simply quantum gate) is a basic quantum circuit operating on a small number of qubits. They are the building blocks of quantum circuits like classical logic gates for conventional digital circuits.

Visualizing operators (QuTip)

Sometimes, it may also be useful to directly visualize the underlying matrix representation of an operator. The density matrix, for example, is an operator whose elements can give *insights about the state it represents*, but one might also be interested in plotting the matrix of a Hamiltonian to inspect the structure and relative importance of various elements.

```
# Visualizing operators
# author: Johansson, J.R.
# license: CC BY 3.0
# code: qutip.org/.../guide/guide-visualization
# activity: active (2018)    # index: 92-3

from qutip import *
import matplotlib as mpl
from matplotlib import cm

a = tensor(destroy(N), qeye(2))
b = tensor(qeye(N), destroy(2))
sx = tensor(qeye(N), sigmax())
H = a.dag() * a + sx - 0.5 * (a * b.dag() + a.dag() * b)

rho_ss = steadystate(H, [np.sqrt(0.1) * a, np.sqrt(0.4) * b.dag()])
fig, ax = hinton(rho_ss)
plt.show()
```

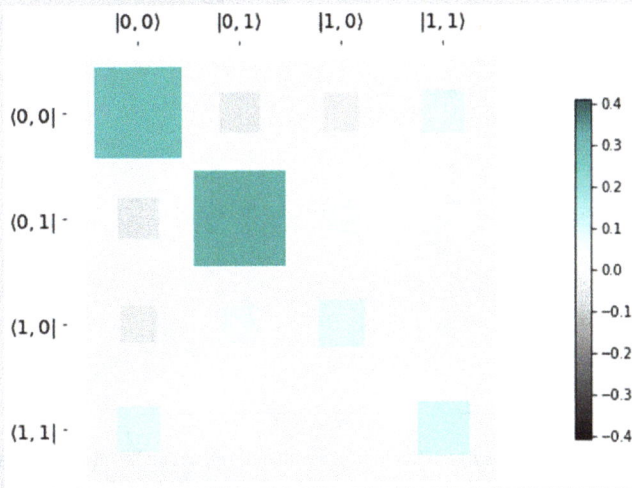

Figure 92.2: Operator visualizations *(Johansson).*

Quantum logic

Any quantum computation can be represented as a network of quantum logic gates from a fairly small family of gates. A choice of gate family that enables this construction is known as a universal gate set. One such common set includes all single-qubit gates as well as the CNOT gate from above. This means any quantum computation can be performed by executing a sequence of single-qubit gates together with CNOT gates.

Operator	Gate(s)	Matrix
Pauli-X (X)	—X— —⊕—	$\begin{bmatrix} 0 & 1 \\ 1 & 0 \end{bmatrix}$
Pauli-Y (Y)	—Y—	$\begin{bmatrix} 0 & -i \\ i & 0 \end{bmatrix}$
Pauli-Z (Z)	—Z—	$\begin{bmatrix} 1 & 0 \\ 0 & -1 \end{bmatrix}$
Hadamard (H)	—H—	$\frac{1}{\sqrt{2}}\begin{bmatrix} 1 & 1 \\ 1 & -1 \end{bmatrix}$
Phase (S, P)	—S—	$\begin{bmatrix} 1 & 0 \\ 0 & i \end{bmatrix}$
$\pi/8$ (T)	—T—	$\begin{bmatrix} 1 & 0 \\ 0 & e^{i\pi/4} \end{bmatrix}$
Controlled Not (CNOT, CX)		$\begin{bmatrix} 1 & 0 & 0 & 0 \\ 0 & 1 & 0 & 0 \\ 0 & 0 & 0 & 1 \\ 0 & 0 & 1 & 0 \end{bmatrix}$
Controlled Z (CZ)		$\begin{bmatrix} 1 & 0 & 0 & 0 \\ 0 & 1 & 0 & 0 \\ 0 & 0 & 1 & 0 \\ 0 & 0 & 0 & -1 \end{bmatrix}$
SWAP		$\begin{bmatrix} 1 & 0 & 0 & 0 \\ 0 & 0 & 1 & 0 \\ 0 & 1 & 0 & 0 \\ 0 & 0 & 0 & 1 \end{bmatrix}$
Toffoli (CCNOT, CCX, TOFF)		$\begin{bmatrix} 1 & 0 & 0 & 0 & 0 & 0 & 0 & 0 \\ 0 & 1 & 0 & 0 & 0 & 0 & 0 & 0 \\ 0 & 0 & 1 & 0 & 0 & 0 & 0 & 0 \\ 0 & 0 & 0 & 1 & 0 & 0 & 0 & 0 \\ 0 & 0 & 0 & 0 & 1 & 0 & 0 & 0 \\ 0 & 0 & 0 & 0 & 0 & 1 & 0 & 0 \\ 0 & 0 & 0 & 0 & 0 & 0 & 0 & 1 \\ 0 & 0 & 0 & 0 & 0 & 0 & 1 & 0 \end{bmatrix}$

Figure 92.3: Common quantum logic gates by name (including abbreviation), circuit form(s), and the corresponding unitary matrices *(Rxtreme[9], Wikipedia)*.

QBit

A qubit[10] is the basic unit of quantum information – the quantum version of the classical binary bit physically realized with a two-state device. A qubit is a two-state (or two-level) quantum-mechanical system, one of the simplest quantum systems displaying the peculiarity of quantum mechanics.

Bloch sphere

The Bloch sphere is a representation of a qubit, the fundamental building block of quantum computers. The prevailing model of quantum computation describes the computation in terms of a network of quantum logic gates.

The sphere[11] is a geometrical representation of the pure state space of a two-level quantum mechanical system (qubit). Quantum mechanics is mathematically formulated in Hilbert space. The space of states of a quantum system is given by the one-dimensional subspaces of the corresponding Hilbert space (or the "points" of the projective Hilbert space).

Example: plotting the Bloch sphere [12]

```
# Plotting the Bloch sphere
# author: Johansson, J.R.
# license: CC BY 3.0  # code: qutip.org/docs/3.1.0/guide/guide-bloch
# activity: active (2018)   # index: 92-4


from qutip import *

b = Bloch()

# Add a single data point
pnt = [1/np.sqrt(3), 1/np.sqrt(3), 1/np.sqrt(3)]
b.add_points(pnt)
vec = [0,1,0]
b.add_vectors(vec)
b.show()
```

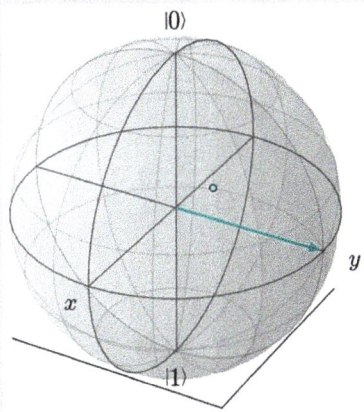

Figure 92.4: Bloch sphere (Johansson).

Example of a logic gate

If the Hadamard gate[9] acts on a single qubit, it maps the basis state, which means that measurement will have equal probabilities to become 1 or 0 (i.e., creates a superposition). It is a one-qubit rotation, mapping the qubit-basis states |0> and |1> to two superposition states with an equal weight of the computational basis states.

References

1. Quantum computing https://en.wikipedia.org/wiki/Quantum_computing.
2. Quantum chemistry (Wikipedia) https://en.wikipedia.org/wiki/Quantum_chemistry.
3. Uncertainty principle (Wikipedia) https://en.wikipedia.org/wiki/Uncertainty_principle.
4. Energy level (Wikipedia) https://en.wikipedia.org/wiki/Energy_level.
5. Wave function (Wikipedia) https://en.wikipedia.org/wiki/Wave_function.
6. Hartree–Fock method (Wikipedia) https://en.wikipedia.org/wiki/Hartree%E2%80%93Fock_method.
7. Computational chemistry (Wikipedia) https://en.wikipedia.org/wiki/Computational_chemistry.
8. Pauli exclusion principle (Wikipedia) https://en.wikipedia.org/wiki/Pauli_exclusion_principle.
9. Quantum logic gate (Wikipedia) https://en.wikipedia.org/wiki/Quantum_logic_gate.
10. Qubit (Wikipedia) https://en.wikipedia.org/wiki/Qubit.
11. Bloch sphere (Wikipedia) https://en.wikipedia.org/wiki/Bloch_sphere.
12. Plotting on the Bloch Sphere – QuTiP Documentation http://qutip.org/docs/3.1.0/guide/guide-bloch.html.

93 QComp: technology vendors

Q-concepts — Q-computing — Q-Algorithms ⊤ Q-Chemistry on Q-computers
QChemistry software ⌐

A quantum processing unit[1] (QPU), also referred to as a quantum chip, is a physical (fabricated) chip that contains a number of interconnected qubits. It is the foundational component of a full quantum computer, which includes the housing environment for the QPU, the control electronics, and many other components.[2]

This chapter contains a selection of quantum ecosystems where Goggle, D-Wave, IBM, and Rigetti have also quantum chips. The cover of this book shows a part of an IBM quantum computer.

Rigetti

Rigetti computing[3] is a Berkeley, California-based developer of quantum integrated circuits used for quantum computers. The company also develops a cloud platform called Forest that enables programmers to write quantum algorithms.

Example

Instantiate a program, declare memory with one single bit of memory, and apply an X gate on qubit 0. Finally, measure qubit 0 into the zeroth index of the memory space (ro[0]). Then, let it run on a locally installed QVM.

```
# Act an X gate on qubit (pyQuil)
# author: Rigetti computing
# license: Apache 2.0
# code: docs.rigetti.com/en/stable SDK
        and setup: qcs.rigetti.com/sdk-downloads
# activity: active (2020)      # index: 93-1

from pyquil import Program
from pyquil.gates import *

p = Program()
ro = p.declare('ro', 'BIT', 1)
p += X(0)
p += MEASURE(0, ro[0])
```

https://doi.org/10.1515/9783110629453-093

```
# print(p) # list the program

# run this program on the Quantum Virtual Machine (QVM)
# There must be a QVM server running at http://127.0.0.1:5000

from pyquil import get_qc
...
qc = get_qc('1q-qvm')
executable = qc.compile(p)
result = qc.run(executable)
print(result)
>    [[1]]
```

Xanadu

Xanadu[4,5] designs quantum silicon photonic chips and integrates them into existing hardware. PennyLane is Xanadu's own cross-platform Python library for differentiable programming of quantum computers.

Via plugin, they also support many other platforms and their QPU's like qisquit, Cirq, and Rigetti Forrest. Also supported is ProjectQ which is an open-source software framework for quantum computing (see Chapter 95) and the Microsoft Quantum Toolkit.

Example

Strawberry fields are a full-stack Python library for designing, simulating, and optimizing quantum optical circuits, which includes a suite of three simulators for execution on CPU or GPU.

```
# Pauli-Z phase shift gate (Xanadu)
# author: Xanadu
# License: Apache 2.0
# code: pennylane.readthedocs.io/.../introduction/circuits
# activity: active (2020)   # index: 93-2

import pennylane as qml

dev = qml.device('default.qubit', wires=2, shots=1000, analytic=False)

def my_quantum_function(x, y):
    qml.RZ(x, wires=0)
    qml.CNOT(wires=[0,1])
```

```
    qml.RY(y, wires=1)
    return qml.expval(qml.PauliZ(1))

circuit = qml.QNode(my_quantum_function, dev)

circuit(np.pi/4, 0.7)
>   0.7648421872844883


print(circuit.draw())
>   0: ──RZ(0.785)── ┌C────────────┐
>   1: ──────────────└X──RY(0.7)──┤ <Z>
```

D-Wave

D-Wave Systems, Inc. is a Canadian quantum computing company, based in Burnaby, British Columbia, Canada. D-Wave was the world's first company to sell computers to exploit quantum effects in their operation. The programming framework supporting the QPU's is named DWave Ocean. D-Wave processors have a different architecture than the other vendors, based on quantum annealing.

Example Boolean NOT gate

This solves a simple problem of a Boolean NOT gate to demonstrate the mathematical formulation of a problem as a binary quadratic model (BQM) and using Ocean tools (submission through an API).

```
# Boolean NOT Gate
# author: DWaveSys Team
# License: Apache 2.0      # code: docs.ocean.dwavesys.com/.../examples/not
# activity: active (2020)  # index: 93-3

from dwave.cloud import Client
client = Client.from_config(token='<<<_YOUR_TOKEN_>>>')
client.get_solvers()
> [Solver(id='2000Q_ONLINE_SOLVER1'), UnstructuredSolver(id='hybrid_v1')]

from dwave.system import DWaveSampler, EmbeddingComposite
sampler = EmbeddingComposite(
                DWaveSampler(solver={'2000Q_ONLINE_SOLVER1': True}))
# 5000 samples
Q = {('x', 'x'): -1, ('x', 'z'): 2, ('z', 'x'): 0, ('z', 'z'): -1}
sampleset = sampler.sample_qubo(Q, num_reads=5000)
print(sampleset)
```

```
>    x  z energy num_oc. chain_.
> 0  0  1   -1.0    2266     0.0
> 1  1  0   -1.0    2732     0.0
> 2  0  0    0.0       1     0.0
> 3  1  1    0.0       1     0.0
> ['BINARY', 4 rows, 5000 samples, 2 variables]

sampleset.first.sample["x"] != sampleset.first.sample["z"]
>  True
# Almost all the returned samples represent valid value assignments for a
NOT gate
```

IBM QISKit

QISKit is a publicly available[6,7] open-source quantum computing framework for programming quantum processors and conducting research. The IBM Q Experience is a platform that gives users in the general public access to a set of IBM's prototype quantum processors via the Cloud.

IBM has collected[8] a reference set of notebooks:

- Basics
- Terra (study circuits)
- Aer (simulate)
- Ignis (study noise)
- Aqua is for application development on NISQ computers

IBMQ example

Running a program on the IBM Quantum Experience[9] chips[10] the IBMBackend and the corresponding compiler can be chosen (an account for IBM's Quantum Experience is needed):

```
# Invoke the IBMQ engine (IBM)
# author: IBM
# license: Apache          # code: github.com/gressling/examples
# activity: active (2020)   # index: 93-4


import projectq.setups.ibm
from projectq.backends import IBMBackend


# create a IBM main compiler engine
eng = MainEngine(IBMBackend(),
                 engine_list=projectq.setups.ibm.get_engine_list())
```

Google

The Cirq library was developed by the Google AI Quantum Team, but it is not an official Google project. Cirq[11] is designed for writing, manipulating, and optimizing quantum circuits and then running them against quantum computers and simulators. Cirq attempts to expose the details of hardware, instead of abstracting them away, because, in the Noisy Intermediate-Scale Quantum (NISQ) regime, these details determine whether or not it is possible to execute a circuit at all.

Cirq example

```
# Square root of NOT (Cirq)
# author: The Cirq Developers, Sung, Kevin J.
# License: Apache 2.0        # code: github.com/quantumlib/cirq
# activity: active (2020)    # index: 93-5


import cirq
# Pick a qubit
qubit = cirq.GridQubit(0, 0)

# Create a circuit
circuit = cirq.Circuit(
    cirq.X(qubit)**0.5,  # Square root of NOT.
    cirq.measure(qubit, key='m')  # Measurement.
)
print("Circuit:")
print(circuit)


simulator = cirq.Simulator()
result = simulator.run(circuit, repetitions=20)
print("Results:")
print(result)

> Circuit:
> (0, 0): ——X^0.5——M(&m&)——

> Results:
> m=11000111111011001000
```

Microsoft QDK

Quantum programs on Microsoft Azure[12] are written in a language called Q# (*"Q-Sharp"*). There is a python wrapper to call Q# programs. Microsoft has no QPU hardware but supports many device types like the superconducting class or devices based on ion traps like Honywell.[13]

References

1. The Quantum Processing Unit (QPU) – pyQuil (Rigetti) https://pyquil-docs.rigetti.com/en/1.9/qpu.html.
2. Introduction to the D-Wave Quantum Hardware https://www.dwavesys.com/tutorials/background-reading-series/introduction-d-wave-quantum-hardware.
3. Rigetti Computing (Wikipedia) https://en.wikipedia.org/wiki/Rigetti_Computing.
4. Software – Xanadu Quantum Technologies https://www.xanadu.ai/.
5. Xanadu (GitHub) https://github.com/XanaduAI.
6. Grant, E.; Benedetti, M.; Cao, S.; Hallam, A.; Lockhart, J.; Stojevic, V.; Green, A. G.; Severini, S. Hierarchical Quantum Classifiers. *arXiv [quant-ph]*, 2018.
7. IBM. Qiskit IQX Tutorials https://github.com/Qiskit/qiskit-iqx-tutorials.
8. qiskit-tutorials (QISKit) https://github.com/Qiskit/qiskit-tutorials.
9. IBM Q Experience (Wikipedia) https://en.wikipedia.org/wiki/IBM_Q_Experience.
10. List of quantum processors (Wikipedia) https://en.wikipedia.org/wiki/List_of_quantum_processors.
11. Cirq documentation https://cirq.readthedocs.io/en/stable/.
12. Azure Quantum | Microsoft Azure https://azure.microsoft.com/en-us/services/quantum/.
13. Honeywell Quantum Computing https://www.honeywell.com/en-us/company/quantum.

94 Quantum computing simulators

Q-concepts − *Q-computing* − Q-Algorithms ⊤ Q-Chemistry on Q-computers
QChemistry software ⌟

Quantum *simulators* permit the study of quantum systems that are difficult to study in the laboratory and impossible to model with a supercomputer. The simulators may be contrasted with generally programmable "digital" quantum computers, which would be capable of solving a wider class of quantum problems.

The simulators can be used to implement and test quantum algorithms like the Variational-Quantum-Eigensolver (VQE).

Quantum simulators may not be confused with software packages for *computing quantum chemistry* like Gaussian™, PYSCF or PSI4, discussed in Chapter 96.

The Quantiki list[1] contains 100+ programs for discovering quantum machines, most of them coded in C++. Also, different languages and API exist. In this chapter, Python-related software is discussed.

ProjectQ (ETH)

ProjectQ[2,3] is an open-source software framework for quantum computing. It was started at ETH Zurich and allows users to implement their quantum programs in Python. ProjectQ can then translate these programs to any type of back-end, be it a simulator run on a classical computer or an actual quantum chip.

```
# Quantum Random Numbers (ProjectQ)
# authors: Steiger, Damian; Haener, Thomas
# License: Apache 2.0       # code: projectq.readthedocs.io
# activity: active (2020)   # index: 94-1

from projectq import MainEngine
# import the operations to perform (Hadamard and measurement)
from projectq.ops import H, Measure

# create a default compiler (the back-end is a simulator)
eng = MainEngine()
# allocate a quantum register with 1 qubit
qubit = eng.allocate_qubit()


H | qubit   # apply a Hadamard gate
Measure | qubit   # measure the qubit
```

https://doi.org/10.1515/9783110629453-094

```
eng.flush()  # flush all gates (and execute measurements)
# converting a qubit to int or bool gives access to the measurement result
print("Measured {}".format(int(qubit)))
> Measured: 1 # creates random bits (0 or 1)
> Measured: 1 # creates random bits (0 or 1)
> Measured: 0 # creates random bits (0 or 1)
```

QuTIP

QuTiP is open-source software for simulating the dynamics of open quantum systems. The library depends on the excellent Numpy, Scipy, and Cython.[4-7] QuTiP is open-source software for simulating the dynamics of open quantum systems.[7] Examples were listed in Chapter 92.

QCircuits

An open-source quantum circuit programming library in Python is QCircuits,[8] with a simple API designed:

```
# Producing Bell states (QCircuits)
# author: Webb, Andrew
# license: MIT License        # code: www.awebb.info/qcircuits
# activity: active (2020)     # index: 94-2

import qcircuits as qc
from itertools import product

# Creates each of the four Bell states
def bell_state(x, y):
    H = qc.Hadamard()
    CNOT = qc.CNOT()
    phi = qc.bitstring(x, y)
    phi = H(phi, qubit_indices=[0])
    return CNOT(phi)

for x, y in product([0, 1], repeat=2):
    print('\nInput: {} {}'.format(x, y))
    print('Bell state:')
    print(bell_state(x, y))

> Input: 0 0
> Bell state:
```

```
> 2-qubit state. Tensor:
> [[0.70710678+0.j 0.        +0.j]
>  [0.        +0.j 0.70710678+0.j]]
… (three more)
```

References

1. Miszczak J. List of QC simulators https://www.quantiki.org/wiki/list-qc-simulators.
2. ProjectQ Website https://projectq.ch/.
3. ProjectQ (GitHub) https://github.com/ProjectQ-Framework/ProjectQ.
4. Johansson, J. R.; Nation, P. D.; Nori, F. QuTiP: An Open-Source Python Framework for the Dynamics of Open Quantum Systems. *arXiv [quant-ph]* **2011**.
5. Johansson, J. R.; Nation, P. D.; Nori, F. QuTiP 2: A Python Framework for the Dynamics of Open Quantum Systems. *arXiv [quant-ph]*, 2012.
6. QuTiP: Quantum Toolbox in Python – QuTiP Documentation http://qutip.org/docs/latest/index.html.
7. Nation, P. D.; Johansson, J. R. QuTiP – Quantum Toolbox in Python http://qutip.org/.
8. QCircuits's documentation http://www.awebb.info/qcircuits/index.html.

95 Quantum algorithms

Q-concepts — Q-computing — Q-Algorithms ⊤ Q-Chemistry on Q-computers
QChemistry software ⌋

Richard Feynman and Yuri Manin suggested that a quantum computer[1] had the potential to simulate things that a classical computer could not. In 1994, Peter Shor developed a quantum algorithm for factoring integers that had the potential to decrypt RSA-encrypted communications.

Quantum supremacy was demonstrated[2] by a programmable quantum device that solved a problem that no classical computer can calculate (irrespective of the usefulness of the problem). An example is generating random numbers in a large solution space ($>2^{50}$) that cannot be computed near time with classic CPUs.

Shor

Shor's algorithm[3] is a quantum algorithm for factoring a number N in O((log N)3) time and O(log N) space, named after Peter Shor. The algorithm is significant because it implies that public-key cryptography might be easily broken, given a sufficiently large quantum computer.

```
# Running Shor's algorithm (IBM qiskit, on QPU)
# author: IBM
# license: MIT License        # code: github.com/gressling/examples
# activity: single example    # index: 95-1

# access: github.com/Qiskit/qiskit-ibmq-provider
# access: https://quantum-computing.ibm.com/

from qiskit import IBMQ
from qiskit.aqua import QuantumInstance
from qiskit.aqua.algorithms import Shor

IBMQ.enable_account('<<API TOKEN>>')
provider = IBMQ.get_provider(hub='ibm-q')
backend = provider.get_backend('ibmq_qasm_simulator')

factors = Shor(21)
result_dict = factors.run(QuantumInstance(backend,
            shots=1, skip_qobj_validation=False))
print(result_dict['factors'])
# Shor(21) will find the prime factors for 21
> 3 and 7
```

https://doi.org/10.1515/9783110629453-095

Oracles

A quantum oracle O is a "black box" operation that is used as input to another algorithm. Choosing the best way to implement an oracle[A] depends heavily on how this oracle will be used within a given algorithm. For example, the Deutsch-Jozsa algorithm relies on the oracle implemented in a first way, while Grover's algorithm relies on the oracle implemented in a second way.

Example (using Rigetti Forest SDK, see Chapter 93)

```
# Quantum oracle (Rigetti Forest, simulated)
# author: Kopczyk, Dawid
# license: "Feel free to use full code"
# code: dkopczyk.quantee.co.uk/grover-search/
# activity: active (2020)              # index: 95-2

from pyquil import Program, get_qc
from pyquil.gates import H, I
from pyquil.api import WavefunctionSimulator

# Construct quantum oracle (not a part of algorithm)
SEARCHED_STRING = "10"
N = len(SEARCHED_STRING)
oracle = np.zeros(shape=(2 ** N, 2 ** N))
for b in range(2 ** N):
    if np.binary_repr(b, N) == SEARCHED_STRING:
        oracle[b, b] = -1
    else:
        oracle[b, b] = 1
print(oracle)

> [[ 1.  0.  0.  0.]
>  [ 0.  1.  0.  0.]
>  [ 0.  0. -1.  0.]
>  [ 0.  0.  0.  1.]]
```

Grover

Grover's algorithm[5] solves the problem of unstructured search: Given a set of N elements forming a set X = {x1,x2,...,xN } and given a boolean function f : X → {0,1}, the goal is to find an element x* in X such that f(x*) = 1.

```
# Grover algorithm with (sympy⁶, simulated)
# author:SymPy Development Team
# License: BSD        # code: docs.sympy.org/.../physics/quantum/grover
# activity: active (2020)        # index: 95-3

from sympy.physics.quantum.qubit import IntQubit
from sympy.physics.quantum.qapply import qapply
from sympy.physics.quantum.grover import OracleGate

f = lambda qubits: qubits == IntQubit(2)
v = OracleGate(2, f)
qapply(v*IntQubit(2))
>  -|2>


qapply(v*IntQubit(3))
>  |3>
```

Deutsch–Jozsa

The Deutsch–Jozsa algorithm was one of the first examples of a quantum algorithm, which is a class of algorithms designed for execution on Quantum computers and has the potential to be more efficient than conventional algorithms by taking advantage of the quantum superposition and entanglement principles.

In the Deutsch–Jozsa problem, a black box is given computing a 0–1 valued function f(x1, x2, ..., xn). The black box takes n bits x1, x2, ..., xn and outputs the value f(x1, x2, ..., xn). The function in the black box is either constant (0 on all inputs or 1 on all inputs) or balanced (returns 1 for half the domain and 0 for the other half). The task is to determine whether f is constant or balanced.

```
# Deutsch-Jozsa
# author: Malykh, Egor
# License: MIT License
# code: github.com/meownoid/...example-deutsch-jozsa
# activity: active (2020)  # index: 95-4

I = QGate([[1, 0], [0, 1]])
H = QGate(np.array([[1, 1], [1, -1]]) / np.sqrt(2))
X = QGate([[0, 1], [1, 0]])
Y = QGate([[0, -1j], [1j, 0]])
Z = QGate([[1, 0], [0, -1]])

from quantum import QRegister, H, I, U
```

```
def is_constant(f, n):
    q = QRegister(n + 1, '0' * n + '1')
    q.apply(H ** (n + 1))
    q.apply(U(f, n))
    q.apply(H ** n @ I)
    return q.measure()[:~0] == '0' * n

def f1(x):
    return x
def f2(x):
    return 1
def f3(x, y):
    return x ^ y
def f4(x, y, z):
    return 0

print('f(x) = x is {}'.format('constant' if is_constant(f1, 1)
else 'balanced'))
print('f(x) = 1 is {}'.format('constant' if is_constant(f2, 1)
else 'balanced'))
print('f(x, y) = x ^ y is {}'.format('constant' if is_constant(f3, 2)
else 'balanced'))
print('f(x, y, z) = 0 is {}'.format('constant' if is_constant(f4, 3)
else 'balanced'))
```

The variational-quantum-eigensolver (VQE)

VQE is one of the latest quantum algorithms. **It opens the bridge to computational quantum chemistry**. VQE is a quantum/classical hybrid algorithm that can be used to find eigenvalues of a (often large) matrix H. When this algorithm is used in quantum simulations, H is typically the Hamiltonian of some system. In this hybrid algorithm, a quantum subroutine is run inside of a classical optimization loop.

The VQE algorithm[7] can be implemented with quantum gates:

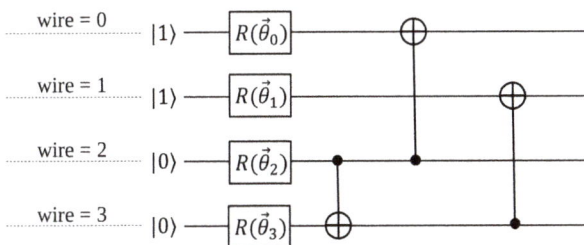

Figure 95.1: A quantum circuit[8] for the construction of VQE on quantum registers (Pennylane[9], Rigetti).

In Chapters 98 and 98, many examples in quantum chemistry make use of the VQE algorithm.

References

1. Quantum computing https://en.wikipedia.org/wiki/Quantum_computing.
2. Arute, F.; Arya, K.; Babbush, R.; Bacon, D.; Bardin, J. C.; Barends, R.; Biswas, R.; Boixo, S.; Brandao, F. G. S. L.; Buell, D. A.; Burkett, B.; Chen, Y.; Chen, Z.; Chiaro, B.; Collins, R.; Courtney, W.; Dunsworth, A.; Farhi, E.; Foxen, B.; Fowler, A.; Gidney, C.; Giustina, M.; Graff, R.; Guerin, K.; Habegger, S.; Harrigan, M. P.; Hartmann, M. J.; Ho, A.; Hoffmann, M.; Huang, T.; Humble, T. S.; Isakov, S. V.; Jeffrey, E.; Jiang, Z.; Kafri, D.; Kechedzhi, K.; Kelly, J.; Klimov, P. V.; Knysh, S.; Korotkov, A.; Kostritsa, F.; Landhuis, D.; Lindmark, M.; Lucero, E.; Lyakh, D.; Mandrà, S.; McClean, J. R.; McEwen, M.; Megrant, A.; Mi, X.; Michielsen, K.; Mohseni, M.; Mutus, J.; Naaman, O.; Neeley, M.; Neill, C.; Niu, M. Y.; Ostby, E.; Petukhov, A.; Platt, J. C.; Quintana, C.; Rieffel, E. G.; Roushan, P.; Rubin, N. C.; Sank, D.; Satzinger, K. J.; Smelyanskiy, V.; Sung, K. J.; Trevithick, M. D.; Vainsencher, A.; Villalonga, B.; White, T.; Yao, Z. J.; Yeh, P.; Zalcman, A.; Neven, H.; Martinis, J. M. Quantum Supremacy Using a Programmable Superconducting Processor. *Nature* **2019**, *574* (7779), 505–510. https://doi.org/10.1038/s41586-019-1666-5.
3. JMiszczak. Shor's factoring algorithm https://www.quantiki.org/wiki/shors-factoring-algorithm.
4. cgranade. Quantum oracles – Microsoft Quantum https://docs.microsoft.com/en-us/quantum/concepts/oracles.
5. Wright, L. J.; Tseng, S. T. Lecture 4: Grover's Algorithm.
6. SymPy documentation https://docs.sympy.org/latest/tutorial/intro.html.
7. Bauer, B.; Bravyi, S.; Motta, M.; Chan, G. K.-L. Quantum Algorithms for Quantum Chemistry and Quantum Materials Science. *arXiv [quant-ph]*, 2020.
8. Sugisaki, K.; Nakazawa, S.; Toyota, K.; Sato, K.; Shiomi, D.; Takui, T. Quantum Chemistry on Quantum Computers: A Method for Preparation of Multiconfigurational Wave Functions on Quantum Computers without Performing Post-Hartree-Fock Calculations. *ACS Cent Sci* **2019**, *5* (1), 167–175. https://doi.org/10.1021/acscentsci.8b00788.
9. A brief overview of VQE – PennyLane (Rigetti) https://pennylane.ai/qml/demos/tutorial_vqe.html.

96 Quantum chemistry software (QChem)

Q-concepts — Q-computing — Q-Algorithms ⊤ Q-Chemistry on Q-computers
QChemistry software ⌋

**Properties of electronically excited states are important in chemistry and con-
densed matter physics. In particular, excited states are a key component of
many chemical reactions and physical processes such as spectral emissions.**

 **By understanding the properties of these states, reaction rates, and reac-
tion pathways, a better understanding of the chemistry of molecules can be
achieved.**

Relevant QM software packages bridging to quantum computing

*Major goals of quantum chemistry include increasing the accuracy of the results for
small molecular systems, and increasing the size of large molecules that can be processed, which is limited by scaling considerations – the computation time increases as
a power of the number of atoms.[1] This is where the use of quantum computing principles
can help.*

 Quantum chemistry computer programs are used in computational chemistry
to implement the methods of quantum chemistry. Most include the Hartree–Fock
(HF) and density functional theory (DFT), molecular mechanics, or semiempirical
quantum chemistry methods.

- Gaussian 16™, a commercial chemistry program (discussed in Chapter xx)
- PySCF, an open-source Python chemistry program
- PyQuante, a pure cross-platform open-source Python chemistry program
- PSI4, a chemistry program that exposes a Python interface allowing for accessing
  internal objects

Gaussian™

Examples with Gaussian – in the context of computational chemistry[2] – were already
discussed in chapters about materials science.

https://doi.org/10.1515/9783110629453-096

pyQuante

PyQuante (Sourceforge Project Page) is an open-source suite of programs for developing quantum chemistry methods. The program is written in the Python programming language but has many "rate-determining" modules also written in C for speed.[3]

pyQuante implements:
– Optimized-effective potential DFT
– Two electron integrals computed
– CI-singles excited states
– DIIS convergence acceleration
– Second-order Moller–Plesset perturbation theory

Example: restricted closed-shell and unrestricted open-shell HF

```
# Hartree Fock and DFT calculation (pyQuante)
# author: Muller, Rick
# License: modified BSD license     # code: pyquante.sourceforge.net
# activity: active (2015)     # index: 96-1

from PyQuante.Molecule import Molecule
h2 = Molecule('h2',[(1,(0,0,0)),(1,(1.4,0,0))])
h2o = Molecule('h2o',[('O',(0,0,0)),
                ('H',(1.0,0,0)),('H',(0,1.0,0))], units = 'Angstrom')

from PyQuante.hartree_fock import rhf
en,orbe,orbs = rhf(h2)
print "HF Energy = ", en

# DFT: LDA (SVWN, Xalpha) and GGA (BLYP) functionals:
from PyQuante.dft import dft
en, orbe, orbs = dft(h2)
# Energy of -1.1353 hartrees.
# The 6-31G** basis set
# In DFT calculations, the functional defaults to SVWN (LDA)

# use different functional:
en, orbe, orbs = dft(h2,functional='BLYP')
# Energy of -1.1665 hartrees.
```

PySCF

PySCF can perform calculations using HF, DFT, perturbation theory, configuration interaction (including full CI), as well as coupled-cluster theory.

Example

A simple example of using a solvent model in the mean-field calculations:

```python
# DFT calculation (PySCF)
# author: Sun, Qiming
# License: Apache
# code: github.com/pyscf/pyscf/.../dftd3/00-hf_with_dftd3
# activity: active (2019)   # index: 96-2
from pyscf import gto
from pyscf import scf
from pyscf import dftd3

mol = gto.Mole()
mol.atom = ''' O    0.00000000    0.00000000   -0.11081188
              H   -0.00000000   -0.84695236    0.59109389
              H   -0.00000000    0.89830571    0.52404783 '''
mol.basis = 'cc-pvdz'
mol.build()

mf = dftd3.dftd3(scf.RHF(mol))
print(mf.kernel())
> -75.99396273778923
```

PSI4

PSI4 is an open-source suite[4] of ab initio quantum chemistry programs designed for efficient, high-accuracy simulations of a variety of molecular properties. It has an optional Python interface.

Python example

```python
# Hartree-Fock computation for water molecule (PSI4)
# author: The PSI4 Project
# License: LGPL-3.0          # code: www.psicode.org/psi4manual/master/psiapi
# activity: active (2020)    # index: 96-3

import psi4

#! Sample HF/cc-pVDZ H2O Computation
psi4.set_memory('500 MB')
```

```
h2o = psi4.geometry("""
O
H 1 0.96
H 1 0.96 2 104.5
""")

psi4.energy('scf/cc-pvdz')
> -76.02663273488399
```

References

1. Quantum chemistry (Wikipedia) https://en.wikipedia.org/wiki/Quantum_chemistry.
2. List of quantum chemistry and solid-state physics software (Wikipedia) https://en.wikipedia.org/wiki/List_of_quantum_chemistry_and_solid-state_physics_software.
3. PyQuante: Python Quantum Chemistry (Sourceforge) http://pyquante.sourceforge.net/.
4. Parrish, R. M.; Burns, L. A.; Smith, D. G. A.; Simmonett, A. C.; DePrince, A. E., 3rd; Hohenstein, E. G.; Bozkaya, U.; Sokolov, A. Y.; Di Remigio, R.; Richard, R. M.; Gonthier, J. F.; James, A. M.; McAlexander, H. R.; Kumar, A.; Saitow, M.; Wang, X.; Pritchard, B. P.; Verma, P.; Schaefer, H. F., 3rd; Patkowski, K.; King, R. A.; Valeev, E. F.; Evangelista, F. A.; Turney, J. M.; Crawford, T. D.; Sherrill, C. D. Psi4 1.1: An Open-Source Electronic Structure Program Emphasizing Automation, Advanced Libraries, and Interoperability. *J. Chem. Theory Comput.* **2017**, *13* (7), 3185–3197. https://doi.org/10.1021/acs.jctc.7b00174.

Quantum Computing Applications

Q-concepts — Q-computing — Q-Algorithms ⊤ Q-Chemistry on Q-computers
QChemistry software ⌐

In computational complexity theory, NP-hardness[1] (nondeterministic polynomial-time hardness) is the defining property of a class of problems that are informally *"at least as hard as the hardest problems in NP."*

Structural network analysis

Networks map relationships on graphs with items as nodes and relationships as edges. This was discussed several times in the book.

Such networks are said to be *structurally balanced* when they can be *cleanly divided* into two sets, with each set containing only entities where all relations between these sets have a directed character (i.e., hydrophilic vs. hydrophobic, friend vs. enemy, or :1 vs. :–1)

The measure of structural imbalance or frustration for a signed network, when it *cannot be cleanly divided*, is the minimum number of edges that violate the rule, "the enemy of my friend is my enemy." Finding a division that minimizes frustration is an NP-hard graph problem. This is an example to use quantum computing.

Example

This example by D-Wave[2,3] simply builds a random sparse graph G using the NetworkX random_geometric_graph() function, which places uniformly at a random specified number of nodes, problem_node_count, in a unit cube, joining edges of any two if the distance is below a given radius – and randomly assigns 0,1 signs to represent friendly and hostile relationships.

```
# Structural Imbalance in a network
# author: D-Wave Ocean
# License: Apache 2
# code: github.com/dwave-examples/structural-imbalance
# activity: active (2020)   # index: 97-1

import networkx as nx
import random
problem_node_count = 300
```

https://doi.org/10.1515/9783110629453-097

```
G = nx.random_geometric_graph(problem_node_count,
                              radius=0.0005*problem_node_count)
G.add_edges_from([(u, v, {'sign': 2*random.randint(0, 1)-1})
                            for u, v in G.edges])

from dwave.system import LeapHybridSampler
sampler = LeapHybridSampler()

import dwave_networkx as dnx
imbalance, bicoloring = dnx.structural_imbalance(G, sampler)
set1 = int(sum(list(bicoloring.values())))
print("One set has {} nodes; the other has {} nodes.".format(
            set1, problem_node_count-set1))
print("The network has {} frustrated
            relationships.".format(len(list(imbalance.keys()))))
> One set has 143 nodes; the other has 157 nodes.
> The network has 904 frustrated relationships.
```

Knapsack

The 0–1 Knapsack[4,5] problem is a well-known combinatorial optimization problem. In a given set of items, each of them has the two properties of *weight* w_i and a *profit* p_i. The problem is to select a subset such that the overall profit sum(p) is maximized without exceeding a given weight capacity C given by sum(w).

The Knapsack problem is also an NP-hard problem that plays an important role in computing theory and in many real-life applications, also in chemistry.[4]

Python example

```
# Knapsack problem
# author: Lucas, Andrew; D-Wave Ocean
# From Andrew Lucas, NP-hard combinatorial problems as Ising spin glasses
# Workshop on Classical and Quantum Optimization; ETH Zürich - August 20, 2014
# based on Lucas, Frontiers in Physics _2, 5 (2014)

# License: Apache 2          # code: github.com/dwave-examples/knapsack
# activity: active (2020)    # index: 97-2

import sys
from dwave.system import LeapHybridSampler
from math import log, ceil
import dimod
```

```python
def knapsack_bqm(costs, weights, weight_capacity):
    costs = costs
    bqm = dimod.AdjVectorBQM(dimod.Vartype.BINARY)
    lagrange = max(costs)
    x_size = len(costs)
    max_y_index = ceil(log(weight_capacity))

    # Slack variable list for Lucas's algorithm. The last variable has
    # a special value because it terminates the sequence.
    y = [2**n for n in range(max_y_index - 1)]
    y.append(weight_capacity + 1 - 2**(max_y_index - 1))

    # Hamiltonian xi-xi terms
    for k in range(x_size):
        bqm.set_linear('x' + str(k),
                              lagrange * (weights[k]**2) - costs[k]))
    # Hamiltonian xi-xj terms
    for i in range(x_size):
        for j in range(i + 1, x_size):
            key = ('x' + str(i), 'x' + str(j))
            bqm.quadratic[key] = 2 * lagrange * weights[i] * weights[j]
    # Hamiltonian y-y terms
    for k in range(max_y_index):
        bqm.set_linear('y' + str(k), lagrange * (y[k]**2))
    # Hamiltonian yi-yj terms
    for i in range(max_y_index):
        for j in range(i + 1, max_y_index):
            key = ('y' + str(i), 'y' + str(j))
            bqm.quadratic[key] = 2 * lagrange * y[i] * y[j]
    # Hamiltonian x-y terms
    for i in range(x_size):
        for j in range(max_y_index):
            key = ('x' + str(i), 'y' + str(j))
            bqm.quadratic[key] = -2 * lagrange * weights[i] * y[j]
    return bqm

# Ingest data
weight_capacity = float(<<VALUE, i.e. 55>>)
df = pd.read_csv(data_file_name, header=None)
df.columns = ['cost', 'weight']
bqm = knapsack_bqm(df['cost'], df['weight'], weight_capacity)

# Calculate
sampler = LeapHybridSampler()
sampleset = sampler.sample(bqm)
for sample, energy in zip(sampleset.record.sample, sampleset.record.energy):
    solution = []
```

```
    for this_bit_index, this_bit in enumerate(sample):
        this_var = sampleset.variables[this_bit_index]
        if this_bit and this_var.startswith('x'):
            solution.append(df['weight'][int(this_var[1:])])

print("Found solution {} at energy {}.".format(solution, energy))
```

References

1. NP-hardness (Wikipedia) https://en.wikipedia.org/wiki/NP-hardness.
2. Getting Started – D-Wave Ocean Documentation documentation https://docs.ocean.dwavesys.com/en/stable/getting_started.html.
3. D-Wave Systems https://en.wikipedia.org/wiki/D-Wave_Systems (Wikipedia).
4. Truong, T. K.; Li, K.; Xu, Y. Chemical Reaction Optimization with Greedy Strategy for the 0–1 Knapsack Problem. *Appl. Soft Comput.* **2013**, *13* (4), 1774–1780. https://doi.org/10.1016/j.asoc.2012.11.048.
5. knapsack – D-Wave example (GitHub) https://github.com/dwave-examples/knapsack.

98 Simulating molecules using VQE

Q-concepts − Q-computing − Q-Algorithms ⊤ Q-Chemistry on Q-computers
                        QChemistry software ⌋

Qiskit chemistry[1,2] **is a set of tools, algorithms, and software for use with quantum computers to carry out research and investigate how to take advantage of the quantum computational power to solve chemistry problems. The module requires a computational chemistry program or library, accessed via a chemistry driver,**[3] **to be installed on the system for the electronic-structure computation of a given molecule.**

A driver is created with a molecular configuration, passed in the format compatible with that particular driver. Then the structure of this *intermediate code* is independent of the driver that was used to compute it. The values and level of accuracy of such data will depend on the underlying chemistry program or library used by the specific driver:

Qiskit Chemistry[4] (driver) ⇔ Qiskit Aqua (algorithm) ⇔ Qiskit Terra (computation)

There is also the option to serialize the Qmolecule data in a binary format known as Hierarchical Data Format 5.

The software has prebuilt support to interface computational chemistry software programs Gaussian 16™, PSI4, PySCF, and PyQuante where the Gaussian 16 driver contains work licensed under the Gaussian Open-Source Public License. The Pyquante driver contains work licensed under the modified BSD license.

Particle–hole transformation on the fermionic operator circuit

This example carries out a particle–hole transformation[5] on an operator (one of the three symmetries of the Hamiltonian[6]).

Example

```
# Particle hole transformation of FermionicOperator
# author: IBM
# License: Apache
# code: github.com/qiskit-community/.../chemistry/ParticleHole...
# activity: active (2020)   # index: 98-1
```

https://doi.org/10.1515/9783110629453-098

```python
from qiskit import BasicAer
from qiskit.aqua import QuantumInstance
from qiskit.chemistry import FermionicOperator
from qiskit.chemistry.drivers import PySCFDriver, UnitsType

# uses PYSCF chemistry driver
driver = PySCFDriver(atom='H .0 .0 .0; H .0 .0 0.735',
                     unit=UnitsType.ANGSTROM,
                     charge=0, spin=0, basis='sto3g')
molecule = driver.run()

# define calculations
newferOp, energy_shift = ferOp.particle_hole_transformation(
                    [molecule.num_alpha, molecule.num_beta])
print('Energy shift is: {}'.format(energy_shift))

# The Jordan-Wigner transformation, maps spin operators onto
# fermionic creation and annihilation operators
newqubitOp_jw = newferOp.mapping(
                map_type='JORDAN_WIGNER', threshold=0.00000001)

newqubitOp_jw.chop(10**-10)
exact_eigensolver = NumPyMinimumEigensolver(newqubitOp_jw)

# calculate
ret = exact_eigensolver.run()

# output
print('The exact ground state energy in PH basis is
      {}'.format(ret.eigenvalue.real))
print('The exact ground state energy in PH basis is {}
      (with energy_shift)'.format(ret.eigenvalue.real - energy_shift))

> Energy shift is: 1.8369679912029837
> The exact ground state energy in PH basis is -0.020307038999395295
> The exact ground state energy in PH basis
        is -1.857275030202379 (with energy_shift)
```

H2 energy plot using different qubit mappings

Example

This notebook by IBM demonstrates using Qiskit chemistry to plot graphs of the ground state energy of the hydrogen (H2) molecule over a range of interatomic distances with different fermionic mappings to quantum qubits.

```
# H2 energy plot using different qubit mappings
# author: IBM
# license: Apache
# code: github.com/qiskit-community/.../chemistry/h2_mappings
# activity: active (2020)   # index: 98-2

from qiskit.aqua import aqua_globals, QuantumInstance
from qiskit.aqua.algorithms import NumPyMinimumEigensolver, VQE
from qiskit.aqua.components.optimizers import L_BFGS_B
from qiskit.circuit.library import TwoLocal
from qiskit.chemistry.core import Hamiltonian, QubitMappingType
…

molecule = 'H .0 .0 -{0}; H .0 .0 {0}'

algorithms = ['VQE', 'NumPyMinimumEigensolver']
mappings   = [QubitMappingType.JORDAN_WIGNER,
              QubitMappingType.PARITY,
              QubitMappingType.BRAVYI_KITAEV]

start = 0.5  # Start distance
by    = 0.5  # How much to increase distance by
steps = 20   # Number of steps to increase by

energies   = np.empty([len(mappings), len(algorithms), steps+1])
hf_energies = np.empty(steps+1)
distances = np.empty(steps+1)
aqua_globals.random_seed = 50

for i in range(steps+1):
    d = start + i*by/steps
    for j in range(len(algorithms)):
        for k in range(len(mappings)):

            driver = PySCFDriver(molecule.format(d/2), basis='sto3g')
            qmolecule = driver.run()

            operator =  Hamiltonian(qubit_mapping=mappings[k],
                                    two_qubit_reduction=False)
            qubit_op, aux_ops = operator.run(qmolecule)

            if algorithms[j] == 'NumPyMinimumEigensolver':
                result = NumPyMinimumEigensolver(qubit_op).run()
            else:
                optimizer = L_BFGS_B(maxfun=2500)
                var_form = TwoLocal(qubit_op.num_qubits,
                                    ['ry', 'rz'], 'cz', reps=5)
                algo = VQE(qubit_op, var_form, optimizer)
                result = algo.run(QuantumInstance(
```

```
                        BasicAer.get_backend('statevector_simulator'),
                                   seed_simulator=aqua_globals.
                                   random_seed,
                                   seed_transpiler=aqua_globals.
                                   random_seed))

        result = operator.process_algorithm_result(result)
        energies[k][j][i] = result.energy
        hf_energies[i] = result.hartree_fock_energy
        # Independent of algorithm & mapping
    distances[i] = d # closes &for i in range(steps+1)&

pylab.rcParams['figure.figsize'] = (12, 8)
pylab.ylim(-1.14, -1.04)
pylab.plot(distances, hf_energies, label='Hartree-Fock')
for j in range(len(algorithms)):
    for k in range(len(mappings)):
        pylab.plot(distances, energies[k][j], label=algorithms[j] +
                              ", " + mappings[k].value)
pylab.xlabel('Interatomic distance')
pylab.ylabel('Energy')
pylab.title('H2 Ground State Energy in different mappings')
pylab.legend(loc='upper right')
pylab.show()
```

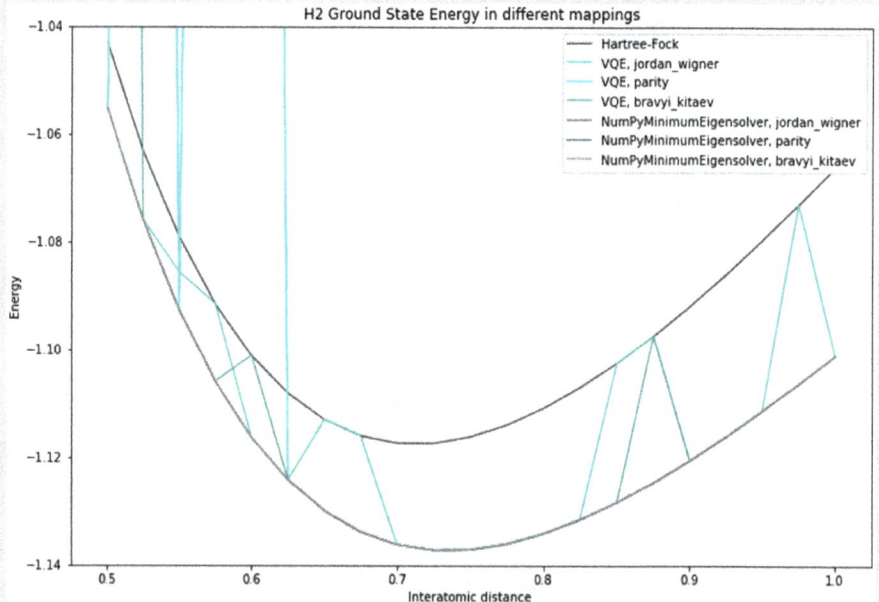

Figure 98.1: H2 ground state energy in different mappings *(IBM)*.

References

1. qiskit-community-tutorials https://github.com/Qiskit/qiskit-community-tutorials/tree/master/chemistry
2. qiskit/advanced/aqua/chemistry (GitHub) **2020**.
3. Chemistry Drivers (qiskit.chemistry.drivers) – Qiskit 0.19.3 documentation https://qiskit.org/documentation/apidoc/qiskit.chemistry.drivers.html.
4. Qiskit's chemistry module (qiskit.chemistry) – Qiskit documentation https://qiskit.org/documentation/apidoc/qiskit_chemistry.html.
5. Wikipedia contributors. Periodic table of topological invariants (Wikipedia) https://en.wikipedia.org/wiki/Periodic_table_of_topological_invariants.
6. Delft, T. U.; course contributors. Topology and symmetry https://topocondmat.org/w1_topointro/0d.html.

99 Studies on small clusters of LiH, BeH$_2$, and NaH

A hydride[1] is the anion of hydrogen, H$^-$, or more commonly it is a compound in which one or more *hydrogen centers* have nucleophilic, reducing, or basic properties. Ionic or saline hydrides are composed of hydride bound to an electropositive metal, generally an alkali metal or alkaline earth metal. In these materials, the hydride is viewed as a pseudohalide. Ionic hydrides are used as bases and as reducing reagents in organic synthesis.

LiH state energy and dipole moments

This code example plots the graphs of the ground state energy and dipole moments of a lithium hydride (LiH) molecule over a range of interatomic distances. The ExactEigensolver is used here. The PYSCF chemistry driver is used.

Example

```
# LiH state energy and dipole moments (IBM)
# author: IBM
# license: Apache
# code: github.com/qiskit-community/.../chemistry/energyplot
# activity: active (2020)   # index: 99-1

import pylab
from qiskit.chemistry import QiskitChemistry

# freeze core orbitals and remove unoccupied
# virtual orbitals to reduce the size of the problem
qiskit_chemistry_dict = {
    'driver': {'name': 'PYSCF'},
    'PYSCF': {'atom': '', 'basis': 'sto3g'},
    'algorithm': {'name': 'ExactEigensolver'},
    'operator': {'name': 'hamiltonian', 'freeze_core': True,
'orbital_reduction': [-3, -2]},
}
molecule = 'Li .0 .0 -{0}; H .0 .0 {0}'

start = 1.25 # Start distance
by    = 0.5  # How much to increase distance by
steps = 20   # Number of steps to increase by
energies  = np.empty(steps+1)
distances = np.empty(steps+1)
dipoles   = np.empty(steps+1)
```

https://doi.org/10.1515/9783110629453-099

```
for i in range(steps+1):
    d = start + i*by/steps
    qiskit_chemistry_dict['PYSCF']['atom'] = molecule.format(d/2)
    solver = QiskitChemistry()
    result = solver.run(qiskit_chemistry_dict) # ExactEigensolver
    distances[i] = d
    energies[i] = result['energy']
    dipoles[i]  = result['total_dipole_moment']


# Plot

pylab.plot(distances, energies)          pylab.plot(distances, dipoles)
pylab.xlabel('Interatomic distance')     pylab.xlabel('Interatomic distance')
pylab.ylabel('Energy')                   pylab.ylabel('Moment a.u')
pylab.title('LiH Ground State Energy');  pylab.title('LiH Dipole Moment');
```

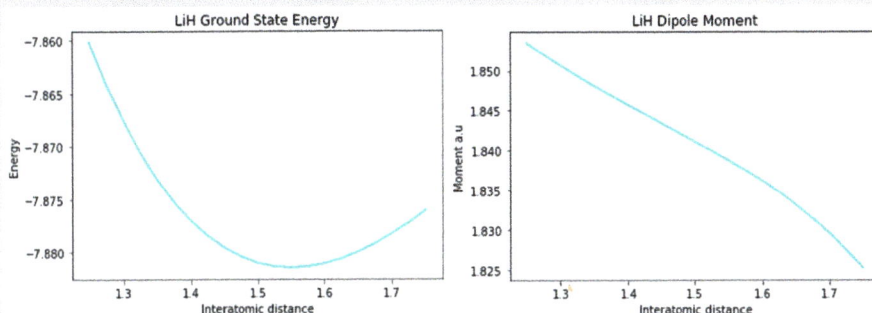

Figure 99.1: LiH ground state and dipole moments *(IBM [2]).*

Orbital reduction comparison for beryllium dihydride (BeH$_2$)

The LiH example was simplified by removing two unoccupied orbitals, along with freezing the core. While "freezing the core electrons" can always be done, discarding unoccupied orbitals might create problems.

Example

In the example for beryllium dihydride (BeH$_2$), the graphs of the ground state energy over a range of interatomic distances are calculated, using NumPyMinimumEigensolver.

Again as in LiH example, the freeze core reduction is true (freeze_core=True) but different virtual orbital removals are tried as a comparison. For better visual understanding in Figure 99.2, *the difference in energy, compared to no reduction*, is plotted.

```python
# Beryllium dihydride (BeH₂) orbital reduction comparison (IBM)
# author: IBM
# License: Apache
# code: github.com/qiskit-community/.../chemistry/beh2_reductions
# activity: active (2020)   # index: 99-2

import pylab
from qiskit.aqua.algorithms import NumPyMinimumEigensolver
from qiskit.chemistry.drivers import PySCFDriver
from qiskit.chemistry.core import Hamiltonian, QubitMappingType

molecule = 'H .0 .0 -{0}; Be .0 .0 .0; H .0 .0 {0}'
reductions = [[], [-2, -1], [-3, -2], [-4, -3], [-1], [-2], [-3], [-4]]

pts  = [x * 0.1 for x in range(6, 20)]
pts += [x * 0.25 for x in range(8, 16)]
pts += [4.0]
energies = np.empty([len(reductions), len(pts)])
distances = np.empty(len(pts))

for i, d in enumerate(pts):
    for j in range(len(reductions)):
        driver = PySCFDriver(molecule.format(d), basis='sto3g')
        qmolecule = driver.run()
        operator =  Hamiltonian(qubit_mapping=QubitMappingType.PARITY,
                                two_qubit_reduction=True,
                                freeze_core=True,
                                orbital_reduction=reductions[j])
        qubit_op, aux_ops = operator.run(qmolecule)
        result = NumPyMinimumEigensolver(qubit_op).run()
        result = operator.process_algorithm_result(result)
        energies[j][i] = result.energy
    distances[i] = d

pylab.rcParams['figure.figsize'] = (12, 8)
for j in range(len(reductions)):
    pylab.plot(distances, np.subtract(energies[j], energies[0]),
label=reductions[j])
pylab.xlabel('Interatomic distance')
pylab.ylabel('Energy')
pylab.title('Energy difference compared to no reduction []')
pylab.legend(loc='upper left');
```

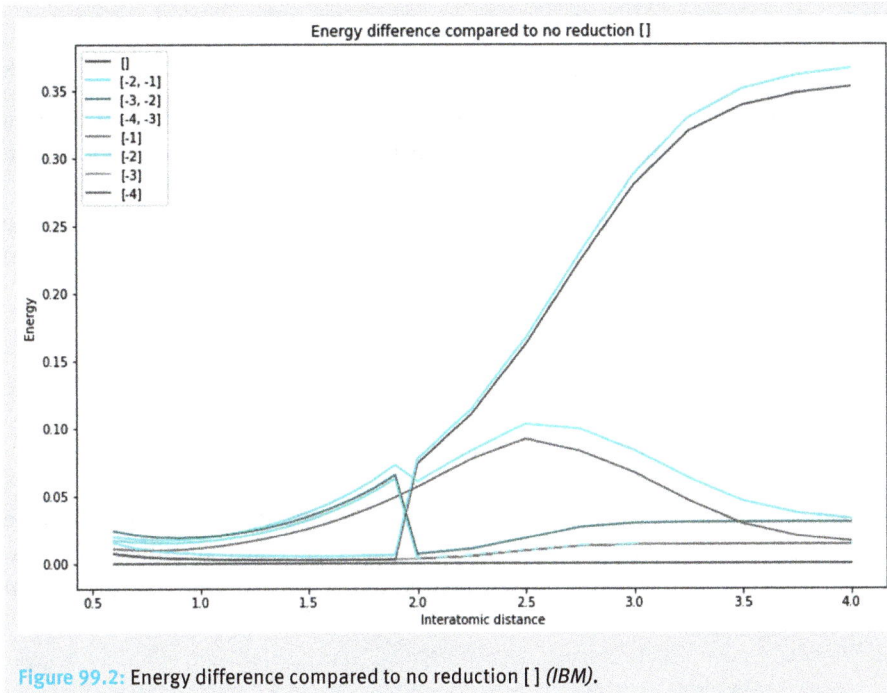

Figure 99.2: Energy difference compared to no reduction [ ] *(IBM).*

References

1. Wikipedia contributors. Hydride (Wikipedia) https://en.wikipedia.org/wiki/Hydride.
2. qiskit-community-tutorials https://github.com/Qiskit/qiskit-community-tutorials/tree/master/chemistry.

100 Quantum machine learning (QAI)

Quantum artificial intelligence[1] (QAI) is an interdisciplinary field that focuses on building quantum algorithms for improving computational tasks within artificial intelligence (AI), including subfields like machine learning (ML).

Quantum mechanics phenomena (superposition and entanglement) allow quantum computing to perform computations more efficiently than classical AI algorithms.

There are two major types of artificial agents:[2]
- classical agents that implement *classical actions* implemented quantum computational;
- *quantum agents* that implement *quantum operations* on a *quantum target*.

In the first case (like in the use case of chemistry domain), the preprocessing is done in a conventional way and the core *optimization* is calculated quantum computational (hybrid neural network). The second case directly invokes the learning step on a quantum level.

To create a quantum-classical neural network,[3] one can implement a hidden layer for a neural network using a parameterized quantum circuit. A "parameterized quantum circuit" is a quantum circuit where the rotation angles for each gate are specified by the components of a classical input vector.

Adding a quantum layer in a neural network does not automatically mean better training performance. To get the advantage of quantum entanglement (the driving force), the code has to be extended to more sophisticated quantum layers than in the following simple code examples.

IBM Qiskit ML

Qiskit ML demonstrates using quantum computers to tackle problems in the ML domain. These include using a quantum-enhanced support vector machine (SVM) to experiment with classification problems on a quantum computer. Resources are on GitHub:
- qiskit-tutorials/legacy_tutorials/aqua/machine_learning/
- qiskit-community / qiskit-community-tutorials

VQC (variational quantum classified)

Experiment using VQC (variational quantum classified) algorithm to train and test samples from a data set to see how accurately the test set can be classified.

https://doi.org/10.1515/9783110629453-100

```
# VQC (Variational Quantum Classified) algorithm
# to train and test samples (IBM)
# author: IBM
# license: Apache
# code: github.com/Qiskit/qiskit-aqua#...learning-programming...
# activity: active (2020)   # index: 100-1


from qiskit import BasicAer
from qiskit.aqua import QuantumInstance, aqua_globals
from qiskit.aqua.algorithms import VQC
from qiskit.aqua.components.optimizers import COBYLA
from qiskit.aqua.components.feature_maps import RawFeatureVector
from qiskit.ml.datasets import wine
from qiskit.circuit.library import TwoLocal


seed = 1376
aqua_globals.random_seed = seed


# Data ingestion
# Wine example data set
feature_dim = 4  # dimension of each data point
_, training_input, test_input, _ = wine(training_size=12,
                                         test_size=4,
                                         n=feature_dim)

# Model definition
feature_map = RawFeatureVector(feature_dimension=feature_dim)

# minimization of the cost function is performed using COBYLA
vqc = VQC(COBYLA(maxiter=100),
          feature_map,
          TwoLocal(feature_map.num_qubits, ['ry', 'rz'], 'cz', reps=3),
          training_input,
          test_input)
# Run
result = vqc.run(QuantumInstance(
            BasicAer.get_backend('statevector_simulator'),
            shots=1024, seed_simulator=seed, seed_transpiler=seed))


result
{'eval_count': 100,
 'eval_time': 82.41949081420898,
 'min_val': 0.49612208701307636,
 'num_optimizer_evals': 100,
 'opt_params': array([ 0.16793776, -2.1698383 ,  0.49180822,  0.8696281 ,
            2.20249542, [...] -1.33564641]),
 'test_success_ratio': 1.0,
 'testing_accuracy': 1.0,
 'testing_loss': 0.3447607173080886,
 'training_loss': 0.49612208701307636}
```

Quantum SVM algorithm: multiclass classifier on chemical analysis

The QSVM algorithm applies to classification problems that require a feature map for which computing the kernel is not efficient classically. This means that the required computational resources are expected to scale exponentially with the size of the problem. QSVM uses a quantum processor to solve this problem by direct estimation of the kernel in the feature space.

Example

Calculated for data are the results of a chemical analysis[4] of wines derived from three different cultivars. The analysis determined the quantities of 13 constituents found in each of the three types.

```
# Quantum SVM algorithm: multiclass classifier
# extension on chemical analysis (IBM)
# author: IBM
# License: Apache  # code: github.com/qiskit-community/...ml.../qsvm_multiclass
# activity: active (2020)   # index: 100-2

from qiskit.ml.datasets import wine
from qiskit import BasicAer
from qiskit.aqua [...]

n = 2  # dimension of each data point
sample_Total, training_input, test_input, class_labels =
        wine(training_size=40,test_size=10, n=n, plot_data=True)
temp = [test_input[k] for k in test_input]
total_array = np.concatenate(temp)
```

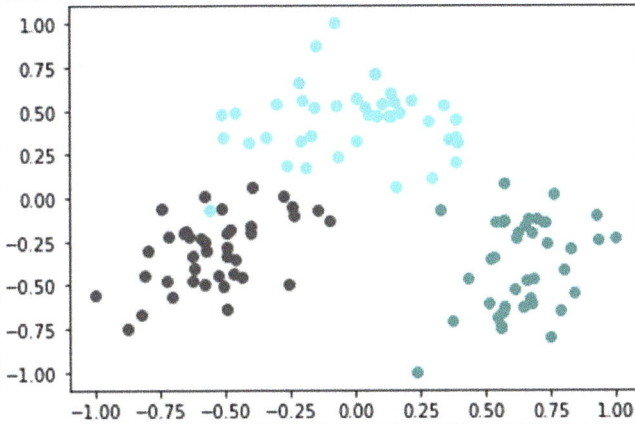

Figure 100.1: PCA dim. reduced dataset.

```
aqua_globals.random_seed = 10598

backend = BasicAer.get_backend('qasm_simulator')
feature_map = SecondOrderExpansion(feature_dimension=
                    get_feature_dimension(training_input),
                    depth=2, entangler_map=[[0, 1]])
svm = QSVM(feature_map, training_input, test_input, total_array,
        multiclass_extension=AllPairs())
quantum_instance = QuantumInstance(backend,
                    shots=1024,
                    seed_simulator=aqua_globals.random_seed,
                    seed_transpiler=aqua_globals.random_seed)

result = svm.run(quantum_instance)
result
>  {'predicted_classes': ['A', 'A', 'B', 'B', 'B', 'C', 'C', 'A', 'C', 'C'],
>   'predicted_labels': array([0, 0, 1, 1, 1, 2, 2, 0, 2, 2]),
>   'test_success_ratio': 0.9,
>   'testing_accuracy': 0.9}
```

Xanadu Pennylane

Pennylane[5] is a cross-platform Python library for quantum ML, automatic differentiation, and optimization of hybrid quantum-classical computations. QML[6] supports
quantum neural networks that have any quantum circuit with trainable continuous
parameters. An example is quantum transfer learning.[7]

Neural net implementation on a quantum computer (Iris dataset)

This code performing a multiclass classification is shortened for better understanding.

```python
# PyTorch on a quantum computer (Xanadu pennylane)
# author: Xanadu
# license: Apache
# code: pennylane.ai/qml/demos/tutorial_multiclass_classification
# activity: active (2020)   # index: 100-3

import pennylane as qml
import torch
from pennylane import numpy as np
from torch.autograd import Variable
import torch.optim as optim
num_classes = 3          train_split = 0.75
margin = 0.15            num_qubits = 2
feature_size = 4         num_layers = 6
batch_size = 10          total_iterations = 100
lr_adam = 0.01

dev = qml.device("default.qubit", wires=num_qubits)


# 1. Quantum circuits definitions
def layer(W):
    qml.Rot(W[0, 0], W[0, 1], W[0, 2], wires=0)
    qml.Rot(W[1, 0], W[1, 1], W[1, 2], wires=1)
    qml.CNOT(wires=[0, 1])


# 3 QNodes, each representing a classifier; uses PyTorch interface
def circuit(weights, feat=None):
    qml.templates.embeddings.AmplitudeEmbedding(feat,
                          [0, 1], pad=0.0, normalize=True)
    for W in weights:
        layer(W)
    return qml.expval(qml.PauliZ(0))


# Quantum circuit is parametrized by the weights
def variational_classifier(q_circuit, params, feat):
    weights = params[0]
    bias = params[1]
    return q_circuit(weights, feat=feat) + bias

qnode1 = qml.QNode(circuit, dev).to_torch()
qnode2 = qml.QNode(circuit, dev).to_torch()
qnode3 = qml.QNode(circuit, dev).to_torch()
```

```python
# 2. Classification, accuracy and loss functions
def multiclass_svm_loss(q_circuits, all_params,
                        feature_vecs, true_labels):
    [...]
    return loss / num_samples

def classify(q_circuits, all_params, feature_vecs, labels):
    [...]
    score = variational_classifier( q_circuits[c],
                (all_params[0][c], all_params[1][c]), feature_vec)
    [...]
    return predicted_labels

def accuracy(labels, hard_predictions):
    [...]
    return loss


# 3. Data ingestion and splitting
def load_and_process_data():
    data = np.loadtxt("multiclass_classification/iris.csv", delimiter=",")
    X = torch.tensor(data[:, 0:feature_size])
    [...]
    Y = torch.tensor(data[:, -1])
    return X, Y
def split_data(feature_vecs, Y):
    [...]
    feat_vecs_train = feature_vecs[index[:num_train]]
    feat_vecs_test = feature_vecs[index[num_train:]]
    return feat_vecs_train, feat_vecs_test, Y_train, Y_test


# 4. Defining training procedure
def training(features, Y):
    [...]
    q_circuits = [qnode1, qnode2, qnode3]
    [...]
    optimizer = optim.Adam(all_weights + all_bias, lr=lr_adam)

    for it in range(total_iterations):
        curr_cost = multiclass_svm_loss(q_circuits,
                    params, feat_vecs_train_batch, Y_train_batch)
        predictions_train = classify(q_circuits, ... feat_vecs_train)
        predictions_test = classify(q_circuits, ... feat_vecs_test)
        acc_train = accuracy(..., predictions_train)
        acc_test = accuracy(..., predictions_test)
```

```
    [...]
    return costs, train_acc, test_acc


# 5. Run
features, Y = load_and_process_data()
costs, train_acc, test_acc = training(features, Y)
```

D-Wave

Also, D-Wave[8,9] implements ML capabilities[10] that can be accessed through Python.

References

1. Quantum artificial intelligence (Wikipedia).
2. dos Santos Gonçalves, Carlos Pedro. Quantum Neural Machine Learning: Theory and Experiments https://www.intechopen.com/books/artificial-intelligence-applications-in-medicine-and-biology/quantum-neural-machine-learning-theory-and-experiments.
3. The Qiskit Team. Hybrid quantum-classical Neural Networks with PyTorch and Qiskit https://community.qiskit.org/textbook/ch-machine-learning/machine-learning-qiskit-pytorch.html.
4. UCI Machine Learning Repository: Wine Data Set https://archive.ics.uci.edu/ml/datasets/wine.
5. Xanadu. Training quantum neural networks with PennyLane, PyTorch, and TensorFlow https://medium.com/xanaduai/training-quantum-neural-networks-with-pennylane-pytorch-and-tensorflow-c669108118cc.
6. QML Demos – PennyLane https://pennylane.ai/qml/demonstrations.html.
7. Quantum transfer learning – PennyLane https://pennylane.ai/qml/demos/tutorial_quantum_transfer_learning.html.
8. Quantum computers can LEARN – Quantum Computing Primer (D-Wave) https://www.dwavesys.com/tutorials/background-reading-series/quantum-computing-primer.
9. Levit, A.; Crawford, D.; Ghadermarzy, N.; Oberoi, J. S.; Zahedinejad, E.; Ronagh, P. Free Energy-Based Reinforcement Learning Using a Quantum Processor. *arXiv [cs.LG]*, 2017.
10. Willsch, D.; Willsch, M.; De Raedt, H.; Michielsen, K. Support Vector Machines on the D-Wave Quantum Annealer. *Comput. Phys. Commun.* **2020**, *248*, 107006. https://doi.org/10.1016/j.cpc.2019.107006.

Code index

4-1	Heavy atom counts code comparison	(various)	CC-BY-SA
5-1	Interact with Apache PredictIO	TappingStone, Inc	Apache
6-1	Pipeline in Kubeflow	Kubeflow team	CC BY 4.0
7-1	Controlling clusters with ipyparallel	IPython Development Team	3-Clause BSD
7-2	Using CUDA in Python	Gressling, Thorsten	MIT
8-1	Make Jupyter a REST Server	Gressling, Thorsten	MIT
8-2	Jupyter Notebook for retrieving from REST APIs	Gressling, Thorsten	MIT
8-3	MQTT subscriber for Jupyter (paho)	Gressling, Thorsten	MIT
8-4	MQTT publisher for Arduino	Gressling, Thorsten	MIT
9-1	Arduino in a Jupyter Notebook Cell	ylabrj	GNU GPL 3
9-2	Plot serial data read from Arduino	Gressling, Thorsten	MIT
9-3	Raspberry GPIO access to log laboratory data	Gressling, Thorsten	MIT
10-1	C++ integration example in Jupyter notebook	Jupyter xeus team	BSD 3
11-1	Using modin instead of pandas	Gressling, Thorsten	MIT
11-2	cuDF / rapid.ai dataframe	Gressling, Thorsten	MIT
11-3	Lambda, filter and map example	Gressling, Thorsten	MIT
13-1	Runtime preparation Google Colab (in the first cell)	Gressling, Thorsten	MIT
14-1	Papermill example	Gressling, Thorsten	MIT
15-1	Notebook interaction examples	Gressling, Thorsten	MIT
15-2	intake example	Gressling, Thorsten	MIT
15-3	debugging with ipdb	Gressling, Thorsten	MIT
15-4	Example hvPlot	holoviz.org	BSD 3
15-5	PlantUML	Gressling, Thorsten	MIT
15-6	tqdm example	Gressling, Thorsten	MIT
17-1	Call KNIME by WebService	Landrum, Greg	CC BY 4.0
17-2	Work with a workflow via KNIME.exe	Gressling, Thorsten	MIT
18-1	Access HDFS file system	Gressling, Thorsten	MIT
18-2	Capture production logs with FLUME	Gressling, Thorsten	MIT
18-3	Massive data broker KAFKA	Gressling, Thorsten	MIT
19-1	PySpark Example	Gressling, Thorsten	MIT

https://doi.org/10.1515/9783110629453-101

19-2	RDD Example on Hadoop	Gressling, Thorsten	MIT
19-3	Spark configuration example	Gressling, Thorsten	MIT
19-4	PySpark and RDKit	Mario Lovrić, José Manuel Molero, Roman Kern	MIT
20-1	Convert SMILES to canonical SMILES comparizson	various	CC-BY-SA
21-1	Parsing data from file Chemical Markup Language (CML)	Gressling, Thorsten	MIT
22-1	Simple ETL example	Gressling, Thorsten	MIT
22-2	ETL with bonobo	Gressling, Thorsten	MIT
23-1	aiida workflow	Aiida Team	MIT
23-2	Build pandas pipelines using pdpipe	Gressling, Thorsten	MIT
23-3	Salt strip in ChEMBL structure pipeline	Gressling, Thorsten	MIT
24-1	Query Pubchem online	Swain, Matt	MIT
24-2	Query ChemSpider online	Gressling, Thorsten	MIT
24-3	Query Wikidata and Wikipedia	Gressling, Thorsten	MIT
24-4	Query ChEBI ontology via Python	Gressling, Thorsten	MIT
24-5	Query ChEBI data via REST	Gressling, Thorsten	MIT
24-6	Query ZINC database	Raschka, Sebastian	GPL
25-1	Example with a relational database	Gressling, Thorsten	MIT
25-2	Neo4J graph example	Gressling, Thorsten	MIT
26-1	Confusion matrix example	Gressling, Thorsten	MIT
29-1	MNIST comparison in major NN frameworks	Gressling, Thorsten	MIT
30-1	Keras to ONNX	ONNX team	MIT
30-2	Running a model in Tensorflow Hub	Google	Apache 2.0
31-1	MLflow Tracking API	Databricks; Teichmann; MLFlow	MIT
31-2	MLFlow environment definition (YAML)	Gressling, Thorsten	MIT
31-3	UBER Ludwig model training	Gressling, Thorsten	MIT
32-1	Symbolic mathematics with Sympy	Gressling, Thorsten	MIT
32-2	General problem solver implementation	Belotti, Jonathon; Connelly, Daniel	(not stated)
33-1	scikit classifier comparison	Varoquaux, Müller, Grobler	BSD 3
34-1	Bayesian Linear Regression with pymc3	Gressling, Thorsten	MIT
35-1	Decision tree visualization	Parr, Terence; Grover, Prince	MIT

36-1	k-nearest neighbors algorithm (k-NN)	Gressling, Thorsten	MIT
36-2	Gaussian and complement Naïve Bayes (scikit)	Gressling, Thorsten	MIT
36-3	Support Vector Machine SVM (scikit)	Gressling, Thorsten	MIT
37-1	Clustering using K-Means and PCA	Roy, Jacques	permission by author
37-2	PCA 2D projection (scikit)	scikit team	BSD
38-1	Sequence classification with 1D convolutions (Keras)	Chollet, François	MIT
38-2	Character-level RNN to classify words (PyTorch)	Robertson, Sean	BSD
38-3	Sequence classification with LSTM (Keras)	Chollet, François	MIT
39-1	SNN in tensorflow	Corvoysier, David	GNU GPL 2
40-1	Explainable AI with LIME	Ribeiro, Marco Tulio (Et al.)	MIT
41-1	Query ICSD crystallographic data (aiida)	aiida team	MIT
42-1	Symmetry analyzer (pymatgen)	materialsvirtuallab	BSD 3
42-2	Crystallographic structure matcher	materialsvirtuallab	BSD 3
42-3	Generate a structure from spacegroup (symmetry)	Shyue Ping Ong	BSD 3
42-4	Data structure for crystallography (crystals)	de Cotret, Laurent P. René	BSD 3
43-1	Thermochemical calculations in metallurgy	Zietsman, Johan	GPL 3
43-2	Chemical kinetics (differential equations, ChemPy)	Dahlgren, Björn	BSD 2
44-1	Air and methane mixed in stoichiometric proportions	Cantera developers	(Cantera)
44-2	Viewing a reaction path diagram	Cantera developers	(Cantera)
45-1	Mendeleev database of elements	Mentel, Lukasz	MIT
45-2	Periodic table in Bokeh	Anaconda	BSD 3
46-1	Isotherms of the van der Waals Equation of State	Segtovich, Iuri	GPL 3
47-1	Quantify Microstructures using 2-Point Statistics (PyMKS)	MATIN materials research group	MIT
48-1	Workflow of a free energy simulation (FeSETUP)	Löffler, Hannes H	GPL 2
48-2	Ovito pipeline / ASE Example	Stukowski; Janssen	GPL 3
49-1	Working with SMIRNOFF parameter sets (OpenForceField)	OpenForceField Team	MIT

49-2	Automate use of AMBER force fields (OpenMM)	OpenMM Team	MIT
50-1	VASP in pymatgen	pymatgen team	MIT
50-2	Plotting the electron localization function (PytLab/VASPy)	Zhengjiang, Shao	MIT
50-3	Single, Threads, Pool and Multiprocessing with vasp.py	Kitchin, John	GNU free documentation
51-1	GAUSSIAN™ Wrapper (ASE), transition state optimization	ASE-developers	GNU
52-1	GROMACS Wrapper	Beckstein, Oliver	GPL 3
53-1	Output energies from NAMD (pyNAMD)	Radak, Brian	MIT
53-2	Analyze coordinates in a trajectory (MDAnalysis)	MDAnalysis team	GPL 2
53-3	Baker-Hubbard Hydrogen Bond Identification (MDTraj)	MDTraj team	LGPL 2
53-4	Energy minimization of a polymer system using LAMMPS	Demidov, Alexander; Fortunato, Michael E.	MIT
53-5	LAMMPS integration with pyiron	Neugebauer, Jörg	BSD 3
54-1	Spectral Neighbor Analysis Potential (SNAP) for Ni-Mo Binary Alloy	Xiang-Guo Li	BSD 3
54-2	Interface reaction of LiCoO2 and Li3PS4	pymatgen team	MIT
54-3	Pourbaix diagram for the Cu-O-H system (pymatgen)	Montoya, Joseph	MIT
55-1	PBE (Perdew–Burke–Ernzerhof) density from NWChem (ASE)	ASE-developers	GNU
55-2	Nudged elastic band calculations (ASE)	ASE-developers	GNU
56-1	NGLView - WebGL Viewer for Molecular Visualization	Rose, Alexander S.	MIT
56-2	py3Dmol - Viewer for Molecular Visualization	Koes, David	MIT
56-3	RDKit Depiction examples	Dahlke, Andrew	CC-BY-SA
57-1	Ingestion: Define, read and write structures (RDKit)	Landrum, Greg; Serizawa; Scalfani (Et al.)	CC BY SA 4.0
57-2	Access atoms and their properties (RDKit)	Landrum, Greg; Serizawa; Scalfani (Et al.)	CC BY SA 4.0
57-3	Substructure Searching (RDKit)	Landrum, Greg; Serizawa; Scalfani (Et al.)	CC BY SA 4.0

57-4	Salt stripping and structure sanitizing (RDKit)	Landrum, Greg; Serizawa; Scalfani (Et al.)	CC BY SA 4.0
58-1	Calculating standard descriptors (RDKit)	yamasakih	MIT
58-2	Structural descriptor examples (RDKit)	Landrum, Greg; Serizawa; Scalfani (Et al.)	CC BY SA 4.0
58-3	Features (RDKit)	Landrum, Greg; Serizawa; Scalfani (Et al.)	CC BY SA 4.0
58-4	Example descriptor: Graph diameter	Gressling, Thorsten	MIT
58-5	Example descriptor: Gasteiger-Marsili atomic partial charges	Gressling, Thorsten	MIT
59-1	Substructure-based transformations (delete, replace) (RDKit)	Landrum, Greg; Serizawa; Scalfani (Et al.)	CC BY SA 4.0
59-2	Break rotatable bonds and depict the fragments (RDKit)	Gressling, Thorsten	CC BY SA 4.0
59-3	Hantzsch thiazole synthesis (RDKit)	Gao, Peng; Jiang, Shilong	MIT
60-1	Benzoic acid Morgan fingerprint visualization (RDKit)	Gressling, Thorsten	MIT
60-2	Fingerprint as input layer in NN	keiserlab.org; Evangelidis, Thomas	MIT
61-1	Fingerprint comparison for three vanillyl-derivatives (RDKit)	Gressling, Thorsten	MIT
62-1	SILA2 Notebook	Gressling, Thorsten	MIT
62-3	labPy laboratory instrument connector (SILA2)	Dörr, Mark	permission by author
63-1	LIMS interaction with LabKey	labkey.com	Apache
63-2	LIMS interaction with eLabFTW	elabftw.net; Carpi, Nicolas	GNU
63-3	Calculations with aqueous electrolyte solutions (pyEQL)	Kingsbury, Ryan S.	GNU
63-4	Calculations with aqueous electrolyte solutions (ionize)	Marshall, Lewis A.	GNU
64-1	Pipette roboting with opentrons	Opentrons	Apache
64-2	Cognitive laboratory assistance (note-taking to ELN, speech)	Gressling, Thorsten	MIT
65-1	Boiling temperature of water (REFPROP)	Bell, Ian	(US public)

65-2	Compute special values in SI units with CoolProp	CoolProp.org	MIT
65-3	Thermophysical Data and psychrometic chart (CoolProp)	CoolProp.org	MIT
66-1	Solubility of CO2 in NaCl brines (reaktoro)	Leal, Allan	LGPL
66-2	Tube Diameter effect on Plug Flow Reactors	Navarro, Franz	MIT
67-1	Read multi vars at Siemens PLC (S7-319 CPU)	Molenaar, Gijs	MIT
67-2	OPC UA Python client example	FreeOPC-UA team	LGPL
68-1	Predictive Maintenance: Turbo-Fan Failure	Gressling, Thorsten	MIT
69-1	Calculation of ph-Value, titration curve (phCalc)	Ryan Nelson; (Johan Hjelm)	MIT
69-2	Calculate Ca2+/EDTA binding (pytc)	Harms, Mike	Unlicense
70-1	Processing 1D Bruker NMR data (nmrglue)	Helmus, Jonathan	BSD 3
71-1	X-Ray Absorption Spectroscopy, working on spectra	Plews, Michael	MIT
71-2	X-Ray diffraction (XRD) and working on spectra	Plews, Michael	MIT
71-3	Calculation of XRD spectrum using MD (ipyMD)	Sewell, Chris	GNU GPL3
71-4	EDS (EDX) spectrum analysis with hyperspy	Burdet, Pierre	GNU GPL3
72-1	MS/MS spectrum processing and visualization (spectrum_utils)	Bittremieux, Wout	Apache
73-1	Thermogravimetric (TGA) plot	Plews, Michael	MIT
73-2	TGA analysis	Gressling, Thorsten	MIT
74-1	Calculate CO infrared equilibrium spectrum (RADIS)	Pannier, Erwan; Laux, Christophe O.	LGPL3
74-2	Raman spectrum fitting with rampy	Le Losq, Charles	GPL2
75-1	Thermogram analysis for activation energy and probability density function	Hemingway, Jordon D.	GPL3
75-2	PFM spectrum unmixing with Singular Value Decomposition (SVD)	Vasudevan, Giridharagopal	MIT
76-1	Multiple GC-MS experiments peak alignment	Davis-Foster, Dominic	LGPL3
77-1	K-matrix Method (SVD) on Beer-Lambert-Bouguer law	Huffman, Scott	permission by author
78-1	Explorative Spectroscopy in Python	Hughes, Adam	FreeBSD

79-1	QSAR multiple linear regression model	Sinyoung, Kim	MIT
80-1	Basic Protein-Ligand Affinity Model (deepchem)	Ramsundar, Bharath; Feinberg, Evan	MIT
81-1	Stochiometry	Dahlgren, Björn	BSD 2
81-2	Balancing reactions and chemical equilibria	Dahlgren, Björn	BSD 2
82-1	Plotting of an amino acid (ALA) substructure vectors	Turk, Samo	BSD 3
83-1	Target prediction: multi-layer perceptron (PyTorch)	Eloy, Felix	MIT
84-1	AutoKeras, auto-sklearn and NeuNetS (IBM)	Gressling, Thorsten	MIT
85-1	Retrosynthesis planner	Tseng, Francis	GPL v3.0
85-2	CNN Layer in 'Predict the outcomes of organic reactions'	Coley, Connor	MIT
86-1	ChemML example workflow	Hachmannlab	BSD 3
86-2	ChemML NN branched layer definition and 'active learning'	Hachmannlab	BSD 3
87-1	Predict DFT trained on QM9 dataset (megnet)	materialsvirtuallab/ crystals.ai	BSD 3
88-1	Create an ontology	Lamy, Jean-Baptiste	LGPL 3
88-2	Adding Python methods to an OWL Class	Lamy, Jean-Baptiste	LGPL 3
88-3	Reasoning example on drug intake	Lamy, Jean-Baptiste	LGPL 3
89-1	Reading Neo4j into network	Gressling, Thorsten	MIT
89-2	Laboratory workplace metric	Gressling, Thorsten	MIT
89-3	M2M Communication analysis chemical production plants	Gressling, Thorsten	MIT
90-1	Printed image to molecule (molvec)	Gressling, Thorsten	MIT
90-2	Structure image to *.mol (imagoPy)	brighteragyemang@ gmail.com	GPL 3.0
90-3	OCR with Tesseract	Gressling, Thorsten	MIT
90-4	NLP of chemical literature (spaCy)	Gressling, Thorsten	MIT
91-1	Patent data ingestion from Google	Gressling, Thorsten	MIT
91-2	Finding key compounds in chemistry patents (SureChEMBL)	Papadatos, George	CC BY 4.0
91-3	Build a Knowledge Graph from Text Data	Gressling, Thorsten	MIT
92-1	General Numerical Solver 1D Time-Dependent Schrodinger's equation	Vanderplas, Jake	BSD
92-2	Dirac notation (sympy)	SymPy Development Team	BSD

92-3	Visualizing operators (QuTip)	Johansson, J.R.	CC BY 3.0
92-4	Plotting the bloch sphere (QuTip)	Johansson, J.R.	CC BY 3.0
93-1	Act an X gate on qubit (Rigetti)	Rigetti Computing	Apache
93-2	Pauli-Z phase shift gate (Xanadu)	Xanadu	Apache
93-3	Boolean NOT Gate (D-Wave)	D-Wave	Apache
93-4	Invoke the IBMQ engine (IBM)	IBM	Apache
93-5	Square root of NOT (Cirq)	The Cirq Developers, Sung, Kevin J.	Apache
94-1	Quantum Random Numbers (ProjectQ)	Steiger, Damian; Haener, Thomas	Apache
94-2	Producing Bell states (QCircuits)	Webb, Andrew	MIT License
95-1	Running Shor's algorithm (IBM qiskit, on QPU)	Gressling, Thorsten	MIT
95-2	Quantum oracle (Rigetti Forest)	Kopczyk, Dawid	"Feel free to use full code"
95-3	Grover algorithm with sympy	SymPy Development Team	BSD
95-4	Deutsch-Jozsa	Malykh, Egor	MIT
96-1	Hartree Fock and DFT calculation (pyQuante)	Muller, Rick	modified BSD license
96-2	DFT calculation (PySCF)	Sun, Qiming	Apache
96-3	Hartree-Fock computation for water molecule (PSI4)	The PSI4 Project	LGPL-3.0
97-1	NP-hard: Structural Imbalance in a network (D-Wave)	D-Wave	Apache
97-2	NP-hard: Knapsack problem (D-Wave)	Lucas, Andrew; D-Wave	Apache
98-1	Particle hole transformation of FermionicOperator (IBM)	IBM	Apache
98-2	H2 energy plot using different qubit mappings	IBM	Apache
99-1	LiH state energy and dipole moments (IBM)	IBM	Apache
99-2	Beryllium dihydride (BeH2) orbital reduction comparison (IBM)	IBM	Apache
100-1	VQC (Variational Quantum Classified) algorithm to train and test samples (IBM)	IBM	Apache
100-2	QSVM Quantum multiclass classifier on chemical analysis	IBM	Apache
100-3	PyTorch on a quantum computer (Xanadu pennylane)	Xanadu.ai	Apache

Index

www.ingramcontent.com/pod-product-compliance
Lightning Source LLC
Chambersburg PA
CBHW060953210326

41598CB00031B/4807